# *Positron Beams*

*and their applications*

# *Positron Beams*

## *and their applications*

*Editor*

*Paul Coleman*

*University of Bath, UK*

*Published by*

World Scientific Publishing Co. Pte. Ltd.

P O Box 128, Farrer Road, Singapore 912805

*USA office:* Suite 1B, 1060 Main Street, River Edge, NJ 07661

*UK office:* 57 Shelton Street, Covent Garden, London WC2H 9HE

**British Library Cataloguing-in-Publication Data**
A catalogue record for this book is available from the British Library.

**POSITRON BEAMS AND THEIR APPLICATIONS**



ISBN 981-02-3394-9

Printed in Singapore by Uto-Print

# PREFACE

Almost thirty years ago the first laboratory-based slow positron beams were developed and used to study simple positron-atom interactions. Beam intensities of less than one particle per second were common, and novel techniques were developed to cope with the resultant low signal-to-background ratios. Since those early days beam intensities have climbed steadily by seven orders of magnitude, and there are plans to produce intense beams of $10^{10}$ positrons per second. This startling increase in intensity has allowed the application of positron beams in a wide variety of fields, both fundamental and applied. Beams can now be produced with characteristics comparable to, if not better than, standard electron beams, but problems associated with their low relative intensity ensure that innovative and imaginative techniques to this day remain a feature of positron beam experimentation.

This volume attempts to provide a coherent and comprehensive overview of the generation and application of mono-energetic positron beams, written by acknowledged experts, at a level accessible to scientists in other spheres who want to know something of the field and to graduate students working, or planning to work, with positron beams.

A brief historical introduction is followed by an overview of the generation and transport of mono-energetic positron beams. A description of the fate of slow positrons in gaseous and condensed matter is followed by discussion of their application in the study of solid surfaces (including microscopy), defect profiling in subsurface regions, interfaces and thin films, and - by acceleration to MeV energies, or by exploiting the positrons' spin polarisation - their use in probing condensed matter in novel ways. The book will end with a look to the future, considering the prospects for intense positron beams and their potential for performing new science.

I am very grateful to all those who have contributed to this book, taking time from their very busy schedules to produce carefully-prepared and comprehensive chapters.

Positron beams will occupy an important place in the future of positron science. The book aims to convey some of the excitement felt by those who work in the field, who look forward both to the routine application of such beams in industrial environments as well as the realisation of many fundamental experimental studies hitherto only dreamed of.

Paul Coleman
Bath, September 1999

## TABLE OF CONTENTS

**CHAPTER 3**
ATOMIC AND MOLECULAR PHYSICS WITH POSITRONS AND POSITRONIUM **41**

G. Laricchia *(University College London, UK)*
and M. Charlton (*University of Swansea, UK)*

**CHAPTER 5**

CHAPTER 1

# INTRODUCTION: A BRIEF HISTORY OF POSITRON BEAMS

P. G. COLEMAN
*Department of Physics*
*University of Bath*
*Claverton Down, Bath BA2 7AY, UK*
*E-mail: p.g.coleman@bath.ac.uk*

The forty-year history of low-energy positron beams is a model example of how an experimental technique can grow and flourish through persistence, ingenuity, and a small measure of good fortune. From modest beginnings the field has witnessed orders of magnitude increase in beam intensities, the development of positron beams which rival or even exceed the specifications of their state-of-the-art electron equivalents, and a burgeoning of applications in almost every field of scientific endeavour. This chapter is a short personal account of some of the milestones in positron beam research since a Princeton graduate student's thesis set the ball rolling - albeit slowly - in 1958.

## 1 Early days

The future for controlled-energy positron experiments did not look too bright when in 1950 Madansky and Rasetti [1] searched without success for the emission of positrons with energies below 150eV from the surfaces of metallic and dielectric materials. It is intriguing to contemplate how the field described in this book would have developed had they been successful; it is particularly ironic that their expectations were perfectly in line with what has been observed since their pioneering efforts, and that they even explained correctly the reasons for their failure.

Eight years after Madansky and Rasetti's results were published Cherry, a graduate student at Princeton, recorded in his PhD thesis on positron and electron backscattering [2] that he had observed the emission of positrons with energies below 10eV from a Cr-covered mica surface bombarded with beta positrons from $^{64}$Cu. This should have been the significant breakthrough; but his results were not published in a journal article and so were not widely known. So it was that the first published experimental evidence for eV positron re-emission from a solid surface did not appear in print in a journal until 1969 [3] when John Madey reported his observation of the re-emission of positrons with a relatively broad energy spectrum (peaked at about 20eV) from the surface of polyethylene. It is interesting to note that Madey's work has largely been overlooked since its publication - perhaps because of the difficulties of working with insulators, with attendant charging effects, and perhaps because of the wide energy spread of the re-emitted positrons. It is almost certain that Madey was the first to observed the emission of hot (or ballistic or epithermal) positrons from a surface. With an increasing interest in applying positron beams to the study of polymer films, and the search for a material which may act as a field-assisted positron emitter, it may well be that Madey's work will be revisited in the near future.

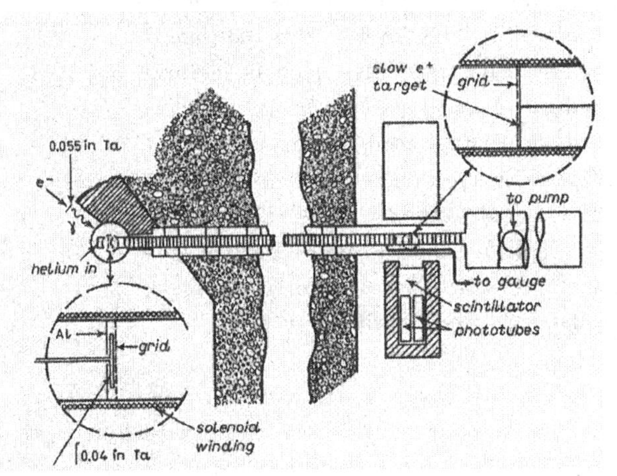


**Figure 1.** Positron beam apparatus used by Costello *et al* to study positron-helium scattering [5].

It was the discovery - or, rather, rediscovery - of the re-emission of slow positrons with narrow energy width from a metallic surface that proved to be the start of today's world-wide activity using positron beams. In the late 1960s, at Gulf Energy and Environment in San Diego, J William McGowan and colleagues were developing experimental apparatus to measure the total scattering cross sections of positrons in gases. They made their positrons by pair production in a Ta target bombarded by bremsstrahlung radiation created by 20 ns bursts of electrons from a 55MeV LINAC (see figure 1). They passed the energetic positrons thus created through a mica sheet (coated with a 150Å-thick gold deposit, primarily for applying an accelerating potential) in order to slow a fraction of them down to energies of the order of $10^2$eV. The positron energies were measured by a time-of-flight technique; the start timing (birth) pulses were obtained from the LINAC, and the stop pulses from annihilation gamma rays emitted from a target at the end of an evacuated, 3m-long, solenoid-surrounded flight tube. The group's original search for $10^2$eV positron proved unsuccessful; fortunately for us, however, they examined the time-of-flight region corresponding to eV positrons, and found signal due to positrons with an energy distribution peaked at 1-2eV. They reported their success at a meeting of the American Physical Society in 1968 [4]. Four years later the San Diego group published their work [5]. They had observed no shift in the peak of the positron energy distribution after repeating their experiments with moderators other than mica; they concluded that the mechanism of enhanced emission was not, as they had originally supposed, a bulk process, but involved emission from the surface of the gold deposit which covered each of their 'moderators'. Tong [6] correctly explained the re-emission phenomenon theoretically in terms of the negative positron-surface affinity (the 'negative work function' for positrons). Costello *et al* also made the first successful measurements of positron-helium scattering cross sections using their beam [7]. They estimated that the efficiency of their positron moderator ( = useable slow positrons per fast positron) was of the order of $10^{-6}$. Unfortunately the project ended and the group members went their different ways, but this time their discovery stimulated others to take up the baton and, eventually, to run with it.

The research which followed Costello *et al*'s 1968 announcement concentrated on efforts to reproduce their result in the laboratory using a radioactive positron source. Mao and Paul [8] looked for free positron emission from very fine plastic fibres, and Daniel *et al* searched for emission from a metallic surface using a magnetic energy spectrometer [9], both initially without success. In 1972 Jaduszliwer *et al* [10] reported the observation of emission of low-energy positrons from a gold-coated mica source assembly, using a 20mCi (740 MBq) source and an energy spectrometer.

## 2 Positron beams in the laboratory

At about the same time, the newly-formed positron group at University College London were in 1969-71 applying the time-of-flight technique used by Costello *et al* in their lab-based system using a weak (50μCi, 200 kBq) $^{22}$Na source. The same gold-covered mica moderator was tried and, assuming the same $10^{-6}$ moderation efficiency, less than one eV positron per second was expected. To cope with this low absolute signal rate every one of the fast positrons was tagged on its way to the moderator by passing it through a thin disc of plastic scintillator. Start pulses were derived from the detector recording annihilation quanta from the target at the end of the flight tube; stop pulses the delayed signals from the plastic scintillator disc. Two years of fruitless searching, accompanied by several system modifications, were ended by the decision to replace the mica moderator with a short tube, the inner surface of which could be lined with a series of foils or films. The adoption of this backscattering geometry led to immediate success; the first-ever time spectrum of low-energy positrons recorded in the laboratory is reproduced in Figure 2 [11]. The signal rate was ~ $10^{-1}$ $s^{-1}$.

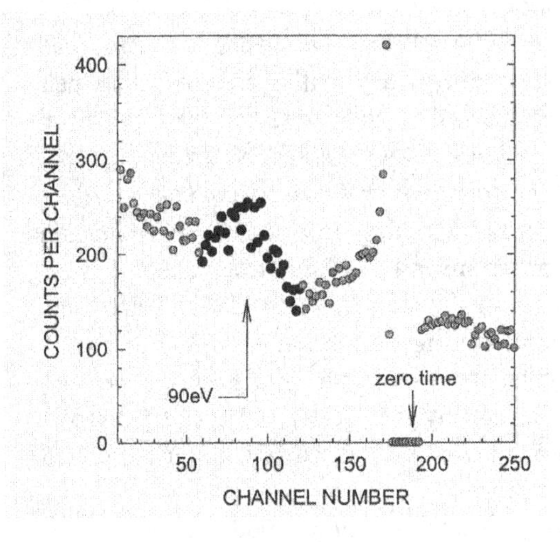


**Figure 2.** The first multichannel analyser time spectrum obtained at University College, London to show a signal peak corresponding to the detection of slow (here 90eV) positrons. 60,000s run; time scale reversed; prompt peak vetoed to reduce background under signal [11].

This meagre signal rate was infinitely greater than had been observed earlier, but was still practically too low for the planned atomic scattering measurements. It was then that the research was blessed by one of the strokes of good fortune which only visit laboratories once or twice in a lifetime, if ever. A paper by Curry and Schawlow (the latter being the laser pioneer) reported than one positronium atom was formed per $10^3$ high-energy positrons entering MgO powder from a radioactive source [12]. It occurred to the University College group that the existence of some mechanism for the dissociation of only 1% of the positronium atoms might result in the emission of slow positrons with considerably higher intensities than that observed from the solid surfaces used earlier. The moderator tube had been replaced by a set of gold-coated vanes in venetian-blind geometry, and this assembly was coated with MgO powder simply by burning Mg ribbon in close proximity. The desired effect was realised; signal rates of up to 1.5 $s^{-1}$ were recorded and, with the signal-to-background enhancement afforded by time-of-flight spectrometry, this was more than sufficient to measure total scattering cross sections for positrons in a range of atomic and molecular gases [13,14].

The London group was joined in the 1970's by a small number of others interested in positron-atom scattering. They all used commercial radioactive sources with the notable exception of the group at Wayne State University, Detroit, who bombarded boron with protons from their 4.75 MeV Van de Graaff accelerator to form the positron emitter $^{11}C$ *in situ*. This source doubled as a moderator and produced beams of only 80 meV energy width, allowing high-resolution low-energy scattering measurements including the first observation of the Ramsauer-Townsend minimum in the cross section for noble gases [15]. A significantly narrower energy spread has yet to be achieved.

## 3 Branching out

Karl Canter, who had been a postdoctoral fellow at University College during the events described above, joined Stephan Berko at Brandeis University in 1972. There, with Berko and Alan Mills, he set up experiments to study positronium formation at solid surfaces bombarded by slow positrons [16], the success of which led to the first major fundamental scientific achievement made possible by positron beams - the observation of positronium in its first excited state in vacuo [17], which had been pursued by many research groups for decades.

Measurement of the natural, or vacuum, lifetime $\lambda_0$ of triplet positronium provides a powerful test of QED, and in the early 1970's these experiments were performed in gaseous or powder media and the results extrapolated to zero density [18,19]. In 1976 Gidley *et al*, at Michigan, measured $\lambda_0$ for the first time truly in vacuo by bombarding a solid surface with a slow positron beam [20].

Allen Mills was responsible for the first true positron-surface measurements, performed after his move to Bell Laboratories from Brandeis. By performing experiments in ultra-high vacuum conditions and with the sample preparation and characterisation facilities essential in surface science, he demonstrated that free positrons and positronium atoms were ejected, with roughly equal probability, from atomically clean, single crystal surfaces of many metals bombarded by keV positrons [21-23]. The other third (or so) of the positrons diffusing back to the surface were shown to be trapped there, with ~2eV binding

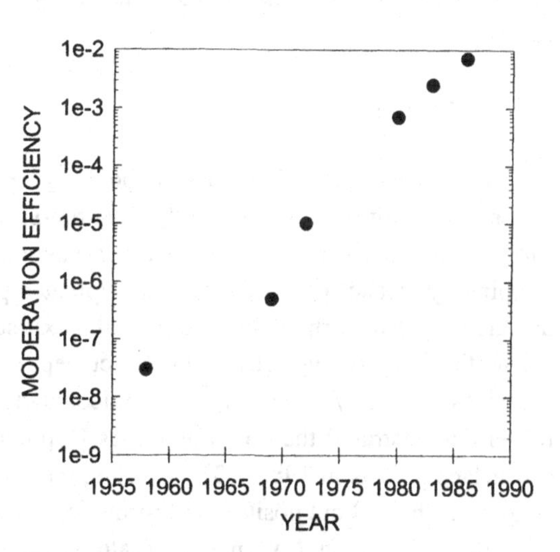


**Figure 3.** Growth in measured positron moderation efficiencies since 1958.

energy, by observing their desorption as thermal positronium atoms upon heating the sample target [24]. These groundbreaking experiments opened the door to an almost bewildering array of novel and exciting experiments in the late seventies and early eighties. The smoked MgO moderator gave way to the annealed metallic moderator (first copper [25] then tungsten [26]), with attendant increases in positron moderation efficiency of two orders of magnitude. Coupled with more intense (~100mCi or 3.4 GBq $^{22}$Na) primary radioactive sources and using single particle counting rather than time-of-flight techniques, absolute useable positron beam intensities in the laboratory approaching $10^6$ $s^{-1}$ were being achieved - a long way from the $1s^{-1}$ characteristic of the first positron scattering experiments a decade earlier.

It is quite remarkable that, in the few years after the discovery that positrons incident on a surface can be re-emitted into the vacuum as positronium (Ps) with a high probability, not only was excited state Ps observed and the ground state lifetime measured in vacuo, but the positronium negative ion was observed and its decay rate measured [27,28]; and, in a triumph of scientific ingenuity and meticulous experimentation, Doppler-free two-photon excitation of the $2^3S$ state of positronium was performed in 1982 and 1984 (and later in 1993 to 3 parts in $10^9$) [29,30].

## 4 Applications

At Brookhaven National Laboratory Kelvin Lynn was *applying* the information gained from the early positron-surface experiments to the study of surface and near-surface phenomena. For example, he looked at subsurface positron trapping by measuring the fraction of implanted positrons which diffused through a defected layer to form positronium on the exit surface [31], in experiments which led to the burgeoning activity in defect depth profiling in solids and thin films which exists today, especially in semiconductor systems [32]. Triftshäuser and Kögel demonstrated the now ubiquitous Doppler broadening method for studying subsurface defects in 1982 [33], obviating the need for constant (and clean) surface conditions throughout positronium formation measurements.

In Brookhaven National Laboratory Lynn used the High Flux Reactor to make highly active $^{64}$Cu sources, which were deposited on to single-crystal tungsten substrates at the end of a beam line constructed in the reactor housing. This was the first operational intense positron beam, producing eV positron beams of absolute intensity ~ $10^7$ $s^{-1}$ [34]. This intensity made possible a series of experiments not possible in a laboratory-based radionuclide system, including two-dimensional angular correlation studies of the surface state and of positronium emitted into the vacuum [35,36] and positronium diffraction [37].

Rich Howell, at the Lawrence Livermore Laboratory, was the first to return to the LINAC method of generating positrons by pair production, to create a pulsed intense positron beam [38]. The pulsed nature of such a beam allows timing experiments such as the measurement of the velocity spectra of positronium atoms leaving metallic surfaces [39].

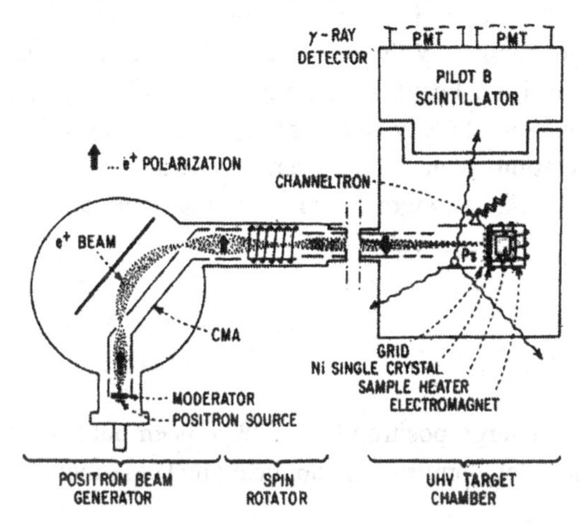


**Figure 4.** Polarised positron beam apparatus used by Gidley *et al* [42] to study surface magnetism. The positron beam is transversely polarised; the spin of the positrons is rotated by ± 90° by an axial magnetic field before entering the target chamber.

Timing measurements with laboratory-based positron beams have been made possible by the development of bunching/chopping techniques, now developed to a degree which enables lifetime measurements to be performed (i.e. with timing resolution of $\sim 10^2$ps) which compare well with traditional bulk measurements. [40].

In the late 1970's the Michigan group found experimentally that thermalised and re-emitted positrons retain a high degree of the longitudinal polarisation which they possess after beta emission [41]. While this natural property of positron beams has been exploited in the study of magnetic surfaces (see figure 4) [42] and in a number of other studies, principally in Michigan, it has been relatively out of the limelight in the past few years. It is very likely that polarised positron beams will play an important role again in the future.

The last dozen years have seen an expansion of research activity involving slow positron beams all over the world. Much of it will be reviewed and discussed in the chapters which follow. There have been inspiring developments in positron surface studies, especially in low-energy positron diffraction at Brandeis [43] and positron-annihilation-induced Auger electron spectroscopy at Arlington [44]. Positronium beams have been generated and used at University College London and at Brookhaven [37,45]. Positron microprobes and microscopes have been built and used [46,47]; brightness enhancement [48] is achieved by repeated focusing and remoderation..

Research using positron beams is today being performed all over the world in many different scientific areas. There is still research into fundamental positron and positronium-related phenomena; but there is much more in which the positron is used as a novel *probe* of matter - in the study of surface, subsurface and interface properties of materials, and of thin films. With the advent of solid rare gas moderators [49], laboratory-based positron beam intensities now stand at seven orders of magnitude above those available twenty five years ago. Encouragingly, there is hope, and indeed expectation, that further developments in beam production technology will lead to even stronger beam intensities and a new generation of positron experiments.

## 5 References

Some excellent review papers on low-energy positron beams have been published at various stages of the history of their development and application. These include the following:

Griffith, T.C. and Heyland, G.R., 1978, *Phys. Rep.* C **39**, 171.
Mills, A.P., Jr, 1982, in *Positron Annihilation*, edited by P.G. Coleman, S.C. Sharma and L.M. Diana (Amsterdam: North Holland) p. 121.
Dale J.M., Hulett, L.D., and Pendyala, S., 1983, *Appl. Spectrosc. Rev.* **19**, 105.
Mills, A.P. Jr., 1983, in *Positron Solid State Physics,* edited by A. Dupasquier and W. Brandt (Amsterdam: North Holland), p. 432.
Charlton, M., 1985, *Rep. Prog. Phys.* **48**, 73.
Dupasquier, A. and Zecca, A., 1985, *Riv. Nuovo Cimento* **8**, 3.
Schultz, P.J. and Lynn, K.G., 1988, *Rev. Mod. Phys.* **60**, 701.
Articles by various authors in *Positron Spectroscopy of Solids,* edited by A. Dupasquier and A.P. Mills, Jr., 1995 (Amsterdam: IOS).

[1] Madansky, L. and Rasetti, F., 1950, *Phys. Rev.* **79**, 397.
[2] Cherry, W., 1958, *Ph.D. dissertation* (Princeton University)
[3] Madey, J.W., 1969, *Phys. Rev. Lett.* **22**, 784.
[4] Groce, D.E., Costello, D.G., McGowan, J.Wm., and Herring, D.F., 1968, *Bull Am Phys Soc* **13**, 1397.

[5] Costello, D.G., Groce,D.E., Herring, D.F., and McGowan, J.Wm., 1972, *Phys Rev* B **5**, 1433.
[6] Tong, B.Y., 1972, *Phys. Rev.* B **5**, 1436.
[7] Costello D.G., Groce, D.E., Herring, D.F. and McGowan, J. Wm., 1972, *Can.J.Phys.* **50**, 23.
[8] Mao, A., and Paul, D.A.L., 1971, reported at *Second Int'l Conf. on Positron Annihilation*, Kingston, Ontario
[9] Daniel, T.N., Dutton, J., Harris, F.M., and Lewis, D.M., 1969, reported at the *2nd International Conference on Atomic Physics,* Amsterdam
[10] Jaduszliwer, B., Keever, W.C., and Paul, D.A.L., 1972, *Procs. Third Intl. Conf. on Atomic Physics,* Boulder, Colorado
[11] Coleman, P.G., 1972, *Ph.D. thesis* (University College London)
[12] Curry, S.M. and Schawlow, A.L., 1971, *Phys. Lett.* A **37**, 5.
[13] Canter, K.F., Coleman, P.G., Griffith, T.C. and Heyland, G.R., 1972, *J Phys* B **5**, L167.
[14] Coleman, P.G., Griffith, T.C. and Heyland, G.R., 1974, *Appl. Phys.* **4**, 89.
[15] Kauppila, W.E., Stein, T.S. and Jesion, G., 1976, *Phys. Rev. Lett.* **36**, 580.
[16] Canter, K.F., Mills, A.P. Jr., and Berko, S., 1974, *Phys. Rev. Lett.* **33**, 7.
[17] Canter, K.F., Mills, A.P. Jr., and Berko, S., 1975, *Phys. Rev. Lett.* **34**, 177 and 1541
[18] Gidley, D.W., Marko, K.A. and Rich, A., 1976, *Phys. Rev. Lett.* **36**, 395.
[19] Gidley, D.W., Rich, A., Zitzewitz, P.W. and Paul, D.A.L., 1978, *Phys. Rev. Lett.* **40**, 737 and Griffith, T.C., Heyland, G.R., Lines, K.S. and Twomey, T.R., 1978, *J. Phys. B* **11**, 743.
[20] Gidley, D.W. and Zitzewitz, P.W., 1978, *Phys. Lett. A* **69**, 97.
[21] Mills, A.P., Jr., Platzman, P.M. and Brown, B.L., 1978, *Phys. Rev. Lett.* **41**, 1076.
[22] Murray, C.A. and Mills, A.P., Jr., 1980, *Solid State Commun.* **34**, 789.
[23] Mills , A.P., Jr., 1978, *Phys. Rev. Lett.* **41**, 1828
[24] Mills, A.P., Jr. and Pfeiffer, L.N., 1979, *Phys. Rev. Lett.* **43**, 1961
[25] Mills, A.P., Jr., 1979, *Appl. Phys. Lett.* **35**, 427.
[26] Vehanen, A., Lynn, K.G., Schultz, P.J. and Eldrup, M., 1983, *Appl. Phys. A.* **32**, 163.
[27] Mills, A.P., Jr., 1981, *Phys. Rev. Lett.* **46**, 717.
[28] Mills, Allen Paine, Jr., 1983, *Phys. Rev. Lett.,* **50**, 671.
[29] Chu, S., and Mills, A.P., Jr., 1982, *Phys. Rev. Lett.* **48,** 1333 and
Chu, S., Mills, A.P., Jr., and Hall, J.L., 1984, *Phys. Rev. Lett.* **52**, 1589.

[30] Fee, M.S., Mills, A.P.,Jr., Chu, S., Shaw, E.D., Danzmann, K., Cichester, R.J. and Zuckerman, D.M., 1993, *Phys. Rev. Lett.* **70**, 1397.
[31] Lynn, K.G., 1980, *Phys. Rev. Lett.* **44**, 1330.
[32] Asoka-Kumar, P., Lynn, K. G. and Welch, D. O., 1994, *J. Appl. Phys.* **76**, 4935.
[33] Triftshäuser, W. and Kögel, G., 1982, *Phys. Rev. Lett.* **48**, 1741.
[34] Weber, M., Berko, S., Canter, K.F., Lynn, K.G., Mills, A.P., Jr., Moodenbaugh, A.R., and Roellig, L.O., 1988, *Intense Positron Beams,* eds. E.H. Ottewitte and W. Kells (Singapore: World Scientific) p.51, and
Lynn, K.G., Mills, A.P.,Jr., West, R.N., Berko, S., Canter, K.F., and Roellig, L.O., *Phys. Rev. Lett.,* **54**, 1702.
[35] Lynn, K.G., Mills, A.P., Jr., West, R.N., Berko, S., Canter, K.F. and Roellig, L.O., 1985, *Phys. Rev. Lett.* **54**, 1702.
[36] Chen, D.M., Berko, S., Canter, K.F., Lynn, K.G., Mills, A.P.,Jr., Roellig, L.O., Sferlazzo, P., Weinert, M. and West, R.N., 1987, *Phys. Rev. Lett.,* **58**, 921.
[37] Weber, M.H., Tang, S., Berko, S., Brown, B.L., Canter, K.F., Lynn, K.G., Mills, A.P., Roellig, L.O. and Viescas, A.J., 1988, *Phys. Rev. Lett.* **61**, 2542.
[38] Howell, R.H., Alvarez, R.A. and Stanek, M., 1982, *Appl. Phys. Lett.* **40**, 751.
[39] Howell, R.H., Rosenberg, I.J., Fluss, M.J., Goldberg, R.E. and Laughlin, R.B., 1987, *Phys. Rev. B* **35**, 5303.
[40] Sperr, P., Kögel, Willutzki and Triftshäuser, W., 1997, *Appl. Surf. Sci.* **116**, 78.
[41] Zitzewitz, P.W., Van House, J., Rich, A. and Gidley, D.W., 1979, *Phys. Rev. Lett.* **43**, 1281
[42] Gidley, D.W., Köymen and Capehart, T.W., 1982, *Phys. Rev. Lett.* **49**, 1779.
[43] Rosenberg, I.J.,Weiss, A.H. and Canter, K.F., 1980, *Phys. Rev. Lett.* **44**, 1139.
[44] Weiss, A.H., in *Positron Spectroscopy of Solids,* eds. A. Dupasquier and A.P. Mills, Jr., 1995 (Amsterdam: IOS), p. 259.
[45] Zafar, Z., Laricchia, G., Charlton, M. and Griffith, T.C., 1991, *J. Phys. B.* **24**, 4461.
[46] Brandes, G.R., Canter, K.F. and Mills, A.P., Jr., 1988, *Phys. Rev. Lett.* **61**, 492.
[47] Brandes, G.R., Canter, K.F., Horsky, T.N. and Lippel, P.H., 1988, *Rev. Sci. Instrum.* **59**, 228.
[48] Mills, A.P., Jr., 1980, *Appl. Phys.* **23**, 189.
[49] Mills, A.P., Jr. and Gullikson, E.M., 1986, *Appl. Phys. Lett.* **49**, 1121.

CHAPTER 2

# THE GENERATION AND TRANSPORT OF POSITRON BEAMS

P. G. COLEMAN
*Department of Physics*
*University of Bath*
*Claverton Down, Bath BA2 7AY, UK*
*E-mail: p.g.coleman@bath.ac.uk*

This chapter will follow the positron from source to target. The variety of primary positron sources and moderation methods will be reviewed. Magnetic and electrostatic transport and focusing of positron beams, including velocity filtering, acceleration, brightness enhancement, bunching, and timing, are described and discussed. The chapter ends with the detection of positrons at the end of their flight path, both directly and via their annihilation radiation.

## 1 Positron sources

### *1.1 Radioactive sources*

The choice of radioactive source for positron beam experiments is governed by the nature of the measurement to be made tempered by cost effectiveness. For most long-term experiments $^{22}$Na is the source of preference, offering an acceptable compromise between cost per Bq and half-life (2.6 y). (The energy spectrum of beta positrons from $^{22}$Na is peaked at 178keV and its end point is 545 keV.) $^{22}$Na positron sources are available commercially in activities up to 3.7 GBq in sealed capsules designed to maximise beta positron output; a high-Z backing material maximises backscattering into the forward direction and a thin low-Z foil cover (usually 5-10μm Ti) ensures safe sealing and low absorption. Moderated positrons retain the spin polarisation of the beta positrons if their direction is unchanged; therefore, to create a highly polarised positron beam it is desirable to use a *low*-Z backing material to minimise backscattering in the source capsule, with consequent unavoidable loss of intensity. $^{22}$Na has been a favourite source for conventional positron studies of bulk material properties (using beta positrons) because of the 1.28 MeV gamma ray emitted essentially immediately after the positron; $^{22}$Na $\rightarrow$

$(\beta^+)\rightarrow{}^{22}Ne^*\rightarrow(\gamma)\rightarrow{}^{22}Ne$. This gamma ray is used as a start signal in high-resolution positron lifetime experiments. In beam experiments, however, this method is not used for timing (see below) and the gamma ray is superfluous. Indeed, if gamma ray spectroscopy is being performed at the sample then all gamma radiation from the source has to be efficiently shielded from the gamma detector to reduce background, and this is one reason why the typical length of positron beam apparatus used for such measurements is between a few and several metres in length.

$^{58}Co$ has been used widely for experiments (such as positron diffraction and microscopy) in which high beam intensity over short periods, rather than long-term count rate stability, is paramount. $^{58}Co$ has a short half-life of 71 days and much higher activity sources may be purchased for the same price when compared with $^{22}Na$.

The in-house production of positron sources has been undertaken by Weng *et al.* [1], who have constructed an apparatus for drop-depositing up to 740MBq $^{22}Na$ from $^{22}NaCl$ solution and sealing with 3μm Ti; Britton *et al.* [2] have produced $^{22}NaCO_3$ at the National Accelerator Centre, South Africa.

In the first reactor-based slow positron beam at Brookhaven National Laboratory [3] highly-active $^{64}Cu$ (~ 3TBq) was created by irradiating a copper ball in the core of the research reactor via the reaction $^{63}Cu(n,\gamma)^{64}Cu$, dropped automatically into a crucible in the source chamber and evaporated on to a tungsten backing. The strong $^{64}Cu$ source acted as a self-moderator for the production of eV positrons, with the high-*Z* backing for efficient backscattering. The half-life of $^{64}Cu$, 12.7 h, meant that the source inevitably required regular replacement.

The development of table-top cyclotrons for medical uses has opened the door to the possibility of creating intense fluxes of fast positrons in the laboratory. Preliminary work on the quasi-continuous production of intense eV positron beams using such machines has been performed (figure 1) [4,5]; an

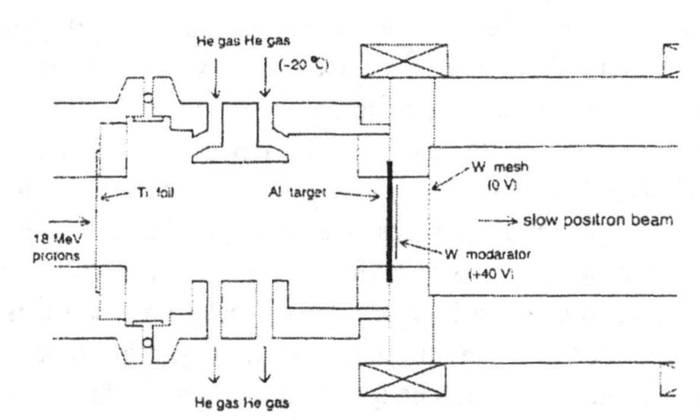


**Fig.1** Moderator section of the compact-cyclotron positron beam of Hirose *et al* [5]

early prototype produced $2x10^6$ eV positrons using 18MeV protons at a current of 30μA. The advantage of such systems is that when the cyclotron is not in operation the strong positron source decays rapidly to safe levels; disadvantages are the need for continual maintenance and the high initial cost. However, the cost per positron compares very favourably with conventional sources when averaged over a period of time.

### *1.2 Pair production*

One of the first successful low-energy positron beam systems, constructed in San Diego in the late 1960's, was based at a LINAC facility [6]. Bremsstrahlung gamma radiation from the pulsed energetic (~$10^1$-$10^2$ MeV) LINAC electrons create electron-positron pairs in a high-*Z* target such as Ta, W or Pt; the fast positrons thus created are then moderated to form the eV positron beam. The pulsed nature of the positron beam created can be exploited in some experiments, or debunching can be achieved to produce a quasi-dc beam. The positron production efficiency depends on the LINAC electron energy $E_{e-}$ and the thickness $d$ of the Ta converter; for example, optimum values of $d$ for producing the maximum numbers of positrons with a small fraction of $E_{e-}$ are ~ 2.5 and 12mm for $E_{e-}$ = 20 and 100 MeV, respectively [7].

Over the past thirty years a number of LINAC-based positron beams have been constructed and used [8-13]. At the reactor in Munich positrons are produced by pair production by gamma rays emitted in the reaction $^{113}Cd(n,\gamma)^{114}Cd$ to produce a continuous beam (figure 2) [14]. In a development parallelling the use of compact cyclotrons to produce intense radioactive positron sources small, relatively low-current accelerators - microtrons - have been used to create positrons by pair production, first by Mills at Bell Laboratories [15] and more recently at the University of Aarhus [16].

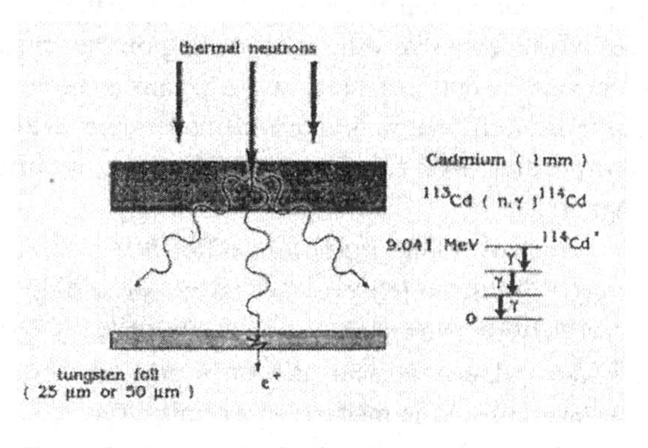


**Fig.2** Positron production by slow neutron reactions with Cd [14].

## 2 Positron moderation

### *2.1 The physics of moderation*

Energetic (primary) positrons entering a solid which are not backscattered at high energies into the vacuum rapidly lose energy by undergoing inelastic collisions, primarily with core and conduction electrons, plasmon excitation becomes important in the $10^1$eV region, as does phonon scattering at eV energies. If implanted into a metal with an energy greater than a few keV a positron will reach thermal equilibrium with its surroundings (*i.e.*, fall to energies ~ 40 meV) in ~ 1ps. If we define the average distance travelled by a diffusing thermalised positron in its lifetime (*i.e.*, before annihilation) as the diffusion length $L_+$, then those positrons which are thermalised within $L_+$ of the surface can diffuse back to it. This fraction is maximised if the moderating material contains no non-equilibrium defects capable of trapping the diffusing positrons; in the case of metallic moderators this means annealing *in situ* to ~ $0.8T_m$ in the best vacuum attainable (say $\leq 10^{-6}$ Pa). This procedure also serves to clean the moderator surface.

Thermalised positrons which reach the surface of a moderator then either (a) fall into the surface well, where they are eventually annihilated, (b) pick up an electron and leave as positronium or, of most interest to us here, (c) leave as free positrons into the vacuum with an energy determined by the positron work function $\phi_+$, as long as $\phi_+$ is negative. ($\phi_+$ for different materials has a distribution about zero, being the small difference between the repulse surface dipole term and the attractive correlation potential; for some materials - such as copper - it has both positive and negative values, depending on the crystal orientation.) The probability of the last event increases as $\phi_+$ becomes more negative [17]; the fast-to-slow conversion efficiency of standard moderators is therefore increased by increasing their positron work functions - for example, by diffusing sulphur to the surface of copper [18].

We shall define moderator efficiency $\varepsilon$ as the number of essentially mono-energetic positrons delivered to a target per unit time divided by the total activity of the primary positron source. (Note of caution: some researchers use an alternative definition which replaces the total primary activity with the numbers of fast positrons striking the moderator per unit time.)

$\varepsilon$ for a radioactive source can be estimated using the expression $\varepsilon = y_0 L_+/R$, where $R$ is the mean range for beta positrons in the moderator, and $y_0$ is the branching ratio for positron re-emission from the surface. For example, for $^{22}$Na positrons bombarding tungsten $L_+ \sim 100$ nm, $y_0 \sim 0.35$ and R ~ 15 μm, so that $\varepsilon \sim 2$

x $10^{-3}$. This illustrates an unfortunate fact of life concerning positron moderation of positrons with primary energies ~ $10^2$keV: only those positrons in the low-energy end of the primary energy spectrum (~$10^1$keV) are stopped within a diffusion length of the surface, and the vast majority penetrate to much deeper depths and can never be re-emitted.

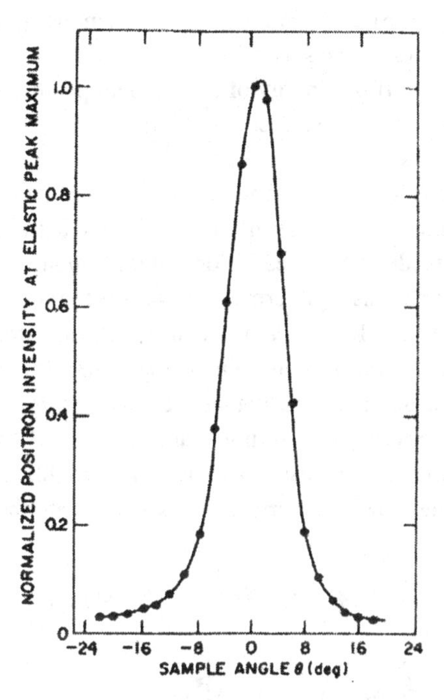


**Fig. 3** Count rate for positrons elastically ejected from W(110) as a function of sample orientation. Angular width at half maximum = 10° [20].

A number of experimental and theoretical investigations have been made of the angles and energies of emission of positrons ejected from solid surfaces. The conjecture that thermalised positrons thrown from the surface by a negative work function are ejected essentially normally to the surface was confirmed by Murray and Mills [19] and later by Fischer *et al* [20]. The angular spread of elastically-emitted positrons was found experimentally to be ~ $2\text{-}[kT/(-\phi_+)]^{1/2}$, as expected from thermal smearing arguments. It is clear from this expression that cooling a moderator will lead to a smaller spread of emission angles; this has been done in electrostatic systems in which the achievement of the best possible beam quality is paramount.

Wilson investigated the probability of inelastic emission and suggested that the fraction of inelastically-emitted positrons increased with work function [21]. Gullikson *et al* [22] showed, by positioning a strong permanent magnet behind the sample target to ensure that elastically-emitted positrons were guided to near-axial paths and could thus be energy-analysed by retarding field analysis, that only a few percent were emitted inelastically from the surfaces studied (nickel and tungsten). This was confirmed, using electrostatic optics, by Fischer *et al* [20]. The probability of energy loss on emission increases when impurities are present on a surface, as indicated by the early measurements on a number of contaminated surfaces by Pendyala *et al* [23] and those performed later on polycrystalline tungsten by Goodyear *et al* [24], both with electrostatic energy analysers. Chen *et al* [25] and Goodyear *et al* demonstrated the diminution of the inelastic tail on cleaning the surface.

If the energy of the incident positrons is below 1 keV it is common for some to reach the surface prior to complete thermalisation. The probability of this occurrence increases rapidly as the energy decreases [26]. The hot, or epithermal, positrons, can escape from most surfaces - cleaned or not. Their energy spectrum is similar to that seen for secondary electrons, being relatively narrower only at higher incident energies. For defected materials, or those with positive positron work functions, epithermal emission is the only process by which positrons are re-emitted [27]. In insulators and semiconductors the maximum energy of epithermal positrons is determined approximately by the band gap; the probability of re-emission at a particular incident positron energy depends on the band gap, inelastic scattering probabilities, and the size of the positron work function. Mills [28] saw epithermal positron emission from the alkali halides, and Knights and Coleman measured epithermal yields from a number of materials (figure 4) [29]. Fits to data

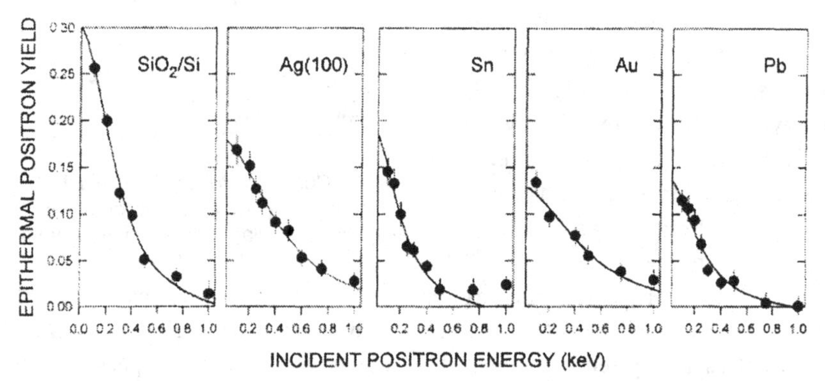


**Fig. 4** Re-emirtted epithermal positron yields for five positive-work-function materials as a function of incident positron energy [29].

in the latter investigation yielded thermalisation lengths - over which positrons are reduced in energy from $\sim 10^1$eV to below the (positive) work function - of between a few Å for metallic samples to 24Å for Si.

Hot positron emitters of great importance are the solid rare gases, to be discussed in the following section, which have very long effective diffusion lengths and, as result, high re-emitted positron yields (approaching 1% of the incident fast positrons). However, although yields are high, the spread of emission angle and energy makes epithermal positron sources less attractive for the generation of bright electrostatic beams unless tandem moderation stages are used.

### *2.2 Moderating materials*

Costello *et al.*'s scheme to moderate energetic positrons created by pair production [6] was to pass them through a thin piece of mica, slowing down positrons by inelastic collisions so that a usable fraction of those starting out with $10^1$-$10^2$ keV energy would emerge in the $10^2$ eV range; a monoenergetic beam was not a necessity because the pulsed nature of the LINAC source meant that positrons were being timed along a flight path several metres in length, so that the interactions of positrons with a range of energies could be studied simultaneously. The reason why their search was fruitless is now clear in hindsight: the number density positrons per eV interval in the $10^2$ eV region was too low. Tong [30] correctly interpreted their observation of a peaked distribution of positrons of eV energies to the ejection of thermalised positrons by the negative work function of the gold layer deposited on the exit surface of the mica sheet, originally intended to provide a planar electrostatic potential. In the early 1970's a number of researchers reported observations of the ejection of eV positrons from a variety of surfaces [e.g., 31-33], none of them atomically clean.

Early moderation efficiencies were $\sim 10^{-6}$, improved by an order of magnitude by the discovery of positron emission from MgO powder [34]. This low efficiency, coupled with the structure seen in early measurements of re-emitted positron energy spectra from metallic surfaces [23], is now understood in terms of trapping of thermalised positrons as they diffuse towards the exit surface and energy loss processes in the contaminant overlayer as they emerge [24 ].

In 1980 Dale *et al.* demonstrated that well-annealed tungsten had $\varepsilon \sim 10^{-3}$, after minimising the concentration of non-equilibrium defects (trapping sites) [35]. The high work function (~3 eV) of tungsten has made it the most widely-used moderator material for many years; its efficiency is not seriously reduced by exposure to air or monolayers of surface contamination (figure 5). The exact procedure for annealing is also not critical; however, heating to as high a temperature as possible (above 2000°C) in as low an ambient pressure as possible (below $10^{-6}$ Pa) achieves the best performance. Both polycrystalline and single crystal tungsten has been successfully used as moderating material [e.g., 36-41]; the geometry employed (see next section) is obviously important in determining $\varepsilon$.

A particularly useful consequence of the ruggedness of the tungsten moderator is that it can be prepared (*ie*, annealed) externally and transported through air to, and installed into, the positron beam system, without significant degradation of performance as a moderator. However, re-annealing *in situ* has been demonstrated to improve moderation efficiency by up to 30%.

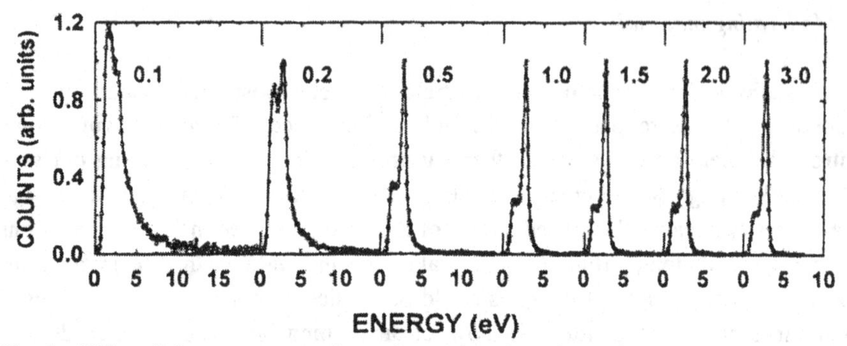


**Fig. 5** Differential energy spectra for positrons re-emitted from polycrystalline tungsten foil. The numbers next to each peak refer to the incident positron energy in keV [24].

Other relatively common metallic moderators include copper and nickel [22,42]. Both these materials have small work functions which can be made effectively positive against positron re-emission by surface contamination. Careful in-situ annealing is therefore required. In the case of copper the (100) crystal face does not re-emit positrons, whereas Cu(110) and (111) have work functions ~ $10^2$ meV. Jibaly *et al* have measured positron work functions for a number of polycrystalline metallic surfaces [43,44]. In some instances surface contamination can increase $\phi_+$ - for example, sulphur on copper [18] (sulphur is present in small quantities even in high-purity copper and migrates to the surface under heating).

Recently a semiconductor re-emitter has been discovered in the form of silicon carbide [45-47]. While having a large band gap (~3eV) which prompts the suspicion that the re-emission may be dominated by epithermal positrons, the yield decreases more slowly with incident energy than would be expected if this were the case, and is affected significantly by the presence of subsurface defects (which would not be very effective at trapping hot positrons). Preliminary results from positron microscopy of SiC also suggest that work-function ejection of thermalised positrons is the dominant re-emission process [48]. The work function for all the polytypes of SiC studied to date appears to be close to -2eV. A second wide band-gap semiconductor - GaN - has recently been shown to re-emit slow positrons with relatively high efficiency [49]. The possibility of using diamond as a positron moderator/remoderator has been studied by Mills and co-workers [50-52], who found $\phi_+ \approx$ -4 eV; because diamond is an insulator it hold promise as a field-assisted moderator (see below).

By definition moderation efficiency $\varepsilon$ is the product of the fraction of beta positrons reaching the exit surface multiplied by the surface transmission probability. The former is dependent on the depth distribution of implanted fast positrons and the effective positron diffusion length $L_{+\mathrm{eff}}$ of the material; in metals and semiconductors $L_{+\mathrm{eff}} \sim 10^2$nm. The highest moderation efficiencies yet recorded are for solid rare gases condensed directly on to the radioactive source capsule. $L_{+\mathrm{eff}}$ is long in the rare gases because of incomplete thermalisation; furthermore, epithermal (hot) positrons efficiently penetrate the surface barrier. Values of ε in the $10^{-3}$ range have been reported, with the value of $7\times10^{-3}$ reported for solid Ne on a $^{22}$Na source [53]. Ar and Kr, easier to prepare in solid form than Ne, have also been successfully used [54,55]. The growth of efficient rare gas moderators requires good starting conditions (base pressure below $10^{-7}$Pa, pure gas, low growth temperature) and careful annealing [56,57]. Such moderators also deteriorate with time so that repeated regrowth is necessary. The energy and angular distributions of epithermal positrons are both significantly broader than for single crystal metallic moderators; high beam intensity is, however, paramount in some applications, and remoderation can still produce a high quality beam of relatively high intensity. Recently the beneficial effect on moderation efficiency of water on the surface of rare gas moderators has been observed (figure 6) [58].

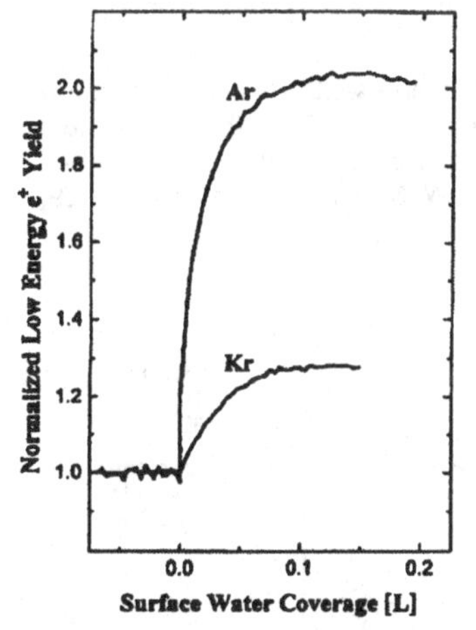


**Fig. 6** eV positron yield from solid Ar and Kr as a function of water coverage [58].

### *2.3 Moderator geometries*

A plethora of moderator geometries have been employed over the past thirty years, and some examples are sketched in Figure 7. The many different geometries may be categorised into three groups: backscattering, transmission, and quasi-transmission.

Backscattering geometries exploit the high efficiencies of well-prepared metallic single crystals. Figure 8 shows a beta source deposited on a thin needle held in front of the moderator; this presents technical difficulties associated with depositing useable source activities on very fine needles (which do not then create significant shadowing problems). The conical moderator geometry of Lynn *et al* [59], which for tungsten achieves an efficiency ε as high as $1.5\times10^{-3}$, is technically sophisticated but is difficult to anneal *in situ*.

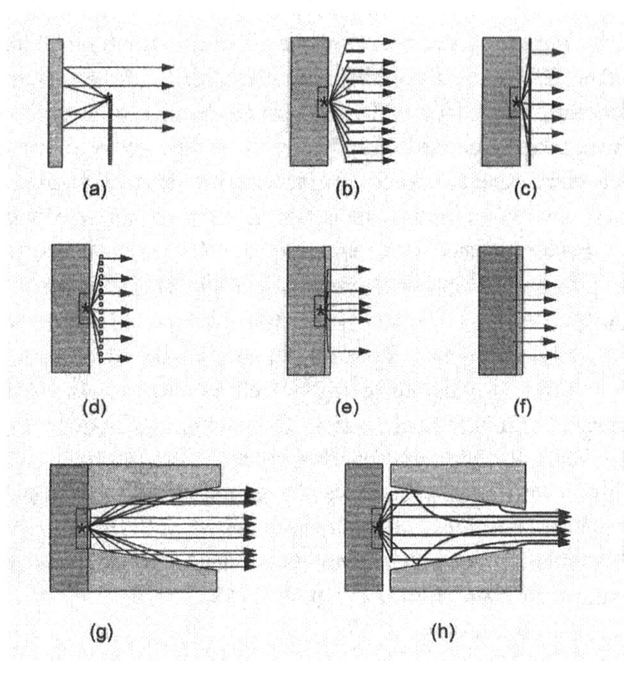


**Fig. 7** A selection of moderator geometries. The asterisk represents the primary positron source. (a) backscattering (see figure 8), (b) venetian blind (eg ref 35), (c) thin film (eg [60-67]), (d) mesh (eg [87]), (e) deposited moderator (eg solid rare gases, [54, 55]), (f) self-moderating source material on high-Z backing (eg $^{64}$Cu on W [3], (g) thin film plus cone [59], and (h) thin film plus reverse cone [60].

Much research into thin-film metallic transmission moderators has been performed and they have been employed in a number of positron beam systems [25, 38, 39, 41, 60-67]. Figure 9 shows re-emission probability as a function of incident positron energy for various Al film thicknesses [62]. Technical problems here are associated with foil quality and difficulties with handling and annealing; however, good single-crystal films provide optimum planar geometry for high-brightness beam generation.

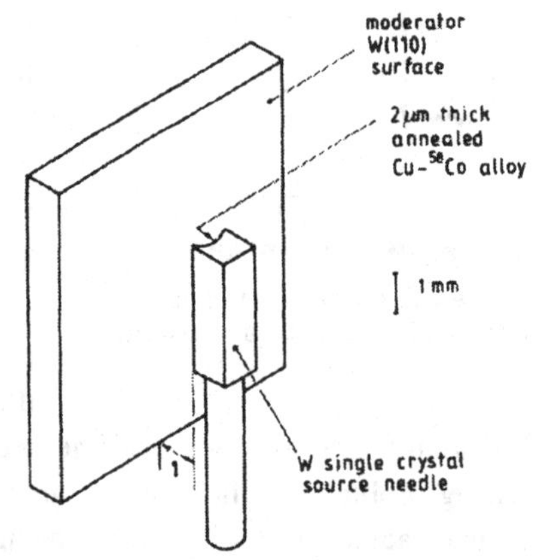


**Fig. 8** Backscattering source-moderator geometry described by Vehanen [37].

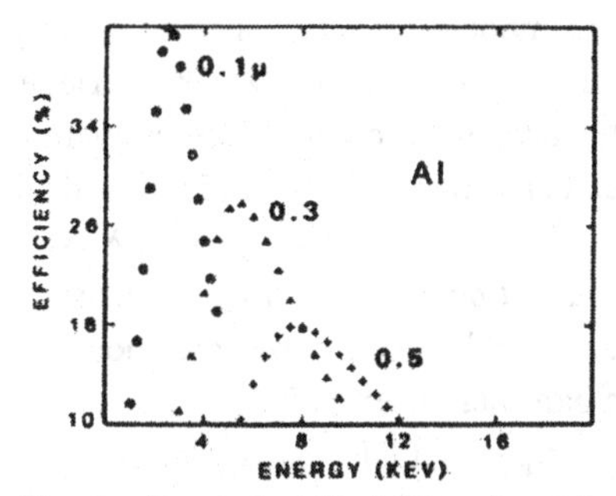


**Fig. 9** Forward yield of eV positrons for three single-crystal Al foils as a function of incident positron energy. The numbers indicate the foil thicknesses in μm.

Quasi-transmission moderators include 'venetian blinds' made of annealed strips of tungsten foil [e.g., 35] and annealed tungsten meshes. A commercial $^{22}$Na source capsule mounted 0.5mm behind a 50%-transmission woven tungsten mesh annealed *in situ* by Joule heating has $\varepsilon \approx 2 \times 10^{-4}$. Efficient moderation by mesh and vane moderators requires application of a small (~ +10V) potential between source and moderator to ensure that positrons leaving the irradiated back side of the moderator are turned away from the source into the beam line.

## 2.4 Field-assisted moderators

Many methods for increasing slow positron yields by modification of known moderator preparation procedures, configurations or geometries have been tried with varying success over the past two decades. For example, these include depositing an epilayer of copper on a (high-*Z*) tungsten backing [3] and the combination of thin foil and cone moderators [60]. Recently Waeber *et al* [68] have proposed increasing moderation efficiency by reducing the mean energy of $\beta^+$ positrons to a few keV, by confinement and repeated passage through a carbon foil, prior to moderation. (As outlined at the beginning of this section, only the first few keV of the $\beta^+$ spectrum is effective in the moderation process.)

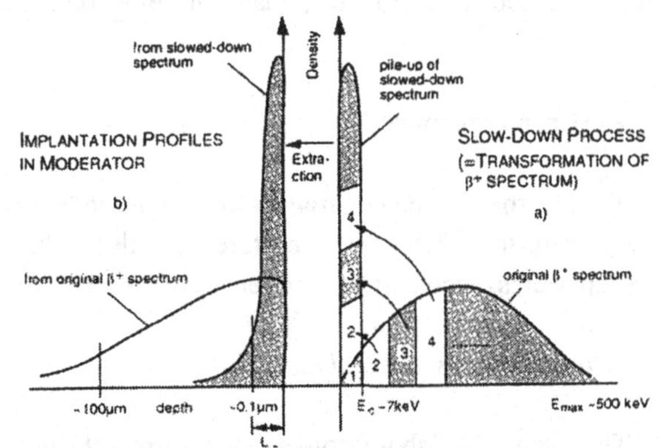


**Fig. 10** Pre-moderation of a beta positron spectrum by repeated passage through a thin C foil (right), and the resulting narrower implantation profile (left) (from Waeber et al [69]).

The most promising method for producing more intense laboratory-based beams, however, rests in field-assisted moderation [69,70]. Here moderation efficiency is greatly enhanced by drifting a larger fraction of thermalised positrons to the exit surface. The electrostatic field can be applied externally or can be created internally in a positron rectifying device, proposed by Debowska *et al* in 1985 [71] and realised by Jørgensen *et al* [72] in 1997. Jørgensen *et al* used a 110nm-thick layered (W-Mo) foil. Coleman *et al* reported observations on insulating polymer films consistent with slow positron ejection by a sub-surface field created by the positron bombardment itself [73].

**Fig. 11** Diamond field-assisted moderator [77]: a = Ti/Au contact, b = high-resistivity diamond, c = B-doped diamond layer, d = Ti/Au ring contact.

Experimental realisation of 'external' field-assisted re-emission has been achieved by charging the surface of a rare gas moderator [74-76] and by applying an electric field across 100 μm diamond [77], both demonstrating the validity of the principle (figure 11). Jacobsen *et al* [78] have recently proposed a new method for growing rare gas moderators across which an electric field could produce a large increase in slow positron emission. A practical field-assisted moderator has yet to be developed, however. One major problem has been the interaction of positrons with the interfaces between the moderator material and surface coatings used to apply a potential; this problem is overcome by growing overlayers of contiguous material such as doped diamond or pure solid rare gas which emit slow positrons. Semi-insulating SiC, GaN or GaAs [79] may, like diamond, provide a suitable candidate for development as field-assisted moderators.

## 3 Positron transport

eV positrons have been transported from moderator to their target by magnetic fields, electrostatic fields, or a mixture of both [80-83] through reflectors, velocity filters, apertures, accelerators, and lenses of various designs.

### *3.1 Magnetic transport and focusing*

The standard laboratory-based positron beam system employs an axial magnetic field for beam guidance, ensuring that positrons which leave the moderator with a small transverse momentum are prevented from straying too far

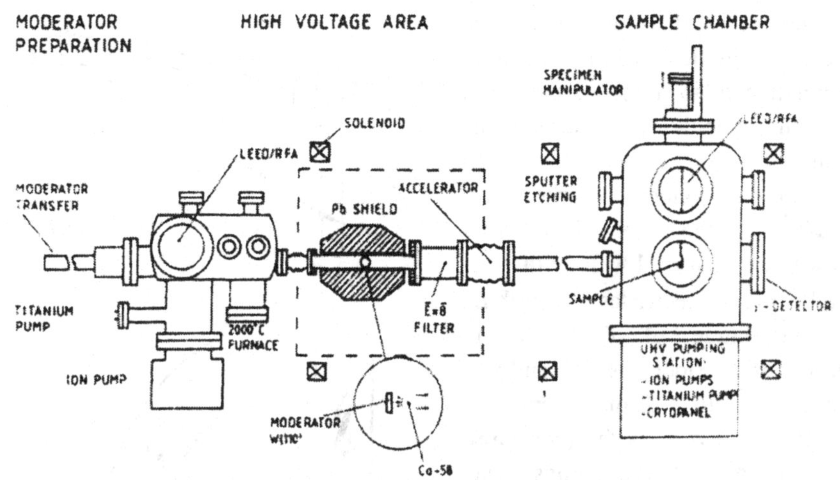


**Fig. 12** Magnetic transport positron beam system [Lahtinen *et al*, 1986, *Nucl. Instrum. Meth.* B **17**, 73]

from the original beam profile, instead undergoing helical motion with a pitch and diameter determined by the field strength and the positrons' momentum. A continuous solenoid around the positron flight tube provides the most uniform magnetic field, but in general prohibits access to the vacuum system and prevents baking of UHV systems; therefore, when solenoids are used they generally only form a section of the transport system (e.g., ref. 83]. More common is the use of quasi-Helmholtz coils which provide uniform magnetic fields to within 5-10%. Such coils are mounted around the evacuated flight tube at intervals approximately equal to their diameter, and typically provide magnetic fields of ~ 100G [84-88].

If annihilation gamma rays are to be detected at or near the target end of a positron beam system then it is essential that there is good radiation shielding between the radiation detector and the primary positron source, and that the number of fast (unmoderated) positrons which annihilate in view of the detector is minimised. To avoid having extremely long systems which rely on the inverse square law to reduce background, or the use of two detectors in coincidence which reduces count rate, most experimenters have opted for bent solenoids [89-93] or ***ExB*** filters [94-96] to remove fast positrons from the beam and allow more efficient shielding of the positron source. ***ExB*** filters, also sometimes called trochoidal monochromators, are widely used to deflect positrons (and electrons) in magnetic transport positron systems for a variety of purposes [e.g., 83,97]. The principle of this device is that an electrostatic field $\boldsymbol{E}$ applied in a direction perpendicular to that of the axial magnetic field $\boldsymbol{B}$ causes the positrons' gyration centres to drift with constant velocity $\boldsymbol{v}$, where $\boldsymbol{v} = \boldsymbol{E}\text{x}\boldsymbol{B}/B^2$. In the common geometry shown in figure 13(a) with two parallel planar electrodes of length $L$, the magnitude of the constant vertical drift velocity is $E/B$ and the vertical beam displacement is $(E/B)L/v_z$, where $v_z$ is the (constant) axial speed of the positrons. The displacement is thus inversely proportional to the positrons' momentum, and eV positrons experience a deflection orders of magnitude greater than unmoderated $\beta^+$ positrons. In this way the eV positrons can be taken through an appropriately offset pair of apertures, while unmoderated positrons are annihilated at a shielded barrier. Because typical positron beam diameters (~several mm) are not negligible compared to the plate separation in $\boldsymbol{E}$x$\boldsymbol{B}$ filters, distortion of the beam cross-sectional profile can be serious (see figure 13(b)) ; Hutchins *et al* [94] employed cylindrical plate geometry to overcome this problem.

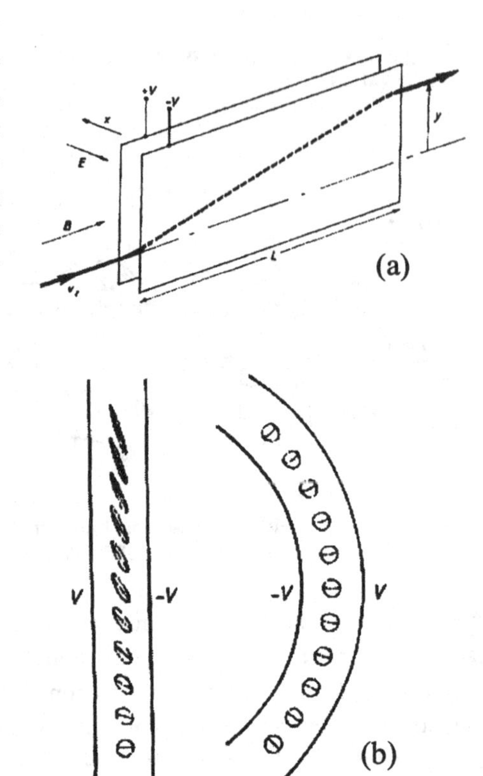


**Fig. 13** (a) Planar-geometry ***ExB*** filter; (b) progressive distortion of positron beam profile passing through planar (left) and cylindrical-geometry (right) ***ExB*** plates [94].

Axial magnetic fields have been used for many years as a simple and reliable method for transporting as many positrons as possible from moderator to target. In recent years, however, magnetic focusing has been developed for positron beams, in which solenoids and appropriately-shaped permanent magnets are employed both to transport and to focus the beam to a small spot size at the target [2, 98, 99].

## *3.2 Electrostatic transport and focusing*

Electrostatic transport and focusing of eV positron beams has been used almost since the early days of experimentation in this field [32,100], but only came to prominence with the development of apparatus for low-energy positron diffraction (see Chapter 5), precise positron-atom scattering measurements (Chapter 3) and the high-resolution measurement of the energy distributions of positrons and electrons ejected from solid surfaces under positron bombardment (Chapter 5). While electrostatic lens systems appear more complex than the magnetic transport beams they offer the means for beam manipulation, reflection, and focusing, together with the ability to measure true energy and angular distributions (*eg* figure 14). Essentially 100% transmission through a system is possible if optimum design is adopted, and high brightness can be achieved if emission from moderating and re-moderating surfaces has small angular and energy dispersion (achieved by using low work-function, cooled, clean, perfectly crystalline, flat surfaces). A number of wholly electrostatic beam systems have been used for atomic and surface research, capable of providing positron beams of energies from a few keV to up to 60 keV [e.g., 101-105]. Canter [106, 107] has discussed the special problems of slow positron optics. He introduced the positron community to the Liouville theorem which states that $\Gamma\sqrt{E}$ is conserved, where $\Gamma$ is the area of the $r$-$\theta$ diagram (figure 15) which

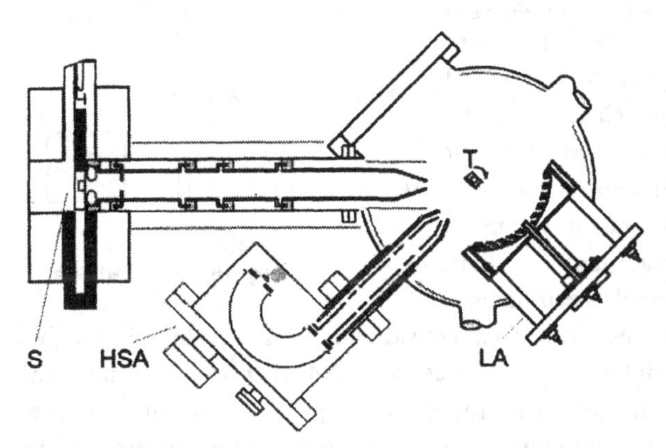


**Fig. 14** Example of an electrostatic beam system. S = source assembly, T = sample target, HSA = hemispherical energy analyser, LA = LEED/Auger head. [102].

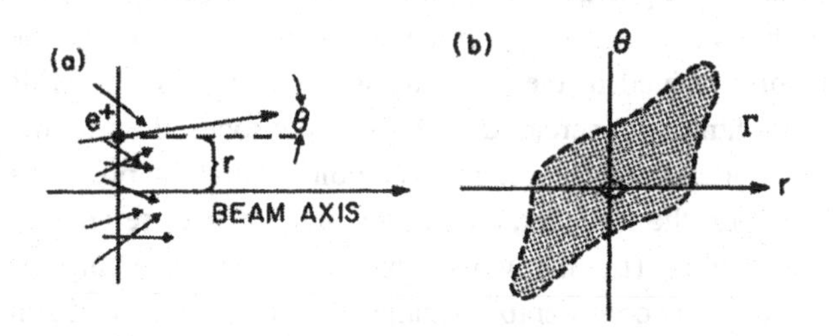


**Fig. 15** Depiction of $r$-$\theta$ phase space for a positron beam [106].

describes the variation of radial positions and directions of an ensemble of positrons, and $E$ is the positron energy. Another important quantity discussed by Canter is the *brightness per volt* of a positron beam - a quantity more important than simple beam intensity or flux when considering beam quality. For a beam flux $I$ Canter defines brightness per volt as $R_V = I(4/\pi)^2/\Gamma^2 E$. The important modification of the Soa immersion lens for positron 'guns' is shown in figure 16 [108]. The reader is referred to refs. [106-108] for guidance on electrostatic positron beam optics.

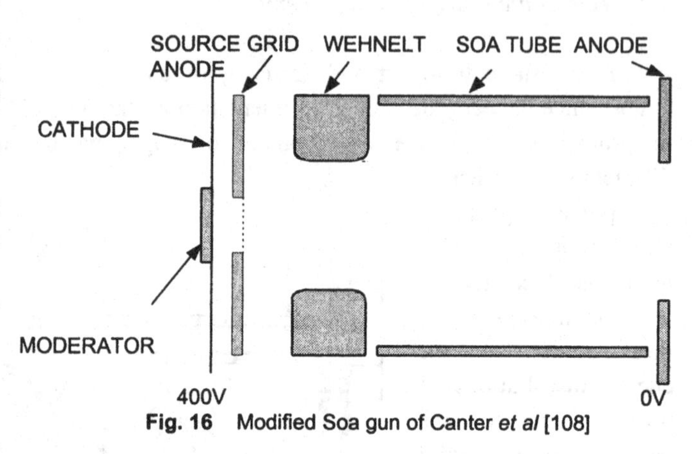

**Fig. 16** Modified Soa gun of Canter *et al* [108]

### 3.2.1 Brightness enhancement

In the previous section we referred to the concept of beam brightness and its importance in many applications of positron beams. Because of the low moderation efficiencies the primary beam (leaving the moderator) is often of relatively large diameter -typically ~ 10mm. This beam may be focused to a spot of less than 1mm diameter in one stage of magnetic or electrostatic focusing, but the *brightness* of the beam is thereby reduced. This is because the area $\Gamma$ of the beam in $r$-$\theta$ space increases as the angular dispersion of the positron paths increases. To enhance the beam brightness - i.e., to obtain more positrons of a particular energy per second per unit cross-sectional area, with a small angular divergence, one has to overcome the limitation of Liouville's theorem stated in the previous section. To do this Mills [109] suggested that beam brightness is increased each time it is focused on to and re-emitted from a secondary moderating surface. The key point is that the positrons are focused on to a small spot on the remoderator surface and are then re-emitted essentially normally to the surface (by the work function) - i.e., the angular divergence of the re-emitted beam is considerably smaller than that of the incident beam, and we have a non-conservative system which thus overcomes Liouville's limitations. Furthermore, the absolute loss of positron intensity is typically only 60-80%, whereas the beam area can be reduced by two to three orders of magnitude, so that the brightness enhancement at each remoderating stage is considerable. Clearly it would be inadvisable to have so many remoderating stages such that the absolute intensity of the final beam was impractically small, and so typical brightness enhancement systems have two or three stages of remoderation at most. A two-stage reflection-geometry brightness enhancement unit, constructed and used by Canter *et al* [107], is shown in figure 17. This geometry has also been used by others to obtain small-diameter, bright beams (e.g., Frieze *et al* [110], Ito *et al* [111], Goodyear and Coleman [112]). Lynn and Wachs suggested that an in-line brightness enhancement geometry could be used by employing thin film remoderators [113]. Zecca and Brusa [114] suggest new methods for obtaining very large reduction in beam area in one stage of demagnification, while Roach *et al* discuss the limits on forming microbeams of positrons with electrostatic optics [115].

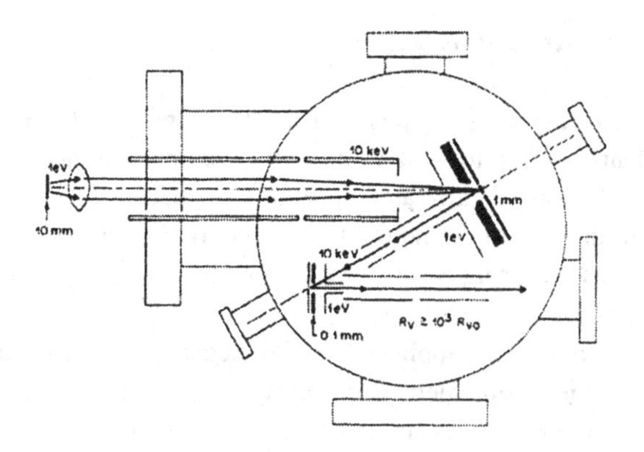


**Fig. 17** Two-stage brightness enhancement stage at Brandeis [107].

### 3.3 *Hybrid positron beams*

In his review of positron optics Canter [106] refers to the problems associated with extracting positrons from a magnetic field into electrostatic lenses. Nevertheless, it is quite common (particularly at intense beam facilities) for a beam to be transported magnetically to the experimental area, creating the need to convert to electrostatic manipulation for a range of experiments. An example of such a system is the JAERI intense beam in Japan [116]. Other hybrid systems have been designed and developed at Brookhaven [117], Hong Kong [118], Kalpakkam [119] and PSI in Switzerland [120].

## 4 Positron acceleration

In the earliest positron beams, and in those which require acceleration potentials of up to ~ $10^3$V, it has been common to apply the accelerating potential to the source/moderator assembly, often with a fine (≥ 90% transmission) mesh held ~ 1mm in front to define the field direction and the acceleration region. However, in systems requiring positrons of energies ~ $10^4$ eV, two acceleration methods have been used.

First is the application of a negative potential to the target. The principal advantage of this method is that for almost all of its flight path the positron is in an environment which is unchanged from run to run and from experiment to experiment. The most important consequence of this is that as the accelerating potential is changed, lateral beam movements associated with ***E*x*B*** forces are very small. The principal disadvantage is that any positrons re-emitted from the target surface are returned to it, distorting the effective positron implantation profile and excluding the possibility of performing any re-emitted positron experiments. A secondary disadvantage may be the lack of manipulability of the sample.

The second method is to accelerate the positrons after they have passed through the velocity filter and before they enter the 'experimental' region. This means that the whole of the source/moderator end of the system is raised to the accelerating potential and that the target environment is at ground. The target is therefore easily moved, and sophisticated multisample holders can be employed. Positron re-emission experiments can be performed and any positrons which do leave the target can be 'lost' - i.e., not returned to it. The main disadvantage of such accelerators in magnetic-transport systems is the ***E*x*B*** effect referred to above; usually the accelerating potential is not perfectly aligned to the guiding magnetic field, and there transverse forces deflect the beam. The deflections at the sample depend upon the beam energy, and unless corrective measures are taken [87] a fraction of the positrons can miss the target and/or strike non-target material, with possibly serious deleterious effects on measured annihilation radiation characteristics. A second feature of pre-accelerated beams is that if defined by a circular aperture then at the target the aperture image will go in and out of focus as the positron energy changes. This can be understood in terms of the positrons going in and out of relative phase with each other; if they are in the same relative phase at the target as they were at the defining aperture then they will appear to be in focus. This occurs each time an integer number of complete helices exist between aperture and target; this number should be inversely proportional to the axial positron velocity. This is confirmed experimentally by recording the positron energies $E$ at

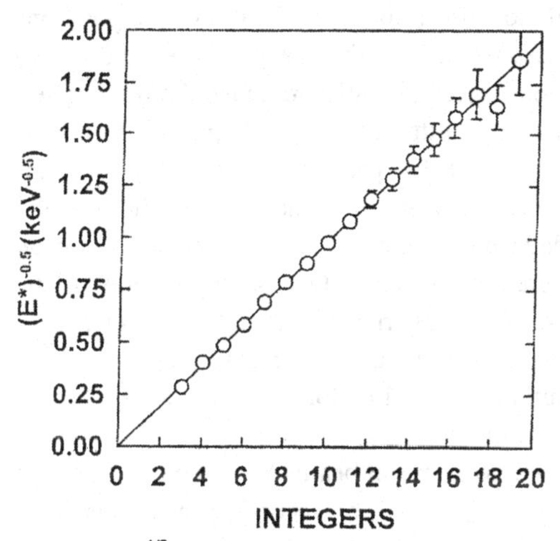


**Fig. 18** $E^{-1/2}$ vs integers, where $E$ is the positron energy at which the beam is in focus at the taget (see text for discussion) (from [87]).

which the beam appears 'in focus' at the target and plotting $E^{-1/2}$ against integers, as shown in figure 18. The combined effects of transverse shifts and defocusing were labelled 'Askolations' by Peter Schultz, after Asko Vehanen of Helsinki [121]; they can be minimised by optimising the uniformity and alignment of the accelerating electric and guiding magnetic fields.

Acceleration to energies of ~ $10^1$keV requires an acceleration unit which provides as uniform a field as possible. This is usually achieved by applying the potential between two plates separated by a number of smooth metallic rings, separated by ceramic tubing to which they are vacuum-sealed. The rings are often terminated externally around their circumferences by smooth hoops of circular cross section, which serve to minimise the possibility of electrical discharge or breakdown. Often the rings are left floating and sit at potentials midway between their two neighbours; sometimes they are resistively connected, internally or externally, to force equipotential differences between adjacent rings. Hutchins *et al* [122] developed a quasi-continuous accelerator in which a commercial ceramic break was internally coated with a thin spiral of resistive carbon.

## 5 Bunching and timing

A wide variety of experiments in all branches of slow positron science benefit from time-tagging incident positrons. One of the earliest experiments in the field - the He scattering measurements of Costello *et al* - used the LINAC pulse to provide a time signal for the birth of the fast, and thus slow, positron at the moderator end of a long flight tube [123]; although the absolute signal:noise ratio in terms of count rate was miniscule, the time-related positron signal events stood measurably above the randomly-related background 'events', which were spread over all times. The same principle was employed by the London group [e.g., 124] in their laboratory-based beam system; the 'birth' timing pulse was obtained by passing the $\beta^+$

positrons through a thin plastic scintillator disc prior to moderation. The positrons which passed through the scintillator deposited enough energy in it (~ 20-30 keV) to create enough photons to give rise to a detectable pulse in a fast photomultiplier. The same timing technique is being used in Stuttgart to tag MeV positrons on their way to the sample under study [125, and see Chapter 8]. In this case, however, every positron whose annihilation radiation is detected has passed through the scintillator and thus has itself been detected (whereas in the former case positron detection is performed *prior* to moderation so that only a small fraction of the scintillator pulses are signal). Because the Stuttgart beam is high (~MeV) energy the timing resolution is good enough to perform lifetime measurements in bulk solids and liquids. The timing resolution of the London system was a few ns, suitable for time-of-flight but not lifetime measurements.

A second method employed to tag slow positron beams has been the detection of secondary electrons from remoderator or sample surfaces. This was used by Lynn *et al* to measure the positron surface state lifetime in Al [126] (figure 19), Van House *et al* to tag polarised positrons [127] and Laricchia and co-workers for time-of-flight measurements of positron and positronium scattering [128, 129]. A recent example, in which sub-500ps timing resolution has been achieved, is the work of Gidley and co-workers on positron lifetimes in thin polymer films [130].

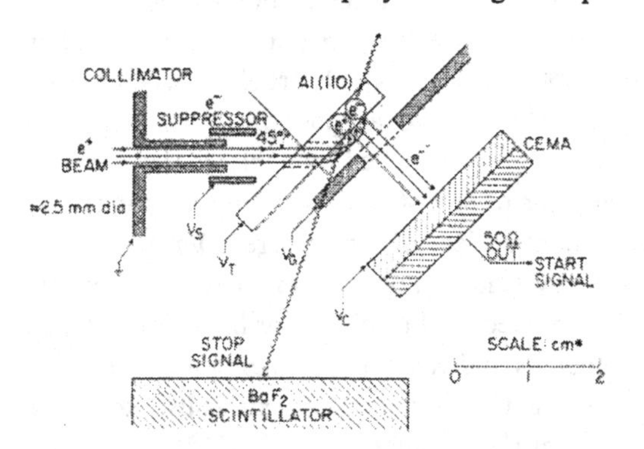


**Fig. 19** Positron tagging by secondary electron detection, used here to provide a start signal in the measurement of the surface state lifetime in Al(110) [126].

LINAC-based beams have built-in pulsing which has been used directly [131] or in conjunction with a chopper/buncher [132] for time-of-flight measurements.

In dc beams, both laboratory and facility-based, one may wish either to ensure that bunches of positrons impact upon a solid surface at essentially the same time, or that each positron is tagged or timed as it impinges on the target. Mills [133] described methods for bunching eV positrons - for example using an harmonic oscillator potential well, which is suddenly turned on (after one quarter of the oscillator period all positrons in the well reach the target at effectively the same time. Another basic method suggested by Mills was the use of a magnetic bottle

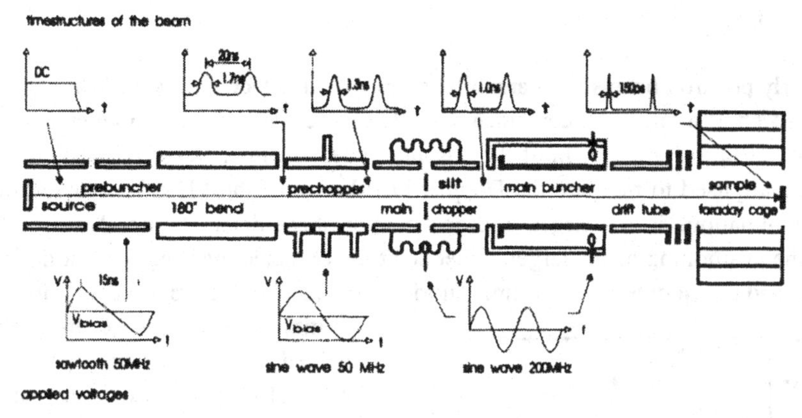


**Fig. 20** Positron pulsing system at Munich [135]. The applied rf voltages are shown below the apparatus, and the beam time structures are shown above. Pulse separation is 20ns, and the pulse FWHM decreases from 1.7 to 0.15 ns from prebuncher to main buncher.

with rf trap; the trapped positrons are released once the trap is saturated. These ideas have been developed by a number of groups world-wide; a notable example is the chopper-buncher system developed in Munich [134, 135], pictured in figure 20, which features pre-buncher and pre-chopper elements and is now capable of sub-200ps timing resolution, putting positron lifetime measurements in thin films and near-surface regions on a par with traditional bulk techniques.

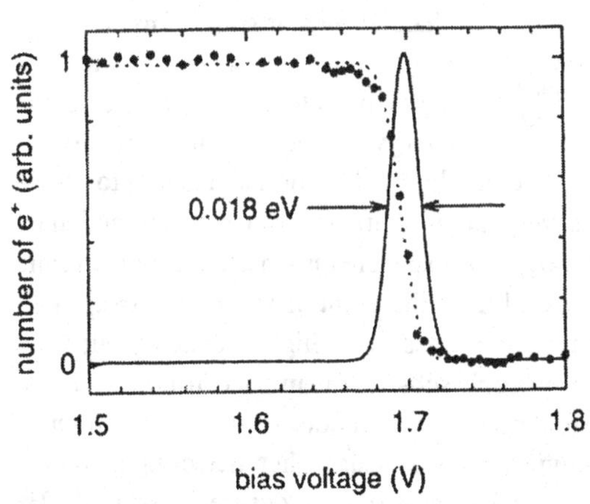


**Fig. 21** Energy distribution of a pulse of positrons extracted from a thermalised room temperature positron plasma stored in a Penning trap [136].

Surko and co-workers [136] have recently developed a pulsed source of very low-energy (~ a few eV) positrons with extremely narrow energy spread (~ 20 meV) by extracting positrons from a room temperature plasma of thermalised positrons stored in a Penning trap (figure 21). Each pulse contains c. $10^5$ positrons and is of 10 μs duration.

## 6 Detectors

Early positron beams used annihilation radiation detection to signal the arrival of a positron at a target; in coincidence mode this can lead to very high signal-to-background ratios, but can limit the absolute detection rate. In today's beams Ge detectors are used to measure the Doppler broadening of the 511keV gamma line from annihilations in the target, and on occasion these detectors are also used to check the positioning of the target. Other scintillation detectors used for fast timing include sodium iodide and barium fluoride; all such detectors used in singles counting mode have to be well-shielded from the primary positron source [137]. Practical issues such as the dependence of Ge detector energy resolution on pulse count rate have to be combated or allowed for (figure 22) [138].

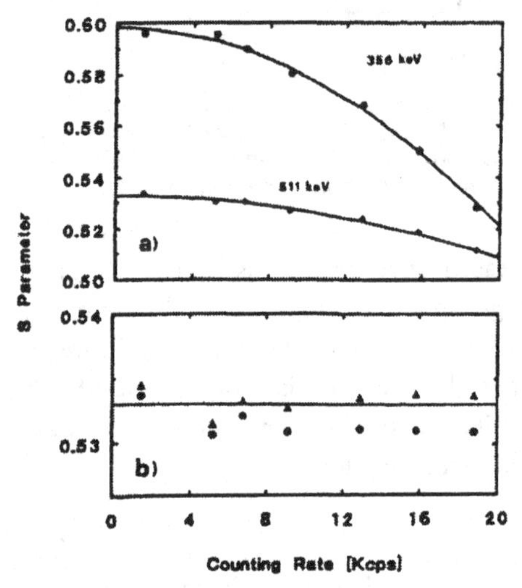


**Fig. 22** The dependence of the measured annihilation lineshape parameter on total detector count rate. Plot (b) shows corrected values [138].

Particle detectors used in positron beams have primarily been of the continuous-dynode electron multiplier type. In 1974 Pendyala *et al* demonstrated that the detection efficiency of single channel electron multipliers (CEMs, also called channeltrons or spiraltrons) for low-energy positrons was very similar to that for electrons of the same energy - being close to 100% between about 20 and 300 eV [139]. The much lower detection efficiency for MeV gamma radiation made these devices very popular in positron systems and they have been used as detectors in a wide variety of experiments - particularly in atomic physics [140]. Secondary electrons released from the point of impact are swept into the channel - a long, thin tube of resistive glass with high secondary electron coefficient - releasing more electrons on each subsequent impact in an avalanche effect which leads to a detectable current pulse at the anode. A potential difference of 2-3000V is maintained across the device, and it is usual when detecting positrons to hold the front end of the single channel at *c.* $-10^2$V to ensure close to 100% detection efficiency. However, a very high negative potential can reduce the detection efficiency by a factor much greater than that expected from the fall due to

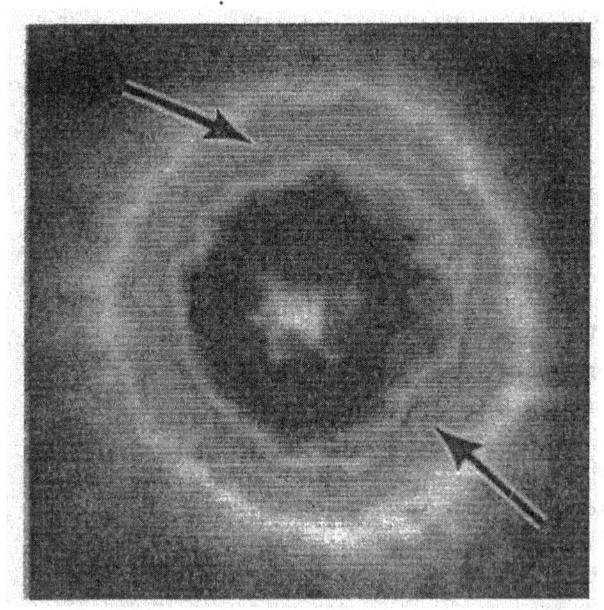

**Fig. 23** Intensity cross-sectional image of a 5keV positron microbeam at Brandeis [143]. The central white peak area contains approx. 465 counts per pixel, and the ring marked with arrows contains c. 180 counts per pixel. The entire image above is 10 μm square.

higher positron energy; the problem is that secondary electrons from the first impact can be swept out of the detector by the electric field (in contrast to the situation with electron detection).

Channel electron multiplier arrays (CEMAs) are commonly used to obtain images of positron beams. They consist of arrays of $10^1$μm-diameter tubes some 500μm long, with a maximum effective open area of ~70%. It is common to mount two CEMA plates, each with $\sim 10^4$ gain, in series, with ~1kV across each. This plate assembly can then be held in front of a phosphor screen (as in ref. 87) or a resistive anode as in positron diffractometers or microscopes (e.g., refs 112, 141).

Several spectroscopic techniques - e.g., retarding field, hemispherical, cylindrical mirror, ***E*x*B*** energy analysers [142] - have been used in positron beam lines. However, these devices are generally (but not always) used after the positron has reached its target, and thus their description belongs more appropriately in the chapters which follow. The reader is also directed to chapters 7-9 for further discussion of microbeams, MeV beams, and polarised beams of positrons.

## 7 References

1. H.M. Weng, Y.F. Hu, C.D. Beling and S. Fung, 1997, *App. Surf. Sci.* **116**, 98.

2. D.T. Britton, M. Härting, C.M. Comrie, S. Mills, F.M. Nortier and T.N. Van der Walt, 1997, *App. Surf. Sci.* **116**, 53.

3. K.G. Lynn, M. Weber, L.O. Roellig, A.P. Mills, Jr. and A.R. Moodenbaugh 1987 *Atomic Physics with Positrons* eds. J.W. Humberston and E.A.G. Armour (New York: Plenum) p. 161.
4. K.F. Canter, 1995, paper at the Workshop on Intense Positron Beams in Europe, University College London.
5. M. Hirose, M. Washio and K. Takahashi, 1995, *App. Surf. Sci.* **85**, 111 and M. Hirose, T. Nakajyo and M. Washio, 1997, *App. Surf. Sci.* **116**, 63.
6. D.G. Costello, D.E. Groce, D.F. Herring and J.Wm. McGowan, 1972, *Phys. Rev.* B**5**, 1433.
7. Richard Ley, 1997, Hyperfine Int. **109**, 167.
8. R.H. Howell, I.J. Rosenberg and M.J. Fluss, 1987, *Appl. Phys.* A**43**, 247.
9. Jun Xu, R. Suzuki, L.D. Hulett, Jr. and T.A. Lewis, 1997, *App. Surf. Sci.* **116**, 34
10. R. Suzuki, T. Ohdaira, T. Mikado, H. Ohgaki, M. Chiwaki and T. Yamazaki, 1997, *App. Surf. Sci.* **116**, 187.
11. G. Graff, R. Ley, A. Osipowitz and G. Werth, 1984, *Appl. Phys.* A**33**, 59.
12. Y. Ito, M. Hirose, S. Takamura, O. Sueoka, I. Kanazawa, K. Mashiko, A. Ichimiya, Y. Murata, S. Okada, M. Hasegawa and T. Hyodo, 1991, *Nucl. Instrum. Methods* A**305**, 269.
13. F. Ebel, W. Faust, C. Hahn, S. Langer, M. Ruckert, H. Schneider, A. Singe and I. Tobehn, 1988, *Nucl. Instrum. Methods* B**272**, 626, and D. Segers, J. Paridaens, M. Dorikens and I. Dorikens-Vanpraet, 1991, *Nucl. Instrum. Methods* B **56-7**, 572.
14. G. Triftshäuser, G. Kögel, W. Triftshäuser, M. Springer, T. Hagner and K. Schreckenbach, 1995, *Mat. Sci. Forum* **175-178**, 221.
15. E.D. Shaw, A.P. Mills, R.J. Chichester, M. Leventhal, D. Zuckerman, E.E. Chaban, T. Kovacs and R.G. Tobin, 1991, *Nucl. Instrum. Methods* B **56-7**,568.
16. Microtron at the University of Aarhus referred to in ref [7].
17. E.M. Gullikson, A.P. Mills, Jr. and C.A. Murray, 1988, *Phys. Rev.* B **38,** 1705.
18. C.A. Murray, A.P. Mills, Jr. and J.E. Rowe, 1980, *Surf. Sci.* **100**, 647.
19. C.A. Murray and A.P. Mills, Jr., 1980, *Sol. State. Commun.* **34**, 789.
20. D.A. Fischer, K.G. Lynn and D.W. Gidley, 1986, *Phys. Rev.* B**33**, 4479.
21. R. J. Wilson, 1983, *Phys. Rev.* B**27**, 6974.
22. E.M. Gullikson, A.P. Mills, Jr., W.S. Crane and B.L. Brown, 1985, *Phys. Rev.* B**32**, 5484.
23. S. Pendyala, D. Bartell, F.E. Girouard and J.W. McGowan, 1974, *Phys. Rev. Lett.* **33**, 1031.
24. A Goodyear, A P Knights and P G Coleman, 1994, *J. Phys.: Condens. Matter* **6**, 9601.
25. D.M. Chen, K.G. Lynn, R. Pareja and B. Nielsen, 1985, *Phys. Rev.* B**31**, 4123.

26. B. Nielsen, K.G. Lynn and Y.-C. Chen, 1986, *Phys. Rev. Lett.* **57**, 1789.
27. A. Goodyear, A.P. Knights and P.G. Coleman, 1996, *Physics Letters* A**212**, 221-226.
28. A.P. Mills, Jr. and W.S. Crane, 1984, *Phys. Rev. Lett.* **53**, 2165.
29. A.P. Knights and P.G. Coleman, 1996 *Surf. Sci.* **367**, 238.
30. B.Y. Tong, 1972, *Phys. Rev.* B**5**, 1436.
31. P.G. Coleman, T.C. Griffith and G.R. Heyland, 1973, *Proc. Roy. Soc.* A**331**, 561.
32. S. Pendyala, P.W. Zitzewitz, J.W. McGowan and P.H.R. Orth, 1973, *Phys. Lett.* A**43**, 298.
33. B. Jaduszliwer, W.C. Keever and D.A.L. Paul, 1972, *Can.J.Phys.* **50**, 1414.
34. K.F. Canter, P.G. Coleman, T.C. Griffith and G.R. Heyland, 1972, *J.Phys.B* **5**, L1617.
35. J.M. Dale, L.D. Hulett and S. Pendyala, 1980, *Surf. Interface Anal.* **2**, 199.
36. A. Vehanen, K.G. Lynn, P.J. Schultz and M. Eldrup, 1983, *Appl. Phys.* A**32**, 163.
37. A. Vehanen, 1987, *Appl. Phys.* A**43**, 269.
38. K.G. Lynn, B. Nielsen and J.H. Quateman, 1985, *Appl. Phys. Lett.* **47**, 239.
39. P.J. Schultz, 1988, *Nucl. Instrum. Methods* B**30**, 94.
40. R.H. Howell, R.A. Alvarez and M. Stanek, 1982, *Appl. Phys. Lett.* **40**, 751.
41. E. Gramsch, J. Throwe and K.G. Lynn, 1987, *Appl. Phys. Lett.* **51**, 1862.
42. A.P. Mills, Jr., 1979, *Appl. Phys. Lett.* **35**, 427.
43. M. Jibaly, A. Weiss, A.R. Koymen, D. Mehl, L. Stiborek and C. Lei, 1991, *Phys. Rev.* B**44**, 12166.
44. M. Jibaly, E.C. Kellogg, A. Weiss, A.R. Koymen, D. Mehl and L. Stiborek, 1992, *Mat. Sci. Forum* **105-110**, 1399.
45. J. Störmer, A. Goodyear, W. Anwand, G. Brauer, P. G. Coleman and W. Triftshäuser, 1996, *J. Phys.: Condens. Matter* **8**, L89-94.
46. L.V. Jørgensen, A. van Veen and H. Schut, 1996, *Nucl. Instrum. Methods* B**119**, 487.
47. A.H. Weiss, E. Jung, J.H. Kim, A. Nangia, R. Venkataraman, S. Starnes and G. Brauer, 1997, *Appl. Surf. Sci.* **116**, 311.
48. C.P. Burrows and P.G. Coleman, 1998, *Proc. Intl. Symposium on the Electronic and Defect Structure of Materials,* ed. M.A. Alam and P.G. Coleman, to be published.
49. R. Suzuki, T. Ohdaira, A. Uedono, S. Ishibashi, A. Matsuda, S. Yoshida, Y. Ishida, S. Niki, P.J. Fons, T. Mikado, T. Yamazaki, S. Tanigawa and Y.K. Cho, 1997, *Mat. Sci. Forum* **255-257**, 714.

50. G.R. Brandes, A.P. Mills, Jr. and D.M. Zuckerman, 1992, *Mat. Sci. Forum* **105-110,** 1363.
51. Y.S. Li, S. Berko and A.P. Mills, Jr., 1992, *Mat. Sci. Forum* **105-110,** 739.
52. A.P. Mills, Jr., G.R. Brandes, D.M. Zuckerman, W. Liu and S. Berko, 1992, *Mat. Sci. Forum* **105-110,** 763.
53. A.P. Mills, Jr. and E.M. Gullikson, 1986, *Appl. Phys. Lett.* **49**, 1121.
54. D. Vasumathi, G. Amarendra, K.F. Canter and A.P. Mills, Jr., 1995, *Appl. Surf. Sci.* **85**, 154.
55. J. Jääskeläinen, T. Laine, K. Fallström, K. Saarinen and P. Hautojärvi, 1997, *Appl. Surf. Sci.* **116**, 73.
56. A.P. Mills, S.S. Voris and T.S. Andrew, 1994, *J. Appl. Phys.* **76,** 2556.
57. M.P. Petkov, K.G. Lynn, L.O. Roellig and T.D. Troev, 1997, *Appl. Surf. Sci.* **116,** 13.
58. M.P. Petkov, K.G. Lynn and L.O. Roellig, 1996, *J. Phys.: Condens. Matter* **8,** l611.
59. K.G. Lynn, E. Gramsch, S.G. Usmar and P. Sferlazzo, 1989, *Appl. Phys. Lett.* 55, 87.
60. X.Y. Wu, P. Dull and K.G. Lynn, 1990, *Appl. Phys. Lett.* **57**, 998.
61. F.M. Jacobsen, K.G. Lynn. 1996, *Phys. Rev. Lett.* **76**, 4262.
62. P. Sferlazzo, 1985, *Appl. Phys.* A**36**, 93.
63. A. Vehanen and J. Mäkinen, 1985, *Appl. Phys.* A**36**, 97.
64. P.J. Schultz, E.M. Gullikson and A.P. Mills, Jr., 1986, *Phys. Rev.* B**34**, 442.
65. N. Zafar, J. Chevallier, G. Laricchia and M. Charlton, 1989, *J.Phys.D:Appl.Phys.* **22,** 868.
66. B. Jin, O. Sueoka and A. Hamada, 1994, *Jpn.J. Appl. Phys.* **33**, L1493.
67. G. Amarendra, K.F. Canter and D.C. Schoepf, 1996, *J.Appl. Phys.* **80**, 4660.
68. W.B. Waeber, M. Shi and D. Gerola, 1995, *Mat. Sci. Forum* **175-178**, 115.
69. K.G. Lynn and B.T.A. McKee, 1979, *Appl. Phys. Lett.* **19**, 10.
70. C.D. Beling, R.I. Simpson, M. Charlton, F.M. Jacobsen, T.C. Griffith, P. Moriarty and S. Fung, 1987, *Appl. Physics* A, **42**, 111.
71. M. Debowska, R. Ewertowski and W. Swiatowski, 1985, *Positron Annihilation,* eds. P.C. Jain, R.M. Singru and K.P. Gopinathan (Singapore: World Scientific) p. 999.
72. L.V. Jørgensen, A. van Veen, H. Schut and J. Chevallier, 1997, *J. Appl. Phys.* **81,** 2725.
73. P.G. Coleman, S. Kuna and R. Grynszpan, 1997, *Mat. Sci. Forum* **255-257**, 668.
74. J.P. Merrison, M. Charlton, M. Deutch and L.V. Jørgensen, 1992, *J. Physics: Condens. Matter* **4,** L207.

75. L.V. Jørgensen, J.P. Merrison, B.I. Deutch, M. Charlton and G.O Jones, 1995, *Phys. Rev.* B **52** 12402.
76. J.P. Merrison, M. Charlton, M. Deutch and L.V. Jørgensen, 1993, *Hyperfine Int.* **76**, 305.
77. G.R. Brandes, K.F. Canter, A. Krupyshev, R. Xie and A.P. Mills, 1997, *Mat. Sci.Forum* **255**, 653.
78. F.M. Jacobsen, M. Petkov and K.G. Lynn, 1998, *Phys.Rev.* B **57**, 6998.
79. Y.Y. Shan, H.L. Au, C.C. Ling, T.C. Lee, B.K. Panda, S.Fung, C.D. Beling, Y.Y. Wang and H.M. Weng, 1994, *Appl. Phys.* A **59**, 259.
80. K.F. Canter and A.P. Mills, Jr., 1982, *Can. J. Phys.* **60**, 551.
81. C.D. Beling and M. Charlton, 1987, *Contemp. Phys.* **28**, 241.
82. A.P. Mills, Jr., 1983, *Positron Solid State Physics,* eds. A. Dupasquier and W. Brandt (Amsterdam: North Holland) p. 432.
83. C. Lei, D. Mehl, A.R. Koymen, F. Gotwald, M. Jibaly and A. Weiss, 1989, *Rev. Sci. Instrum.* **60**, 3656.
84. N. Zafar, G. Laricchia, M. Charlton and T.C. Griffith, 1992, *Hyperfine Int.* **73**, 213.
85. S.M. Hutchins, P.G. Coleman, M.A. Alam and R.N. West, 1985, *Positron Annihilation,* eds. P.C. Jain, R.M. Suingru and K.P. Gopinathan (Singapore: World Scientific) p.983.
86. A. Vehanen, J. Lahtinen, H. Huomo, J. Mäkinen, K. Rytsölä and P. Hautojärvi, 1985, ibid, p. 989.
87. N.B. Chilton and P.G. Coleman, 1995, *Meas. Sci. Technol.* **6**, 53.
88. P.J. Schultz, 1988, *Nucl. Instrum. Methods* B**30**, 94.
89. K.G. Lynn and H. Lutz, 1980, *Rev. Sci. Instrum.* **51**, 977.
90. P.C. Rice-Evans, D.T. Britton and B.P. Cowan, 1987, *Appl. Phys.* A**43**, 283.
91. T. Kumita, M. Chiba, R. Hamatsu, M. Hirose, T. Hirose, H. Iijima, M. Irako, M. Washio and J. Wang, 1997, *Mat. Sci. Forum* **255-257**, 656.
92. K.F. Canter, A.P. Mills, Jr. and S. Berko, 1975, *Phys. Rev. Lett.* **34**, 177.
93. D.T. Britton, P.C. Rice-Evans and J.H. Evans, 1985, *Nucl. Instrum. Methods* B**12**, 426.
94. S.M. Hutchins, P.G. Coleman, R.J. Stone and R.W. West, 1986, *J. Phys. E: Sci. Instrum.* **19**, 282.
95. A. Vehanen, K.G. Lynn, P.J. Schultz and M. Eldrup, 1983, *App. Phys.*A **32**, 163.
96. M. Charlton, T.C. Griffith, G.R. Heyland, K.S. Lines and G.L. Wright, 1980, *J.Phys. B* **13**, L757.

97. P.G. Coleman, L. Albrecht, A.B. Walker and K.O. Jensen, 1992, *J. Phys. Condens. Matter* **4**, 10311.
98. D.T. Britton, K. Uhlmann and G. Kögel, 1995, *Appl. Surf. Sci.* **85**, 158.
99. Y. Shirai, H. Miura and M. Yamaguchi, 1995, ibid, p. 138.
100. P.W. Zitzewitz, J.C. Van House, A. Rich and D.W. Gidley, 1979, *Phys. Rev. Lett.* **43**, 1281.
101. G.R. Massoumi, N. Hozhabri, W.N. Lennard, P.J. Schultz, S.F. Baert, H.H. Jorch and A.H. Weiss, 1991, *Rev. Sci. Instrum.* **62**, 1460.
102. A. Goodyear and P.G. Coleman, 1995, *Meas. Sci. Technol.* **6**, 415.
103. A. Zecca, M. Bettonte, J. Paridaens, G.P. Karwasz and R.S. Brusa, 1998, *Meas. Sci. Technol.* **9**, 409.
104. J. Moxom, G. Larrichia and M. Charlton, 1995, *Appl. Surf. Sci.* **85**, 118.
104a. S. Yang, H.Q. Zhou, E. Jung, A.H. Weiss and P.H. Citrin, 1997, *Rev. Sci. Instrum.* **68**, 3893.
105. T. Yamamoto, Y. Honda and T. Okada, 1995, *Appl. Surf. Sci.* **85**, 132
106. K.F. Canter, 1995, *Positron Spectroscopy of Solids,* eds. A.P. Mills, Jr. and A. Dupasquier (Amsterdam: IOS) p. 361.
107. K.F. Canter, 1986, *Positron Studies of Solids, Surfaces and Atoms,* eds. A.P. Mills, Jr., W.S. Crane and K.F. Canter (Singapore: World Scientific), p. 102.
108. K.F. Canter, P.H. Lippel, W.S. Crane and A.P. Mills, Jr., 1986, ibid, p. 199.
109. A.P. Mills, Jr., 1980, *Appl. Phys.* **23**, 189.
110. W.E. Frieze, D.W. Gidley and K.G. Lynn, 1985, *Phys. Rev.* B**31**, 5628.
111.Y. Ito, M. Hirose, S. Takamura, O. Sueoka, I. Kanazawa, K. Mashiko, A. Ichimiya, Y. Murata, S. Okada, M. Hasegawa and T. Hyodo, 1991, *Nucl. Instrum. Methods* B**305**, 269.
112. A. Goodyear and P.G. Coleman, 1995, *Appl. Surf. Sci.* **85,** 98.
113. K.G. Lynn and A. Wachs, 1982, *Appl. Phys.* A**29**, 93.
114. A. Zecca nd R.S. Brusa, 1992, *Nucl. Instrum. Methods* B**313**, 337.
115. T. Roach, A. Bakshi and K.F. Canter, 1995, *Meas. Sci. Technol.* **6**, 496.
116. I. Kanazawa, Y. Ito, M. Hirose, H. Abe, O. Sueoka, S. Takamura, A. Ichimiya, Y. Murata, F. Komori, K. Fukutani, S. Okada and T. Hattori, 1995, Appl. Surf. Sci. **85**, 124.
117. F.J. Mulligan and M.S. Lubell, 1993, Meas. Sci. Technol **4**,197.
118. W. LiMing, S.B. Melwani, C.D. Beling and S. Fung, 1995, *Appl. Phys.* A**61**, 325.
119. G. Amarendra, B. Viswanthan, G.V. Rao, K.V.T. Kutty, B. Purniah and K.P. Gopinathan, 1994, *Slow Positron Techniques for Solids and Surfaces*, eds. E. Ottewitte and A.H. Weiss (New York: AIP Conf. Series 303), p. 452.

120. W. M. Waeber and M. Shi, 1997, *Appl. Surf. Sci.* **116**, 91.
121. P.J. Schultz, *Proceedings of the Workshop on Positron Beams,* Helsinki University of Technology, 1984.
122. S.M. Hutchins, P.G. Coleman and R.N. West, 1986, *Positron (Electron)-Gas Scattering*, eds. W.E. Kauppila, T.S. Stein and J.M. Wadehra (Singapore: World Scientific) p. 342.
123. D.G. Costello, D.E. Groce, D.F. Herring and J.W. McGowan, 1972, *Can. J. Phys.* **50**, 23.
124. P.G. Coleman, T.C. Griffith and G.R. Heyland, 1973, *Proc. Roy. Soc.* A**331**, 561.
125. P. Wesolowski, K. Maier, J. Major, A. Seeger, H. Stoll, T. Grund and M. Koch, 1992, *Nucl. Instrum. Methods* B**68**, 468.
126. K.G. Lynn, W.E. Frieze and P.J. Schultz, 1984, *Phys. Rev. Lett.* **52**, 1137.
127. J. Van Houase, A. Rich and P.W. Zitzewitz, 1984, *Origins of Life* **14**, 413.
128. G. Laricchia, S.A. Davis, M. Charlton and T.C. Griffith, 1988, *J.Phys.E:Sci.Instrum.* **21**, 886.
129. J.P. Merrison, M. Charlton and G. Laricchia, 1991, *Meas. Sci. Technol.* **2**, 176.
130. G.B. DeMaggio, W.E. Frieze, D.W. Gidley, M. Zhu, H.A. Hristov and A.F. Yee, 1997, *Phys.Rev.Lett.***78**, 1524.
131. R.H. Howell, M. Tuomisaari and Y.C. Jean, 1990, *Phys. Rev.* B**42**, 6921.
132. T. Ohdaira, R. Suzuki, T. Mikado, H. Ohgaki, M. Chiwaki and T. Yamazaki, 1997, *Appl. Surf. Sci.* **116**, 177.
133. A.P. Mills, Jr., 1980, *Appl. Phys.* **22**, 273.
134. D. Schödlbauer, P. Sperr, G. Kögel and W. Triftshäuser, 1988, *Nucl. Instrum. Methods* B**34**, 258.
135. P. Willutski, J Störmer, G. Kögel, P. Sperr, D.T. Britton, R. Steindl and W. Triftshäuser, 1994, *Meas. Sci. Technol.* **5**, 548.
136. S.J. Gilbert, C. Kurz, R.G. Greaves and C.M. Surko, 1997, *Appl. Phys. Lett.* **70**, 1944.
137. For a broad review of radiation and particle detectors see P.G. Coleman, 1998, *Procs. 1997 Positron Summer School (PSS-97),* ed. D.M. Schrader (in press).
138. A. Zecca, R.S. Bruss and M. Duarte Naia, 1994, *Meas. Sci. Technol.* **5**, 61.
139. S. Penyala, J. Wm. McGowan, P.H.R. Orth and P.W. Zitzewitz, 1974, *Rev. Sci. Instrum.* **45**, 1347.
140. T.S. Stein, R.D. Gomez, Y.F. Hsieh, W.E. Kauppila, C.K.Kwan, Y.I. Wan, 1985, *Phys. Rev.Lett.* **55**,488.

141. K.F. Canter, 1995, *Positron Spectroscopy of Solids,* eds. A.P. Mills, Jr. and A. Dupasquier (Amsterdam: IOS) p. 385.
142. A.P. Mills, Jr., 1995, *Positron Spectroscopy of Solids,* eds. A.P. Mills, Jr. and A. Dupasquier (Amsterdam: IOS) p. 211.
143. K.F. Canter, V. Dharmavaram, A.G. Smirnov, S.A. Wesley, K.A. Wong, R. Xie, G.R. Brandes and A.P. Mills, Jr., 1994, *Slow Positron Techniques for Solids and Surfaces*, eds. E. Ottewitte and A.H. Weiss (New York: AIP Conf. Series 303), p.385.

CHAPTER 3

# ATOMIC AND MOLECULAR PHYSICS WITH POSITRONS AND POSITRONIUM

G. LARICCHIA
*Department of Physics and Astronomy*
*University College London, Gower Street, London WC1E 6BT, UK*
*E-mail:g.laricchia@ucl.ac.uk*

M. CHARLTON
*Department of Physics*
*University of Wales Swansea, Singleton Park, Swansea SA2 8PP, UK*
*E-mail: m.charlton@swansea.ac.uk*

**In recent years there have been considerable advances in studies of atomic and molecular physics with positrons and positronium. This chapter is devoted to a discussion of the fundamental properties of these particles, their interactions with atoms and molecules and possibilities for the formation of low-energy antihydrogen.**

## 1 Introduction

In 1951, during a discussion at the first post-war conference in West Berlin, Prof. Carl Ramsauer remarked how interesting it would be if it were possible to repeat the experiments on the scattering of slow electrons ($e^-$) by atoms and molecules with positrons ($e^+$) (Massey 1982), the similarities and differences in the interactions of these two projectiles with matter making this kind of comparison valuable. For example, whilst the polarization field is essentially the same for both particles, the $e^+$ is repelled (rather than attracted) by the mean static atomic field and, although typical positron densities are such as to make exchange non-existent, a non-local interaction does exist through the possibility of positronium (Ps) formation whether virtual or real (Mott and Massey 1965).

Approximately 20 years after Ramsauer's comment, the first measurements of the total cross-section of monoenergetic positrons scattering from helium were reported (Costello et al 1972). Over the following 25 years, a steady development in the techniques for the production of slow $e^+$ beams has enabled the compilation of an extensive data library on $e^+$ scattering from a variety of atomic and molecular targets with the measurement of total and partial cross-sections, the latter both in integral and, more recently, differential form. This steady progress has been periodically reviewed (e.g. Griffith and Heyland 1978, Charlton 1985, Charlton and Laricchia 1990, 1991; Kauppila and Stein 1990, Laricchia 1995a,b; 1996; Zecca et al 1996).

Today the study of positron collisions remains an exciting branch of atomic scattering physics, combining fundamentally unique processes and interactions (e.g. annihilation, lack of exchange, Ps formation) with the opportunity of investigating more general collision effects related to the charge and/or mass of the projectile rather than its antimatter nature. Noteworthy examples in this respect are recent comparative studies of atomic ionization using particle/antiparticle pairs (i.e. electron/positron and proton/antiproton) which have been valuable for an increased understanding of the dynamics of ionization (e.g. Schultz et al 1991, Knudsen and Reading 1992, Paludan et al 1997a).

In positronic physics, a novel branch of investigation has emerged over the last decade with the realization of atomic beams of positronium (e.g. Laricchia 1995c). From a general atomic physics point of view, the study of atomic collisions with positronium offers the prospect of almost isolating effects arising from exchange forces, since the static and first-order polarisation interactions are absent (due to the coincidence of its centres of charge and mass). Such studies have relevance also for QED tests based upon precision measurements of Ps lifetimes (e.g. Al-Ramadhan and Gidley 1994), astrophysical studies (e.g. Drachman 1990) and the production of atomic antimatter (e.g. Charlton et al 1994).

In this chapter, some of these topics will be highlighted, beginning with a discussion of the fundamental properties of positrons and positronium, moving on to their interactions with atoms and molecules and then onto current efforts towards experimentation with antihydrogen.

## 2 Fundamentals

### *2.1 Positronium*

#### *2.1.1 Basic properties*

Positronium is structurally hydrogen-like, although the reduced mass factor of one half in a Bohr analysis decreases the gross features of the energy level separation such that its binding energy ($B_{Ps}$) in a state of principal quantum number $n$ is approximately 6.8eV/$n^2$. The fine and hyperfine separations are markedly different from the corresponding hydrogen values due to the large magnetic moment of the positron

(658 times that of the proton) and the presence of QED effects such as virtual annihilation (see e.g. Berko and Pendleton 1980 and Rich 1981 for summaries).

Positronium can exist in the two spin states, $s$=0, 1. The $s$=0 case with the positron and electron spins anti-parallel is a singlet and is usually referred to as para-positronium (p-Ps). The $s = 1$ triplet is called ortho-positronium (o-Ps).

### 2.1.2 Annihilation modes and lifetimes

The annihilation modes of positronium are governed by the selection rule derived from charge conjugation invariance by Yang (1949) and Wolfenstein and Ravenhall (1952). This states that the number, $n_p$, of photons liberated depends upon the values of the spin and the orbital angular momentum, $\ell$, according to

$$(-1)^{n_p} = (-1)^{\ell+s} \tag{1}$$

Thus free p-Ps decay must occur by the emission of an even number of photons and that of o-Ps via the emission of an odd number. In the p-Ps case, the lowest order process dominates with two gamma-rays being emitted in the p-Ps rest frame back-to-back, each with an energy of 511keV. The most common decay mode for o-Ps is via the emission of three (coplanar) quanta possessing a characteristic distribution of energies (Ore and Powell 1949, Chang et al 1985) whilst the probability for one-quantum annihilation is suppressed by the requirement of a third body to absorb the excess momentum. High-resolution studies of the latter process have been achieved by implanting monoenergetic positrons in condensed matter (Palathingal et al 1991, 1995).

The lowest order contributions to the decay rates were first calculated by Pirenne (1946) ($^1$S) and Ore and Powell (1949) ($^3$S) and are given by

$$\Gamma_0(n^1S) = \frac{mc^2\alpha^5}{2\hbar n^3} \tag{2}$$

and

$$\Gamma_0(n^3S) = \frac{2(\pi^2-9)mc^2\alpha^6}{9\pi\hbar n^3} \tag{3}$$

Inspection of these two expressions reveals that, due both to the extra factor of $\alpha$, the fine structure constant, and the numerical factor in Eq. (3), $\Gamma_0(n^1\text{S}) \gg \Gamma_0(n^3\text{S})$ and inserting values for $n$=1 they are found to be approximately 8GHz and 7MHz respectively, corresponding to lifetimes of 125ps and 142ns.

The lifetimes against annihilation of the $n$=2 states follow the $n^3$ scaling law implicit in Eqs. (2) and (3) for the S states, however when $\ell$ >0 the overlap of the positron and electron wavefunctions is much lower and Alekseev (1958, 1959) finds lifetimes for 2P states to be ~$10^{-4}$s. The actual lifetimes of the excited states may thus be governed by optical de-excitation, as is the case for the 2P-1S transition, which has a characteristic value of 3.2ns, i.e. double that for the corresponding transition in hydrogen.

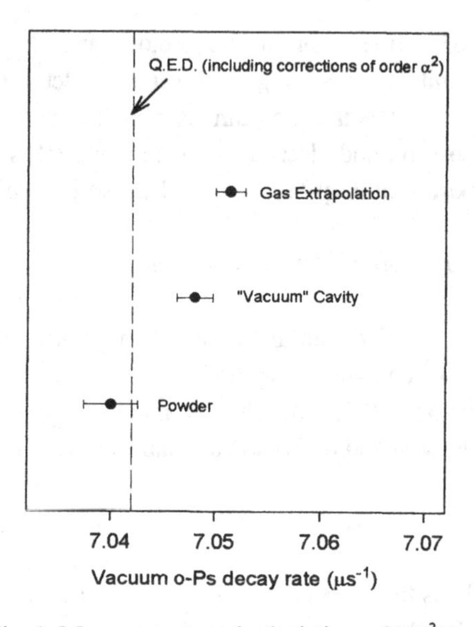


Fig. 1: Measurements and calculations of $\Gamma(1^3S)$.

The decay rate of ground state o-Ps in vacuum has been measured more often than any other property of positronium, since the atom can be produced in abundance and the accurate determination of its relatively long lifetime is well within technological capabilities. By contrast, the 125ps p-Ps ground state lifetime has only been measured twice with a precision better than 1%, by utilising the mixing of the $m$=0 o-Ps and p-Ps states in a uniform magnetic field and assuming a value for the ground state hyperfine splitting. The situation regarding measurements and calculations of $\Gamma(1^3S)$ is summarised in Fig. 1 which emphasises the discrepancy which exists between recent experiments, particularly the "powder" work of Asai et al (1995) and the detailed work of the Michigan group, and between the latter and QED theory. The most recent measurement of $\Gamma(1^1S)$ is by Al-Ramadhan and Gidley (1994) who found a value of 7990.9±1.7$\mu s^{-1}$, in accord the result of 7989.5$\mu s^{-1}$ from the calculation by Khriplovich and Yelkhovsky (1990).

### 2.1.3 *Spectroscopic properties*

The ground state hyperfine structure was the first to be investigated experimentally and the historic measurements of Deutsch and Dulit (1951) first verified the necessity of incorporating the virtual annihilation term in the calculation of the frequency of the ground state hyperfine splitting. The most recent values for this quantity are 203.3875±0.0016GHz (Mills and Bearman 1975) and 203.3890±0.0012GHz (Egan et al 1977) which are to be compared with the theoretical result of 203.400GHz (Caswell

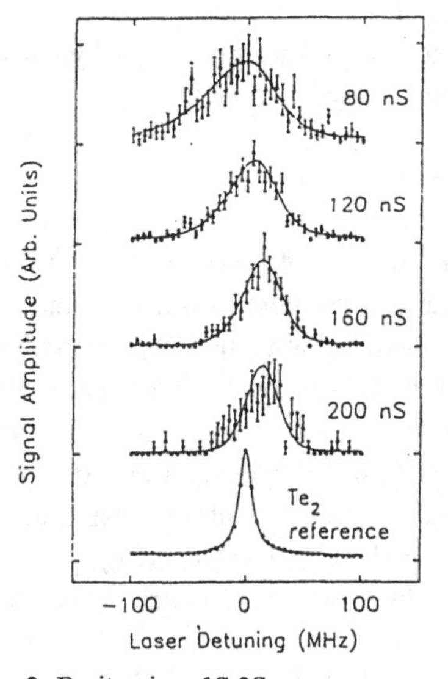


Fig. 2: Positronium 1S-2S resonance curves at various laser pulse time delays along with the $Te_2$ calibration line (Fee et al 1993a).

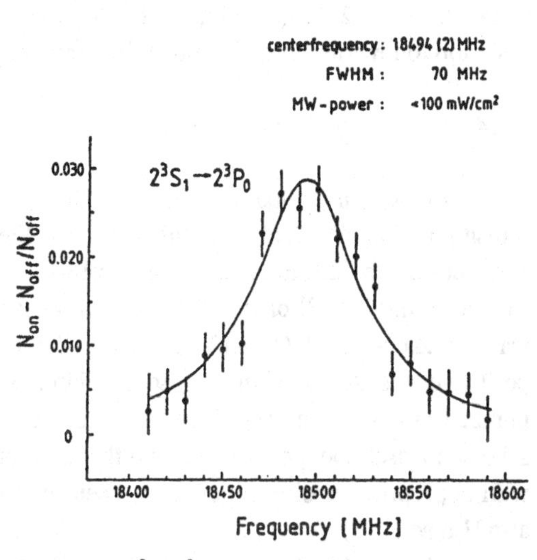


Fig. 3: $2^3S_1$-$2^3P_0$ transition. Note the increase in Lyman-α emission stimulated by microwave power (Hagena et al 1993).

and LePage 1978 and Bodwin and Yennie 1978). Rich (1981) has stated that uncalculated terms of order $\alpha^2$ would contribute around 7MHz to this value if their coefficients were unity. This puts into perspective the apparent discrepancy between theory and experiment.

Spectroscopy of the 1S-2S transition has been performed by Mills and co-workers (Chu et al 1984 and Fee et al 1993a,b). Figure 2 shows a positronium 1S-2S resonance curve, along with the $Te_2$ resonance line which was used for calibration. A fit to these data yielded a value for the frequency interval of 1233607216.4±3.2MHz (Fee et al 1993a,b) in good accord with the latest calculations which give 1233607221.7MHz (Fell 1992 and Khriplovich et al 1992).

The first crude spectroscopic measurement performed on excited state positronium was the identification of the 243nm Lyman-α radiation emitted in the 2P-1S transition (Canter et al 1975). The basic principle of the experiment was to collide low energy positrons with a surface and look for a coincidence between a Lyman-α photon and a delayed gamma-ray due to the subsequent annihilation of the $1^3S$ o-Ps. This is a signature of excited state Ps (Ps*) formation from the surface which, due to work function considerations, is generally only possible for positrons which reach the surface epithermally (see also section 4.2). This simple method of producing $n$=2 positronium in vacuum has allowed measurements of the frequencies for the $2^3S_1$-$2^3P_J$

(J=0,1,2) transitions (Hagena et al 1993, Hatamian et al 1987). As an example, Fig.3 shows the $2^3S_1$-$2^3P_0$ transition at 18494 MHz, recorded by the increase in Lyman-$\alpha$ emission following the application of microwave power (Hagena et al 1993).

### *2.1.4 Non-spectroscopic laser studies and exotic tests*

Ziock et al (1990a) have reported the optical saturation of the one photon $1^3S$-$2^3P$ transition. A notable feature of this work was the use of a dye laser modified to obtain a sufficient bandwidth to cover a significant fraction of the positronium Doppler profile. This also allowed all of the $2^3P$ states to be accessed. This accomplishment paved the way for Ziock et al (1990b) to achieve the resonant excitation of high-$n$, Rydberg, positronium states by shining a second laser, again with a large bandwidth, onto the pumped atoms. The optical saturation of the 1S-2P transition is also the basis for a scheme to laser cool positronium and the desirability and feasibility of so doing, together with appropriate cooling rates, have been discussed by Laing and Dermer (1988) (see also Hirose 1998).

The combination of a readily available purely leptonic system which is also composed of a particle-antiparticle pair has, over the years, attracted the interests of many experimenters as a test for the existence of exotic particles or couplings. The latter may perhaps manifest themselves in the decay properties of positronium, such that attempts to observe forbidden modes have been made. An in-depth survey of all of these measurements is beyond the scope of the present discussion. Instead we present a partial list to which the interested reader may refer. A summary of the situation up to around 1980 has been given by Rich (1981). Later work includes: a CPT test on polarised positronium (Arbic et al 1988); searches for the forbidden two-gamma (Nico et al 1992), and four-gamma annihilations of o-Ps (Yang et al 1996); a search for spatial anisotropy in o-Ps annihilation (Mills and Zuckerman 1990) and various attempts to detect particles other than gamma-rays in the decay of o-Ps (Gninenko et al 1990, Orito et al 1989, Mitsui et al 1993, Skalsey and Conti 1995 and references therein). It is sufficient to say for the present purposes that no evidence for any forbidden decays or new particles have been forthcoming within the limits set by the relevant experiments.

## *2.2 The positronium negative ion*

This entity, denoted $Ps^-$, is formed from a positron together with two electrons in a relative spin singlet state. It was predicted to be bound by Wheeler (1946) and first detected experimentally by Mills (1981). There have been numerous calculations of its binding energy (though no measurements to date), with the most accurate being by

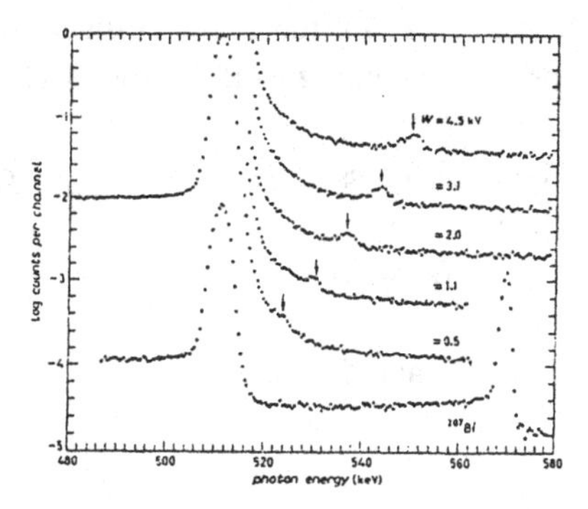


Fig.4: Doppler-shifted two-gamma annihilation of $Ps^-$ moving towards the detector as a function of the accelerating potential (Mills 1981).

Bhatia and Drachman (1983) who found a value of 0.326677eV. (See also Ho 1983, Petelenz and Smith 1988 and Frolov and Yeremin 1989). There are no other bound states, though the system possesses resonances in the electron-positronium continuum (Ho 1984).

The observation of $Ps^-$ by Mills (1981) used a technique analogous to the production of $H^-$ by proton bombardment of thin foils (e.g. Allison 1958). The $Ps^-$ was produced by partial transmission of a $e^+$ beam through a carbon film. The ion was accelerated away from the foil and hence detected by recording the Doppler-shifted two-gamma annihilation. Figure 4 shows the observed $Ps^-$ signature in the form of a small peak which was progressively blue-shifted with respect to the 511keV annihilation line as the accelerating voltage was increased.

The rate of two-gamma annihilation of the positron with one of the electrons has also been calculated by Bhatia and Drachman (1983). They obtained a value of $2.0861 ns^{-1}$, which is in accord with the work of Ho (1983, 1990), and, as expected, close to the weighted annihilation rates for the singlet and triplet rates of free ground state positronium. There has been only one reported measurement of this quantity, again due to Mills (1983) who obtained a value of $2.09 \pm 0.09 ns^{-1}$ for the intrinsic decay rate. This value is in good accord with, though much less precise than, theory.

## 3 Interactions of positrons with atoms and molecules

### *3.1 Introduction*

During the last decade, there has been much progress in the study of positron-atom collisions, including near-threshold and differential measurements. An overview of the current status is given in Table 1.

Table 1. Current status in $e^+$ - atom (molecule) collision experiments. (E denotes the positron incident energy and Q cross-sections).

| Processes | Available cross-sections | Lowest threshold |
|---|---|---|
| $e^+ + A \rightarrow$ all | $Q_t$ for many targets up to O(keV) | none |
| elastic scattering $e^+ + A \rightarrow e^+ + A$ | some $Q_{el}$ and $dQ_{el}/d\Omega$ | none |
| target excitation $e^+ + A \rightarrow e^+ + A^*$ | scarce | $E_{ex}$ |
| electron capture $e^+ + A \rightarrow Ps + A^+$ | $Q_{Ps}$ ($1 \leq (E-E_{Ps}) \leq 10^2$ eV) and some $dQ_{Ps}/d\Omega$ | $E_{Ps}=E_i - B_{Ps}$ |
| target ionization $e^+ + A \rightarrow e^+ + ze^- + A^{z+}$ | $Q_i^{z+}$ ($1 \leq (E-E_i) \leq 10^3$ eV) some SDCS, DDCS and TDCS for $Q_i^+$ | $E_i$ |
| annihilation $e^+ + A \rightarrow 2\gamma + A^+$ | until recently only over a very restricted energy range | none |
| formation of compounds $e^+ + A \rightarrow (eA)^+ + h\nu$ | scarce | $E_c$ |

### 3.2 *Experimental systems*

Examples of the typical positron beams used in atomic collisions are shown in Fig. 5 a,b. The principal difference between them is in the transport field used: the first employs a magnetic field and is suitable for the measurements of angle integrated cross-sections (Kara et al 1997), whilst the second uses an electrostatic field and is used for angle and energy differential studies (Kövér et al 1993, 1994). Both types of beamlines usually employ commercially available isotopes which have a sufficiently long half-life (e.g. $^{22}$Na or $^{58}$Co) to enable experimentation away from their production site. Alternatively, the beams are coupled to accelerators providing fast positrons either from pair creation (e.g. Howell et al 1982, Hulett et al 1987, Ley 1997) or through the production of intense short lived isotopes (e.g. Lynn and Jacobsen 1994, Xie et al 1994). Typical moderators consist of a film of a metal (e.g. Zafar et al 1988, 1989; Jacobsen et al 1990 and references therein) or a solid rare gas (Mills et al 1994 and references therein). The particular choice depends upon operating vacuum conditions and the specific requirements on the beam intensity versus energy spread. Usually, the slow positrons are deflected to separate them from the flux of fast particles and gamma rays originating from the source region and are detected by conventional charge particle detectors (e.g. channeltrons, multi-channel-plates) and/or $\gamma$-ray counters (see Chapters 2 and 9 for further details on positron beam technology). A retarding field analyzer, as shown in Fig. 5a, may be used to decrease the energy spread of the beam whilst a parallel-plate analyzer, as in Fig. 5b, may be employed to determine the energy of the scattered positron or ionized electron. Both systems comprise an ion extraction system for the investigation, in these cases, of ionization or Ps formation. The extraction field is usually pulsed after the collision in order not to perturb the interaction.

### *3.3 Total cross-sections, $Q_t$*

For these, a vast data library is available and extensive reviews on this topic have been written recently by Kauppila and Stein (1990) and Zecca et al (1996). The inert gases were the first atoms to be studied experimentally since they exist in atomic form at room temperature. More recently, investigations have been extended to more difficult targets such as atomic hydrogen and the alkali metals.

The main features of the total cross-sections for $e^{\pm}$ - inert atoms are illustrated in Fig. 6 where the comparison with corresponding $Q_t(e^-)$ serves to highlight some of the interesting differences and similarities in the atomic interactions between the projectiles. Unlike the Ramsauser-Townsend minima which arise in $e^-$ scattering from argon, krypton and xenon due to the strength of the attractive potential at low energies, the troughs in $Q_t(e^+)$ at low energies arise from the partial cancellation of the static (repulsive) and polarization (attractive) interactions. As a result, at low energies, the cross-sections for helium and neon differ by orders of magnitude between the two projectiles. Whilst $Q_t(e^-)$ in general varies smoothly as different inelastic channels open, $Q_t(e^+)$ manifests a pronounced increase at the threshold for Ps formation, $E_{Ps}$ (see Table 1). Both the minimum and the sharp rise at $E_{Ps}$ are features typical also of $e^+$ scattering from many molecular targets. At high kinetic energies, the increasingly dominant static interaction ultimately results in the merging of the cross-sections of the two projectiles.

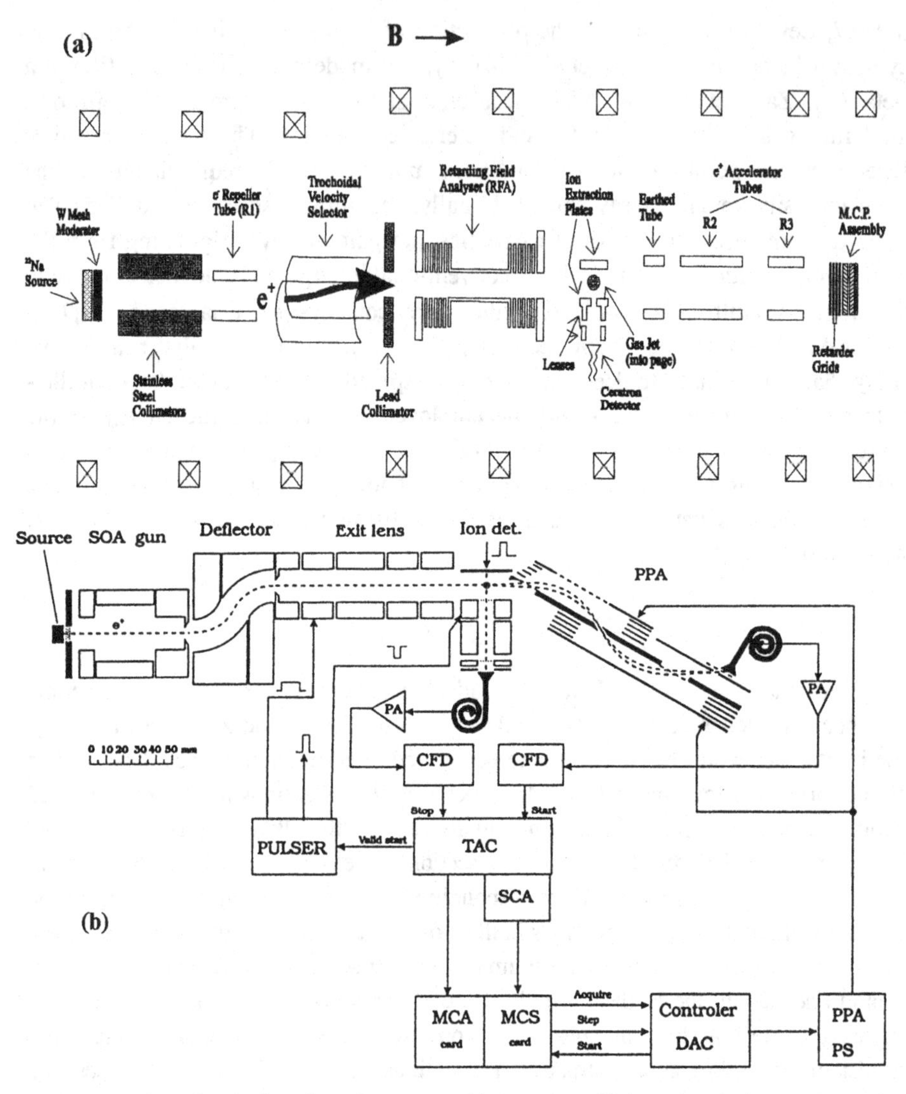


Fig. 5: Examples of typical positron beams used in atomic collisions: a) employs a magnetic field for beam confinement (Kara et al 1997), whilst b) uses an electrostatic field (Kövér et al 1993).

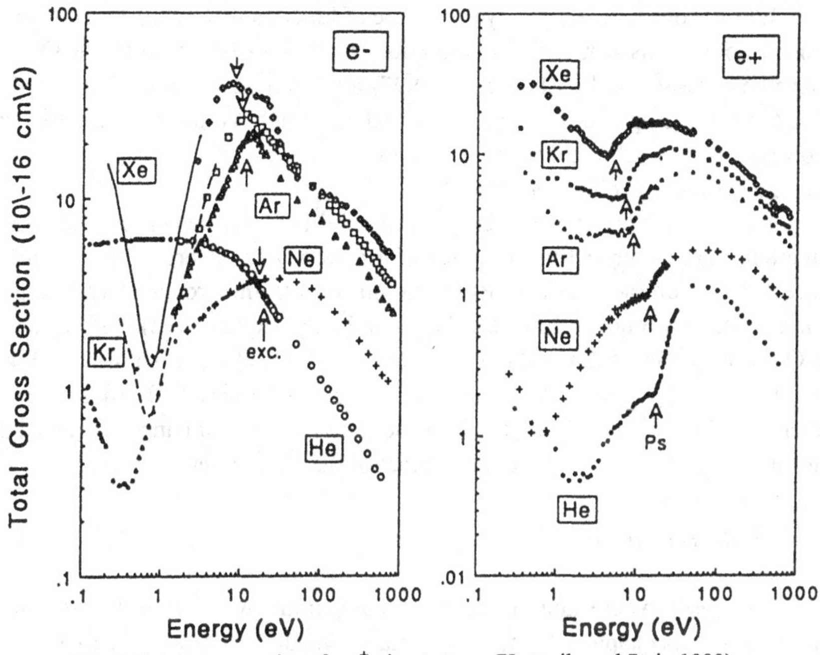


Fig. 6: Total cross-sections for $e^{\pm}$ - inert atoms (Kauppila and Stein 1990).

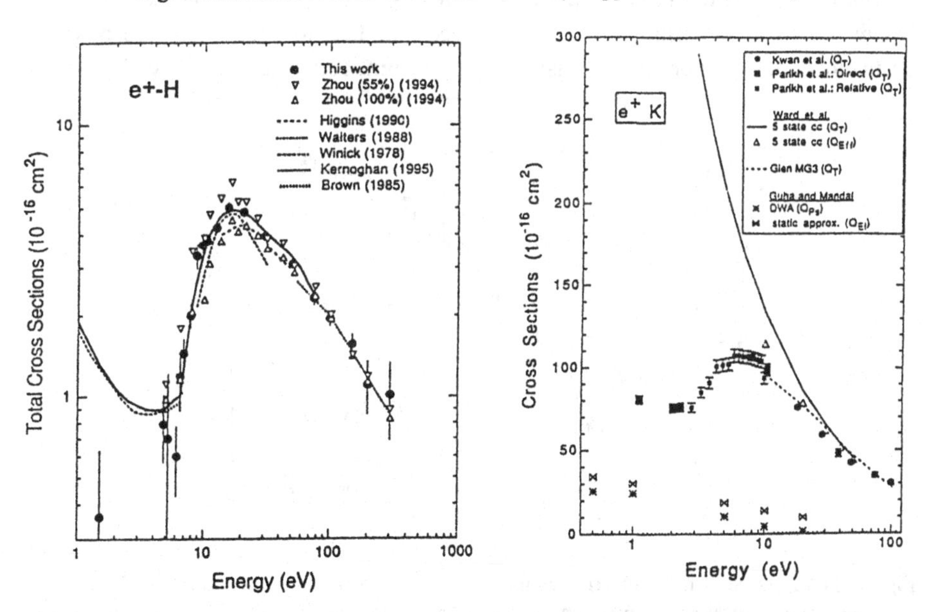


Fig. 7: Total cross-section measurements for $e^+$-H compared to various theories (Zhou et al 1997).

Fig. 8: Total cross-section measurements for potassium compared to various theories (Stein et al 1996).

Interactions with atomic hydrogen are of fundamental interest and, being the simplest atom, it has attracted most attention from theorists. Recently, the $Q_t$ for $e^+$-H have been measured by Zhou et al (1997) and their results are shown in Fig. 7 where they are compared with theories. Good agreement is found with these except at lowest energies where, however, forward scattering effects could significantly affect the experiment (Zhou et al 1997).

The alkali atoms possess large polarizabilities and ionization energies less than 6.8eV. The latter fact means that Ps formation is an exoergic reaction and, for this reason, the cross-section for Ps formation ($Q_{Ps}$) in the ground state diverges as the impact energy tends to zero (e.g. Watts and Humberston 1992). As an example, the $Q_t$ measurements for potassium (Stein et al 1996) are shown in Fig. 8. Interestingly, the broad peak in the total cross-section observed around 6eV has been ascribed by Walters et al (1995) to the progressive dominance of excited state Ps formation with increasing atomic number of the alkali atom.

### 3.4 *Elastic scattering ($Q_{el}$ and $dQ_{el}/d\Omega$)*

In cases where annihilation is a comparatively improbable process (see section 3.8 in this chapter), $Q_{el}$ is approximately equal to $Q_t$ below the first inelastic threshold. For the inert atoms, the behaviour of $Q_t$-$Q_{Ps}$ (approximately equal to $Q_{el}$ below the first excitation threshold) has been studied to investigate the possible occurrence of Wigner cusps arising at $E_{Ps}$ from coupling effects with the Ps

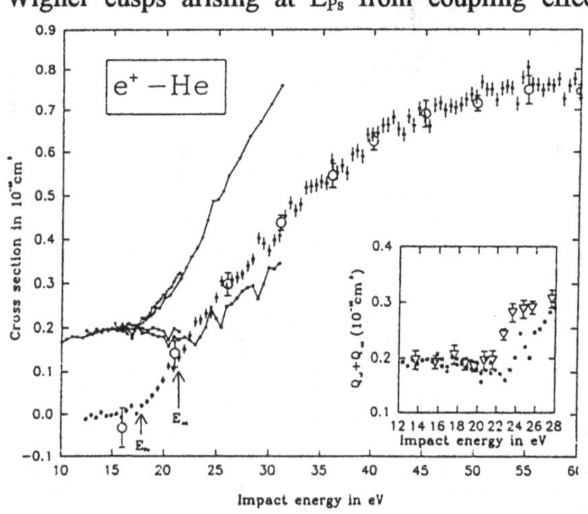


Fig. 9: Total (upper lines), elastic scattering (lower lines) and total ionization (points) cross-sections from He. The inset shows the sum of the elastic scattering and excitation cross-sections. The Ps formation threshold in helium is at 17.79eV (Moxom et al 1993).

formation channel (Coleman et al 1992; Moxom et al 1993). Measurements of $Q_{el}+Q_{ex}$ for helium are given in the inset of Fig. 9 which show, in contrast to previous indications (Campeanu et al 1987), a smooth variation of $Q_{el}$ across $E_{Ps}$=17.79eV for this system. This result implies a minor s-wave contribution to $Q_{Ps}$ and is in agreement with near threshold $Q_{Ps}$ measurements of Moxom et al (1994) and calculations of Van Reeth and Humberston (1997) which will be discussed in more detail in the next section.

A more sensitive (and direct) method of investigating elastic scattering is through the measurement of the differential cross-section ($dQ_{el}/d\Omega$). Data are available for the inert atoms (Coleman and McNutt 1979, Floeder et al 1988, Smith et al 1990, Kauppila et al 1996) and some molecules (Przybyla et al 1997) albeit with an energy resolution insufficient to discriminate, in the latter case, against rotational and vibrational excitations. As an example, the relative measurements of $dQ_{el}/d\Omega$ for the $e^+$-Ar system by Smith et al (1990) are reproduced in Fig. 10 where they have been normalized to and compared with calculations using polarized orbital (McEachran et al 1979) and optical potential (Bartschat et al 1988) methods. The discrepancy between theories and between these and experiment at the higher energies has been attributed to the importance of inelastic channels. Indeed, small deviations which occur in the vicinity of $E_{Ps}$ (both above and below) have also been found to be consistent with the influence of the Ps formation channel (Meyerhof and Laricchia 1997).

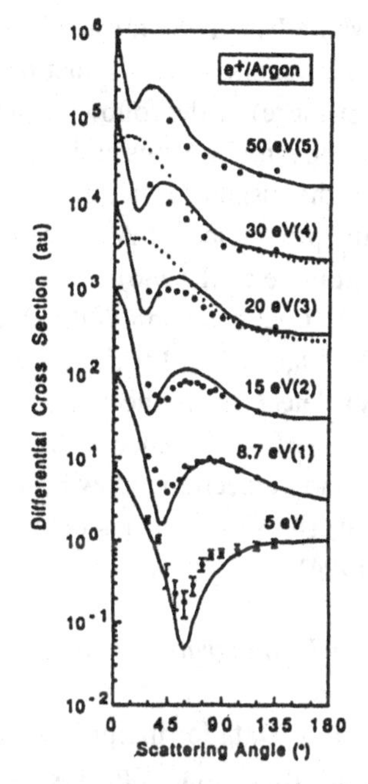


Fig. 10: $dQ_{el}/d\Omega$ for $e^+$-Ar (Smith et al 1990) normalized to theory: -- McEachran et al 1979, .... Bartschat et al 1988.

## 3.5 *Positronium formation*

The striking rise in $Q_t$ above the threshold of Ps formation was an early indication of the importance of this process (Canter et al 1972). For the cases considered in this section, the reaction may be represented as

$$e^+ + A \rightarrow Ps + A^+ \tag{4}$$

where Ps may be formed in the ground or higher states and the ion itself may be left in an excited state. Reaction (4) has been studied experimentally by observing one (or more) or the following signals:
i) the simultaneous emission of three γ-rays (o-Ps) (e.g. Charlton et al 1983),
ii) the disappearance of the $e^+$ in the final state (all Ps) (e.g. Fornari et al 1983),
iii) production of $A^+$ in conjunction with the disappearance of the $e^+$ (all Ps) (e.g. Fromme et al 1986),
iv) direct detection of the Ps atom itself (o-Ps) (e.g. Laricchia et al 1987; Falke et al 1995a,b; 1997; Finch et al 1996),
v) detection of the 5.1eV Lyman-α photon from the (2P-1S) transition of Ps in delayed coincidence with an annihilation photon (Ps*) (Laricchia et al 1985),
vi) ion detection below $E_i$ (Moxom et al 1993),
vii) the simultaneous emission of two γ-rays (p-Ps and quenched o-Ps) (Zhou et al 1994).

### *3.5.1 Integrated cross-section, $Q_{Ps}$*

Data for integrated cross-sections (summed over all final states of the Ps and ion) are available for the inert atoms (e.g. Moxom et al 1993, 1994; Overton et al 1993; Meyerhof and Laricchia 1997 and references therein; also for a compilation of earlier data see Charlton and Laricchia 1990), some molecules (Moxom et al 1993, Laricchia and Moxom 1993, Laricchia et al 1993), atomic hydrogen (Sperber et al 1992, Weber et al 1994, Zhou et al 1997) and the heavier alkali metals (Stein et al 1996, Zhou et al 1994, Surdutovich et al 1996). As shown in Fig. 11, in the case of helium, a degree of agreement is found among the various experiments below

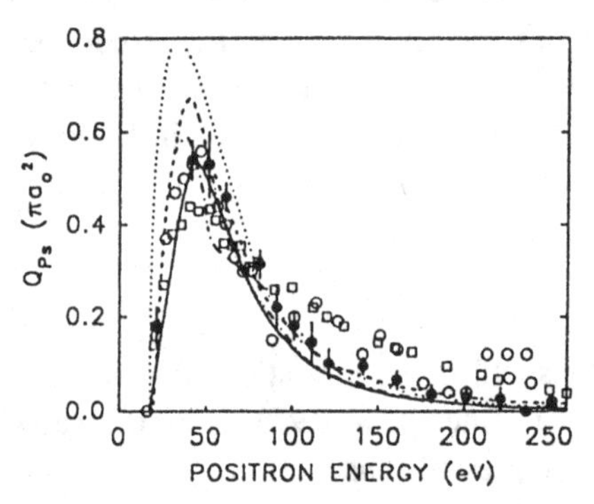


Fig. 11: Positronium formation cross-section for helium. Filled circles: Overton et al (1993) compared to earlier measurements and theories.

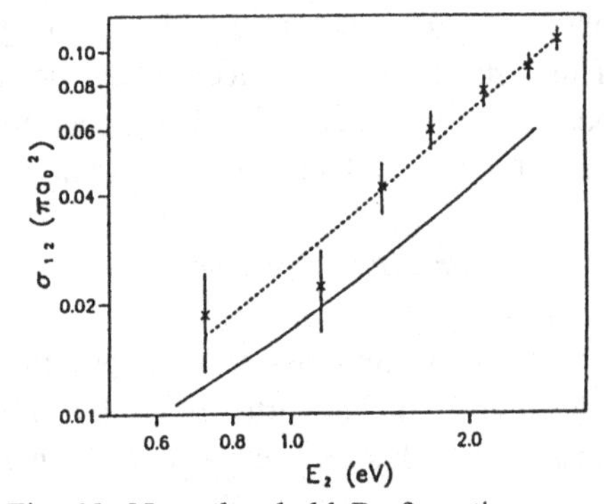


Fig. 12: Near threshold Ps formation cross-section for He: expt (Moxom et al 1993), theory (Van Reeth and Humberston 1997).

~60 eV after which those of Overton et al (1993) begin to decrease approximately as $E^{-2.5}$ in better agreement with calculations, most of which predict exponents in the range (3-5) (Mandal et al 1979; Khan and Ghosh 1983; Deb et al 1987; Schultz and Olson 1988; Tripathi et al 1988; Bransden et al 1992; McAlinden and Walters 1992; Fojon et al 1995).

The near-threshold measurements for helium (Moxom et al 1994) are shown in more detail in Fig. 12 . Here they are compared to recent variational calculations (Van Reeth and Humberston 1997) for the s-, p- and d-wave contributions to $Q_{Ps}$ which have been supplemented by a First Born Approximation (FBA) calculation for higher partial waves. In this way, theory is found to be ~30% lower than experiment. More elaborate calculations for higher partial waves might resolve the discrepancy (Van Reeth 1997).

In the measurements, a gated positron beam and an efficient ion extraction technique were employed (method (vi)) to determine the so-called total ionization cross-section $Q_t^+ = Q_{ann} + Q_{Ps} + Q_i^+ + \Sigma Q_{ho}^+$ where $Q_{Ps}$ includes contributions from excited states, $Q_i^+$ is the cross-section for single direct ionization and $\Sigma Q_{ho}^+$ the sum over higher-order processes. If $Q_{ann}$ is negligible (see section 3.8 below), then below $E_i$ , $Q_t^+ = Q_{Ps}$. In this way, neon, argon, krypton and xenon have also been investigated (Moxom et al 1994, Meyerhof et al 1996, Meyerhof and Laricchia 1997). The results for neon and xenon are shown in Fig. 13 where the measurements are compared to the calculations of a truncated coupled-static approximation (McAlinden and Walters 1992) and to R-matrix partial wave cross-sections (Meyerhof and Laricchia 1997). The results of McAlinden and Walters have been found to agree increasingly better with experiment as the target atomic number increases. The reason for this is not understood. Both sets of calculations indicate that, whilst the s-wave contribution matches the steep rise of the experimental $Q_{Ps}$ from threshold, particularly as the atomic number of the target increases, this component makes a progressively smaller contribution to $Q_{Ps}$ so that inclusion of higher partial waves becomes crucial (Meyerhof and Laricchia 1997).

In section 3.3, we remarked that for the alkali atoms Ps formation is an exoergic reaction. The results for the cross-section for this process in sodium and potassium of Zhou et al (1994) are shown in Fig. 14 together with various calculations. The lower limits (LL) were obtained by measuring two gamma coincidences (method vii) which are related to the self-annihilation of p-Ps and quenched o-Ps, assuming that $Q_{ann}$ is negligible at the energies considered. The upper limits (UL) were obtained by a method similar to (ii) above. In the case of sodium, (LL) and (UL) agree to within 35% whilst the divergence between (UL) and (LL) for potassium at low energies were interpreted as being due to incomplete collection of the scattered positrons and the (LL) data have been taken as good estimates of $Q_{Ps}$. As indicated earlier, coupled eigenstate calculations (Walters et al

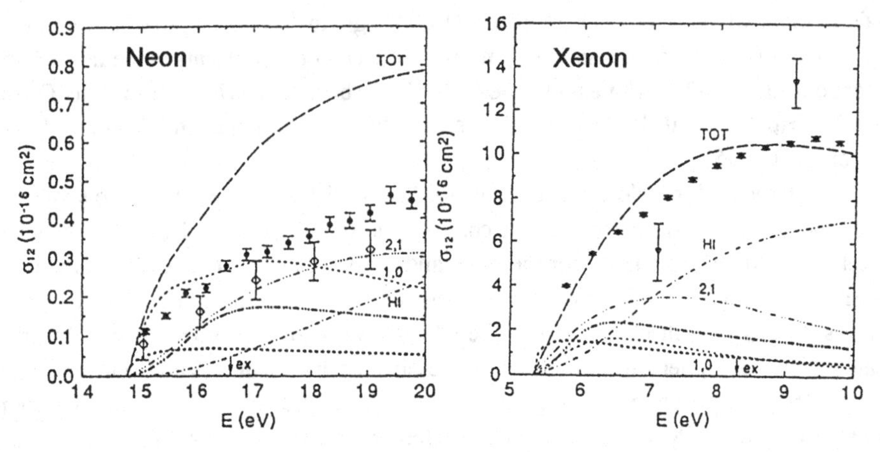


Fig. 13: Positronium formation cross-sections for Ne and Xe (Meyerhof and Laricchia 1997). Measurements (points) are compared to theory. The truncated coupled-static approximation of McAlinden and Walters (1992) is indicated by TOT. The thinner curves are the corresponding partial wave contributions $(l_1,l_2)=(1,0)$, $(2,1)$ and $l_1\geq3$ (HI). The bold curves are the R-matrix partial wave cross-sections of Meyerhof and Laricchia (1997).

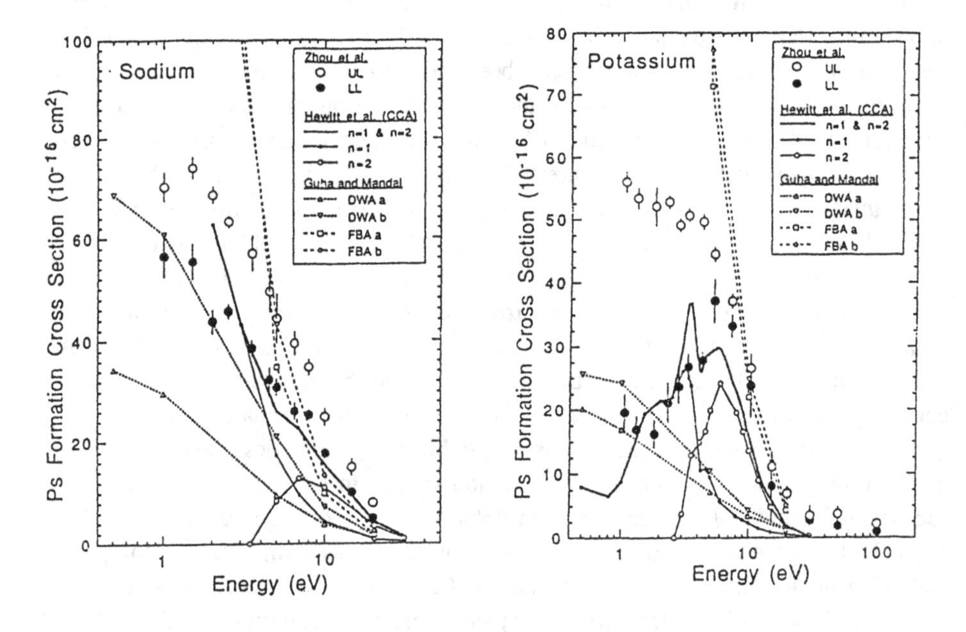


Fig. 14: $Q_{Ps}$ measurements for Na and K of Zhou et al (1994) compared with theories.

1995, Kahali and Ghosh 1995) have ascribed the broad peak in the total cross-section observed around 6eV to an increasing importance of excited state Ps formation as the target is increased from lithium to cesium. This trend is shown in Fig. 15 and awaits direct experimental investigation (Walters et al 1995).

Excited state Ps formed in binary collisions between a positron and an atom (molecule) was first observed by Laricchia et al (1985) in low density neon, argon and molecular hydrogen using method (v). The highest yield was found in molecular hydrogen followed by argon and neon. It might be interesting to investigate whether a trend in $Q_{Ps}(n>1)/Q_{Ps}(n=1)$ with target atomic number or ionization potential exists also for other atomic families.

The first measurement of $Q_{Ps}$ for atomic hydrogen was reported by the Bielefeld group (Sperber et al 1992) using method (iii) above. Their results were later re-evaluated to correct for the incomplete detection of scattered positrons (Weber et al 1994). More recently, the Detroit group has also measured $Q_{Ps}$ for atomic hydrogen by using method (vii) (Zhou et al 1997). Their results together with those from Bielefeld are compared to some theories in Fig. 16 where particularly good agreement is noted with the coupled 33 state calculations of Kernoghan et al (1996) and 28 state close-coupling calculation of Mitroy (1996). More recently, however, following a re-determination of their direct ionization cross-section (see section 3.7.1 below), the $Q_{Ps}$ from Bielefeld group has been lowered by almost a factor of 2 and now disagrees with the Detroit measurements and most theories (Hofmann et al 1997).

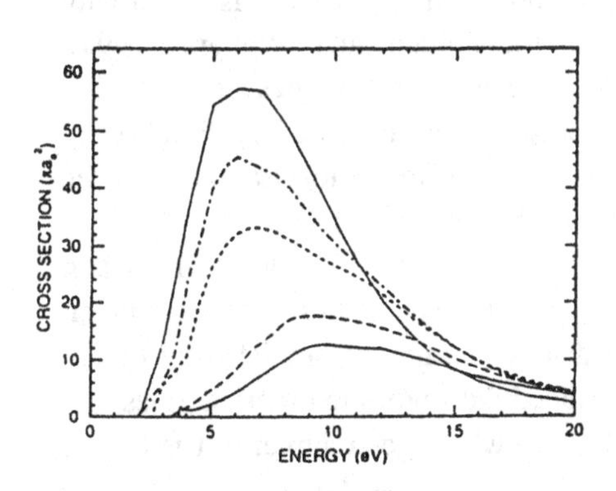


Fig. 15: Ps* formation cross-sections for the alkali atoms. The magnitude of the peak is calculated to increase progressively from Li to Cs (Walters et al 1995).

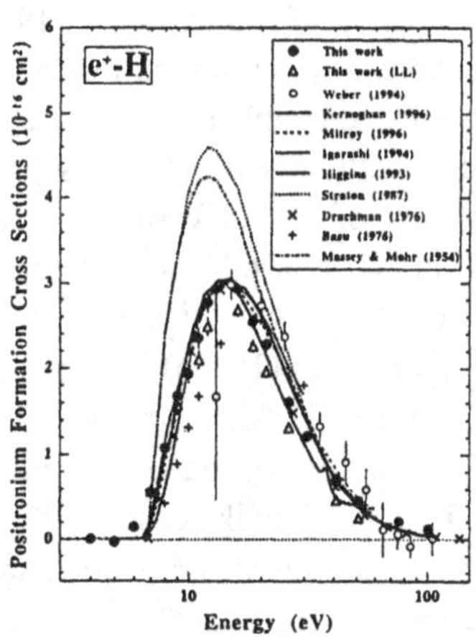


Fig. 16: Positronium formation cross-section measurements for atomic hydrogen by Zhou et al (1997) compared to those of Weber et al (1994) and theories.

### 3.5.2 *Differential cross-section, $dQ_{Ps}/d\Omega$*

The first experimental investigations of $dQ_{Ps}/d\Omega$ were stimulated by theories which indicated a high degree of forward collimation of the outgoing positronium and hence by the prospect of being able to produce positronium beams (see section 4 below). For this reason, the behaviour of $dQ_{Ps}/d\Omega$ was investigated at small forward angles as a function of incident positron energy and target gas. Fig. 17 shows the results of Laricchia et al (1987) for the variation of the fraction ($F_{Ps}$) of $e^+$ scattered by helium and emitted as positronium within an angle $\theta'$ about the incident $e^+$ direction. Also shown in the figure are the corresponding predictions computed from the theory of Mandal et al (1979) according to

$$F_{Ps}(E,\theta') = (2\pi / Q_t)\int_0^{\theta'} dQ_{Ps}(E) / d\Omega \sin\theta d\theta \qquad (5)$$

Further investigations have ensued and their relevance to Ps beam production will be reviewed in section 4 below. The most recent are those of Garner et al (1996) shown in Fig. 18. Here the energy dependence of the measured efficiency $\varepsilon_{Ps}$ for producing positronium from helium, argon and molecular hydrogen within an angle of $\pm 1.5°$ is compared to theories (note that for these measurements, $\varepsilon_{Ps} \approx F_{Ps}$). Fair agreement is again found in the case of helium with the distorted wave approximation of Mandal et al (1979) and, in the case of $H_2$, with the first Born calculations of Biswas et al (1991). For argon, no overall agreement is found with the truncated-coupled static calculation of McAlinden and Walters (1994), although both sets of results suggest a decrease at high energies as $E^{-1.5}$. The dependence of $dQ_{Ps}/d\Omega$ with incident $e^+$ energy has been studied by Finch et al (1996) in argon at 60° and their results have also been found to disagree with the energy dependence predicted by McAlinden and Walters (1994).

Falke et al (1995a) have studied $dQ_{Ps}/d\Omega$ in $e^+$ - argon, krypton collisions at 30 eV from 0°-120°. Their results, compared with the calculations of McAlinden and Walters (1994), did not reproduce the distinct minimum predicted by theory near 15° at 30eV. Falke et al (1997) have extended the studies to other energies and to krypton gas. Their results are shown in Fig. 19 where good agreement is found, in the case of argon at 75eV, with the calculations of McAlinden and Walters (1994) over the whole angular range, whilst for argon at 90eV and krypton at 75eV

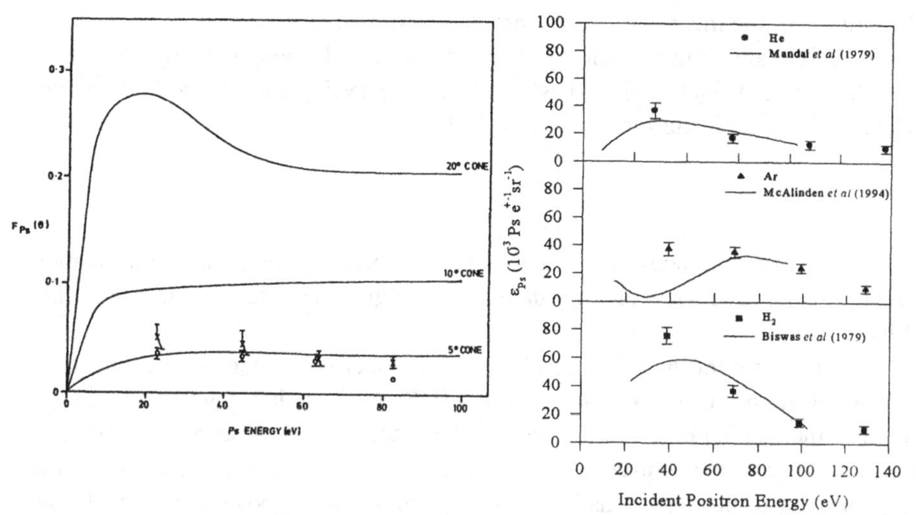


Fig. 17: Fraction ($F_{Ps}$) of $e^+$ scattered by He and emitted as Ps within an angle θ' about the incident $e^+$ direction. Expt: Laricchia et al (1987); theory: Mandal et al (1979).

Fig. 18: Measured energy dependence of the efficiency for producing Ps from He, Ar and $H_2$ within an angle of ±1.5° (Garner et al 1996) compared to theories as indicated.

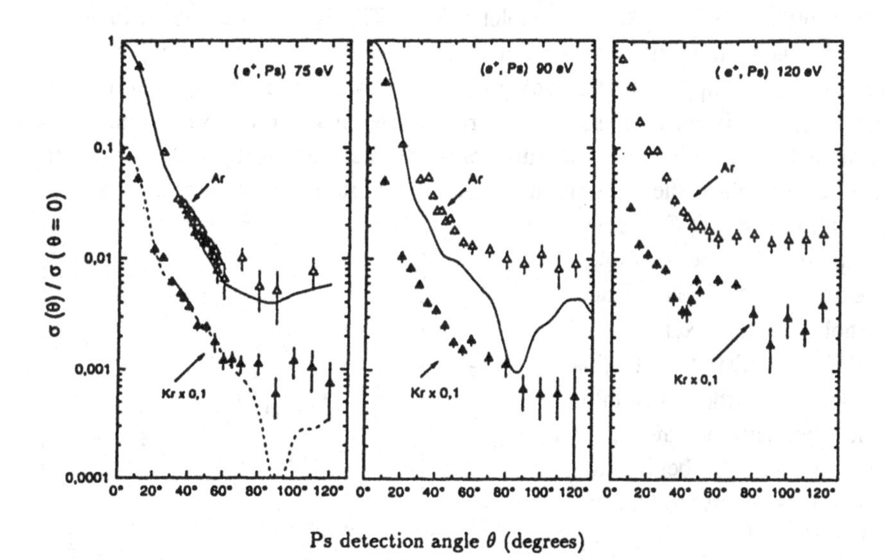


Fig. 19: Differential Ps formation cross-section measurements at 75,90 and 120eV for Ar and Kr (Falke et al 1997) compared to theory (McAlinden and Walters 1994).

the minima in the theory around 85° are not confirmed by experiment. An anomaly observed for transfer ionization ($e^+$, Ps $e^-$) in argon at 45° and 75eV by Falke et al (1995b) has been found not to persist at 90eV suggesting that the structure is not related to Thomas scattering (Falke et al 1997).

### *3.6 Excitation*

Experimental activity in this area has been sporadic and information remains sparse. A review of early data has been given by Charlton and Laricchia (1990).

A more recent investigation of the excitation cross-sections for helium, neon and argon atoms, using a time-of-flight (TOF) method, has found the data for argon in the whole range and for neon in the range 50-100 eV to coincide roughly with that for electron impact (Mori and Sueoka 1994). For helium they found the positron data to be much smaller than the electron data above 30eV and below 35eV to agree with theory (Varracchio 1990).

Effects ascribed to the influence of channel-coupling have been observed (Laricchia et al 1993) in the total ionization cross-section, $Q_t^+$, of $O_2$ in the vicinity of the threshold for excitation to the Schumann-Runge Continuum, a broad photoabsorption band in the ultraviolet region. The results are shown in Fig. 20 where $Q_t^+$ is seen to rise near the threshold for Ps formation at 5.27eV before decreasing above approximately 9eV impact energy. After reaching a minimum at around 11eV, it is seen to rise over the remaining energy range investigated. Also shown in the figure are the measurements for the cross-section $Q_{ex}$ for positron impact excitation to the Schumann-Runge continuum (Katayama et al 1987). The location in the dip of $Q_t^+$ appears to correspond to the peak of $Q_{ex}$. The decrease in $Q_t^+$ is near the threshold for the excitation $X^3\Sigma^-_g \rightarrow B^3\Sigma^-_u$, whilst $Q_{ex}$ begins to decrease near the threshold for single ionization at 12.07eV around which $Q_t^+$ begins to rise again. This correspondence has been attributed to strong coupling between the Ps formation, excitation and direct ionization channels (Laricchia et al 1993).

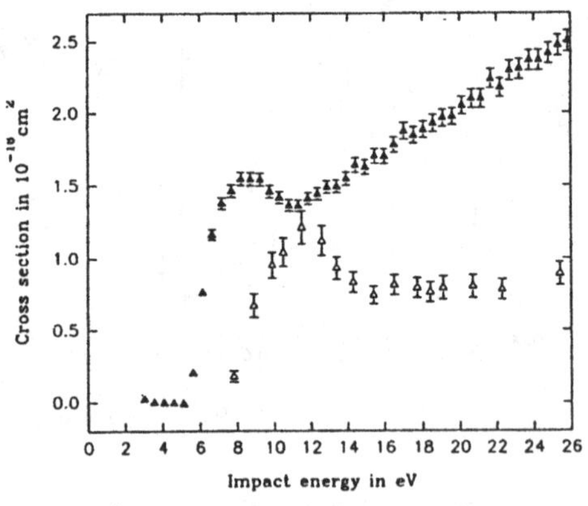


Fig. 20: $Q_t^+$ for $O_2$ (filled symbols, Laricchia et al 1993); $Q_{ex}$ (hollow symbols, Katayama et al 1987).

An enhancement of $Q_t^+$ has been observed near the threshold for formation of positronium simultaneous to excitation of the remnant ion (Laricchia and Moxom 1993). This confirms an earlier observation whereby a photon of energy around 3.5eV was found to be related in time to Ps formation (Laricchia et al 1988a). It was suggested that the photon originated from the $^2\Pi_u$-$^2\Pi_g$ transition of $CO_2^+$, the simultaneous excitation of the product ion resulting in an increased Ps formation probability. The reason for this is thought to be due to the fact that the C-state of $CO_2$ lies at 10.56eV above the ground-state, i.e. only 60meV above the threshold for Ps formation simultaneous to ion excitation. Ps formation would then seem to proceed via a virtual-excitation of the molecule, followed by capture of the excited electron by a quasi-stationary positron, the remnant ion absorbing the difference in binding energies.

### *3.7 Ionization*

Considerable advances in the understanding of the dynamics of ionization by charged particle impact have been effected in recent years through comparative studies using particle/antiparticle pairs which offer the opportunity of identifying collision effects related to the charge and/or mass of the projectile (Schultz et al 1991, Knudsen and Reading 1992, Paludan et al 1997a). An extensive data library is now available for the cases of positron and antiproton impact which complements that for electrons and protons. In the rest of this section comparisons, whenever applicable and appropriate, will be drawn among these projectiles.

Data for all four ($e^{\pm}$, $p^{\pm}$) projectiles are now available for the inert atoms (Paludan et al 1997a,b and references therein), atomic hydrogen (Jones et al 1993, Knudsen et al 1995 and references therein) and some molecules (Jacobsen et al 1995a,b; Moxom et al 1996 and references therein). In the case of positrons, the following reaction

$$e^+ + A \rightarrow e^+ + ze^- + A^{z+} \qquad (6)$$

has been studied with z=1,2,3 by

i) measuring energy loss spectra (e.g. Mori and Sueoka 1994);

ii) detecting an ion in coincidence with a $e^+$ (e.g. Fromme et al 1986; Knudsen et al 1990; Jacobsen et al 1995a,b; Moxom et al 1996; Kövér et al 1993; Falke et al 1997);

iii) detecting an ion in coincidence with the ejected $e^-$ (e.g. Moxom et al 1992; Kövér et al 1993; Falke et al 1997).

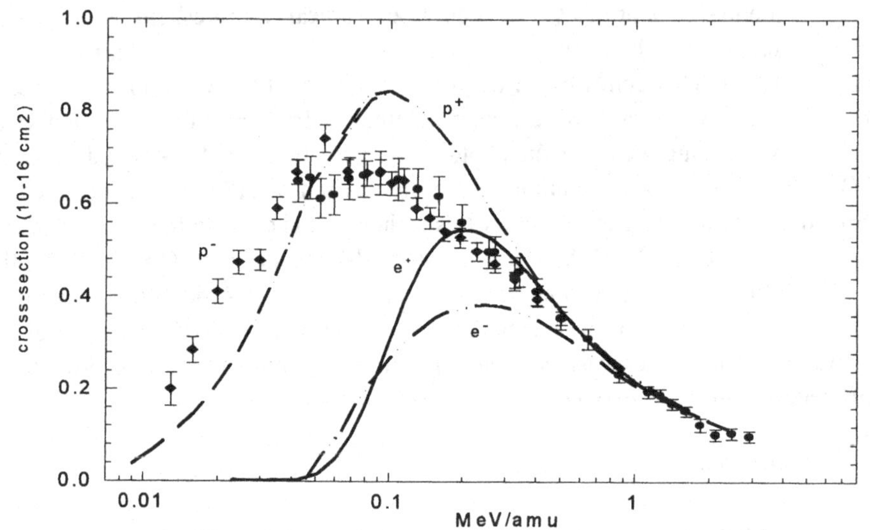


Fig. 21: The direct ionization cross-section with equivelocity ($e^{\pm}$, $p^{\pm}$) projectiles on He (Knudsen and Reading 1992).

### 3.7.1 *Integrated cross-sections, $Q_i^{z+}$*

*Single ionization*

The results for $Q_i^+$ obtained with equivelocity ($e^{\pm}$, $p^{\pm}$) projectiles on He are shown in Fig. 21 (Knudsen and Reading 1992). From comparisons of this type for several targets, certain trends in ionization by charged particles have been discerned. At high velocities, $Q_i^+$ is found to display a similar energy dependence for all four projectiles and the cross-sections are assumed to have merged in accordance with the first Born approximation. At lower velocities, the heavier projectiles have higher $Q_i^+$ (mass effect) whilst, for each particle/antiparticle pair, the positively charged particles have larger $Q_i^+$ (charge effect). The latter has been ascribed to target polarisation during the first part of the collision (Knudsen and Reading 1992). At the lowest velocities, $Q_i^+(e^+, p^+) < Q_i^+(e^-, p^-)$. This is interpreted as being due in part to competition from electron capture for the positive projectiles and, in the case of the heavier projectiles, in part to binding/antibinding effects (Knudsen and Reading 1992).

Notably, however, Moxom et al (1996) observed that, around the peak in the cross-sections, $Q_i^+(e^+)/ Q_i^+(e^-)$ decreased with increasing atomic number of the target in contrast to the expectations based on the influence of the polarisation interaction. It was suggested (Laricchia 1995a,b;1996) that the effect might be a consequence of the increasing static interaction with the undistorted target which results in progressively larger (smaller) impact parameters for the positron (electron) with associated deceleration (acceleration) of the projectile. Stimulated by this hypothesis, studies have been extended to include all inert atoms for both $e^+$ (Kara et al 1997) and $p^-$ (Paludan et al 1997b). These studies have confirmed the importance of trajectory effects for the lighter particles (Paludan et al 1997a). In Fig. 22, $R^+ = Q_i^+(e^+)/Q_i^+(p^+)$ and $R^- = Q_i^+(e^-)/Q_i^+(p^-)$ are plotted versus velocity for various targets showing that, whilst $R^+ \sim R^-$ for H, $H_2$ and helium, as Z increases so does the difference between $R^+$ and $R^-$, with $R^+$ ($R^-$) $\rightarrow$ 1 progressively more slowly (quickly).

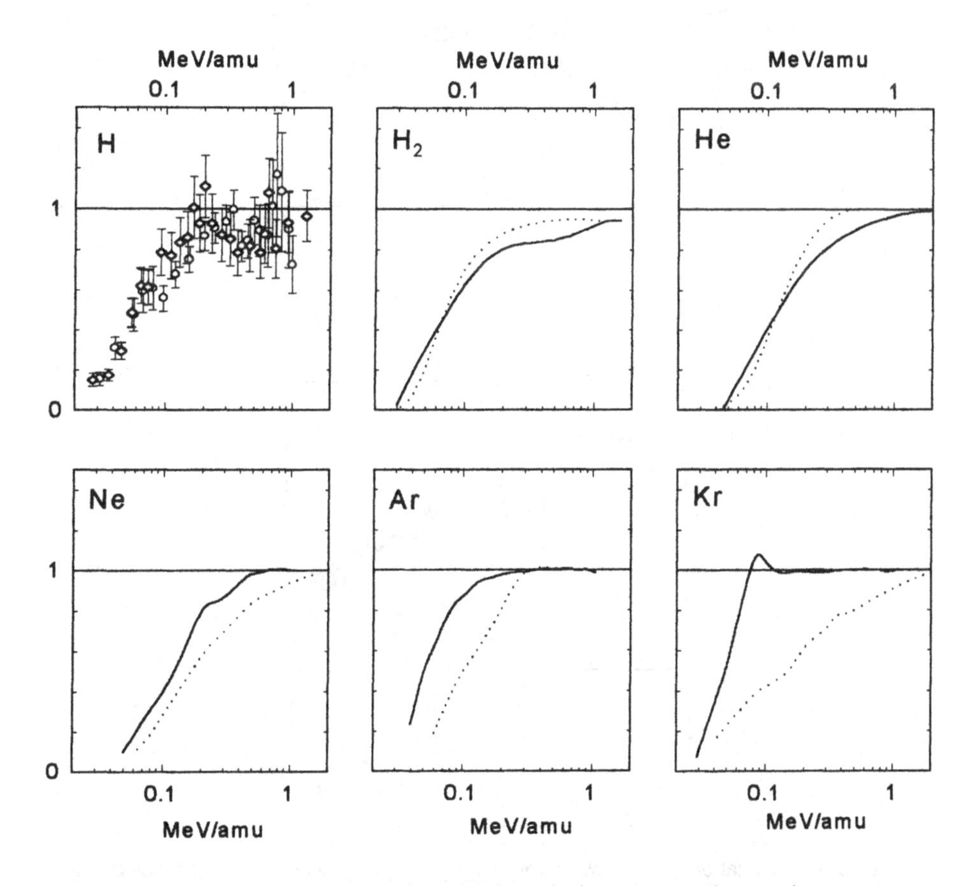


Fig. 22: Renormalized ratios $R^+ = Q_i^+(e^+)/ Q_i^+(p^+)$ and $R^- = Q_i^+(e^-)/ Q_i^+(p^-)$ plotted versus projectile velocity (Paludan et al 1997a).

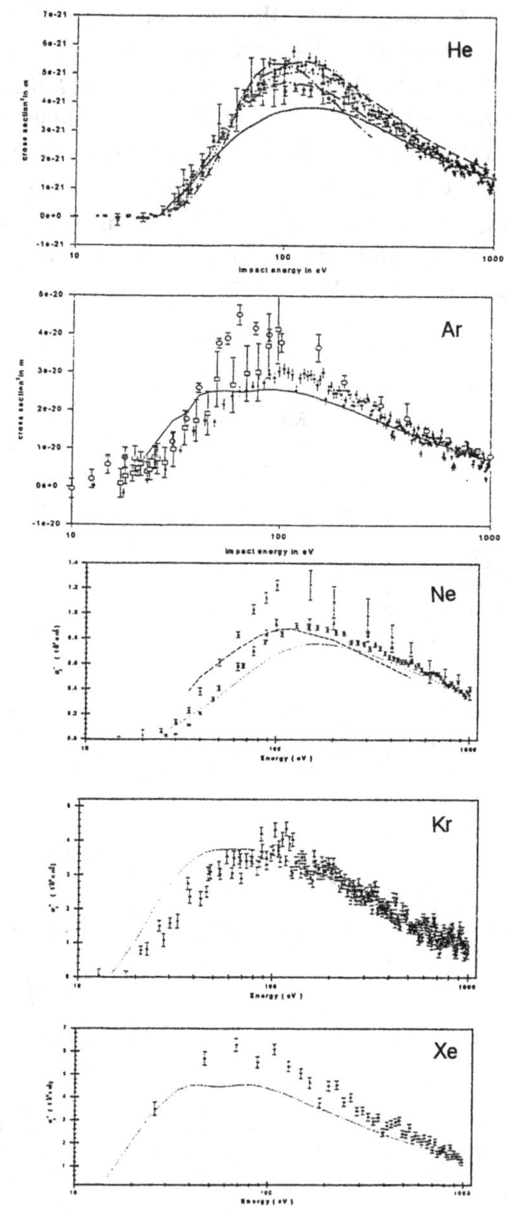

Fig. 23: Direct ionization cross-section for the inert atoms compared to corresponding electron measurements and theories (He and Ar: Moxom et al 1996; Ne, Kr and Xe: Kara et al 1997).

As shown in Fig. 23, fair agreement is generally found among recent experimental determinations of $Q_i^+(e^+)$ for He and theories. Measurements for the other inert atoms are also shown in Fig. 23 where they are compared with corresponding electron measurements and calculations.

The study of the near threshold behaviour of $Q_i^+$ has attracted much interest from both theorists and experimenters since Wannier (1953) proposed that, near the threshold, the energy dependence of the cross-section for double $e^-$ escape from a singly charged positive ion depends purely on the asymptotic configuration of the final state. Using classical arguments, he obtained the following expression for the cross-section $Q_i^+(e^-)$ for single ionization of a neutral atom:

$$Q_i^+(e^-) \propto (E')^{\eta} \qquad (7)$$

where $E'=E-E_i$ and $\eta=1.127$. Much support for this power law has ensued from both experiment (e.g. Cvejanovic and Read 1974; Donahue et al 1982; Hippler et al 1983) and theory (e.g. Peterkop 1971; Rau 1971; Crothers 1986; Rost 1994) although from time to time doubts have been raised (e.g. Crowe et al 1990; Friedman et al 1992) as to whether another model, in which the outer $e^-$ escapes in the Coulomb dipole field set up by the inner $e^-$ and the ion, might provide a better description (Temkin 1984). Theoretically, the problem has been considered also for positron impact and various results are included in Table 2.

Table 2. Near-threshold energy dependence of the single ionization cross-section by $e^{\pm}$ impact.

| Theory | Prediction |
|---|---|
| Wannier (1953), Peterkop (1971), Rau (1971), Crothers (1986), Rost (1994) | $Q_i^+(e^-) \propto (E')^{1.127}$ |
| Klar (1981), Rost and Heller (1994) | $Q_i^+(e^+) \propto (E')^{2.651}$ |
| Wetmore and Olson (1986) | $Q_i^+(e^+) \propto (E')^{3.01}$ |
| Sil and Roy (1996) | $Q_i^+(e^+) \propto (E')^{1.5}$ |
| Ihra et al (1997) | $Q_i^+(e^+) \propto (E')^{2.640} \exp(-0.73\sqrt{E'})$ |
| Temkin (1984) | $Q_i^+(e^{\pm}) \propto E' (\ln E')^{-2} [1 + C \sin(\alpha \ln E' + \mu)]$ |

Threshold measurements of $Q_i^+(e^+)$ are difficult because of the low beam intensities, poor energy resolutions and the large background from Ps formation. Recently, however, $Q_i^+(e^+)$ has been determined in the vicinity of the threshold for helium and molecular hydrogen by Ashley et al (1996). Their results are shown in Fig.

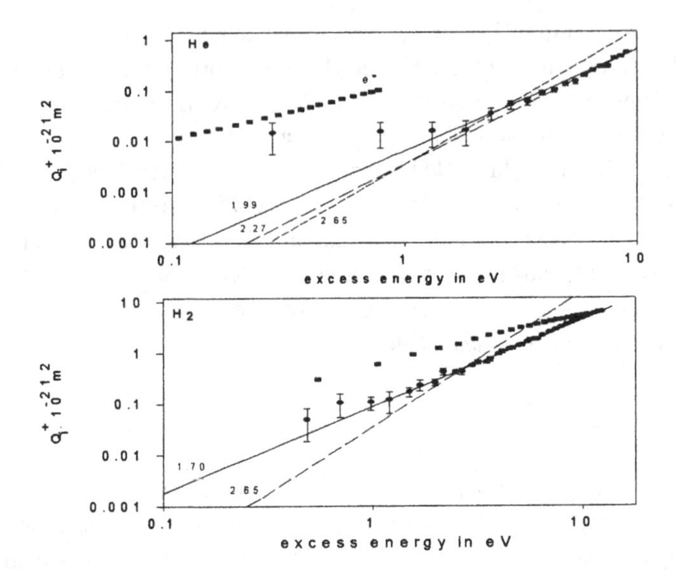


Fig. 24: Near threshold energy dependence of the direct ionization cross-section in He and $H_2$ (Ashley et al 1996). Various fits to Eq. (7) are indicated.

24 where it can be seen that, contrary to earlier suggestions (Knudsen et al 1990), this cross-section has a different energy dependence (as well as magnitude) from its electron counterpart. If fitted to a Wannier-type power law, the data up to E'=3eV are consistent with an exponent of ~2, i.e. significantly smaller than predicted by theory over the same energy range (e.g. Klar 1981). This discrepancy was interpreted by Ashley et al (1996) as suggesting that the Wannier description for positrons might apply over a smaller range of energies above threshold than predicted. This conclusion is now supported by calculations (Ihra et al 1997) where anharmonic corrections to the Wannier law are found to become important at smaller values of the energy above threshold for positron rather than for electron impact.

*Multiple ionization*

In double ionization, electron-electron correlations complicate an accurate theoretical description. At high velocities, three main mechanisms may be distinguished. The first is a two step-process involving the projectile interacting with one electron which then liberates a second electron. In the second, the projectile interacts with both electrons in succession whilst, in the third, the second electron is ejected as the ion relaxes after the collision. McGuire (1982) first suggested that whilst the cross-sections for each of these processes are proportional

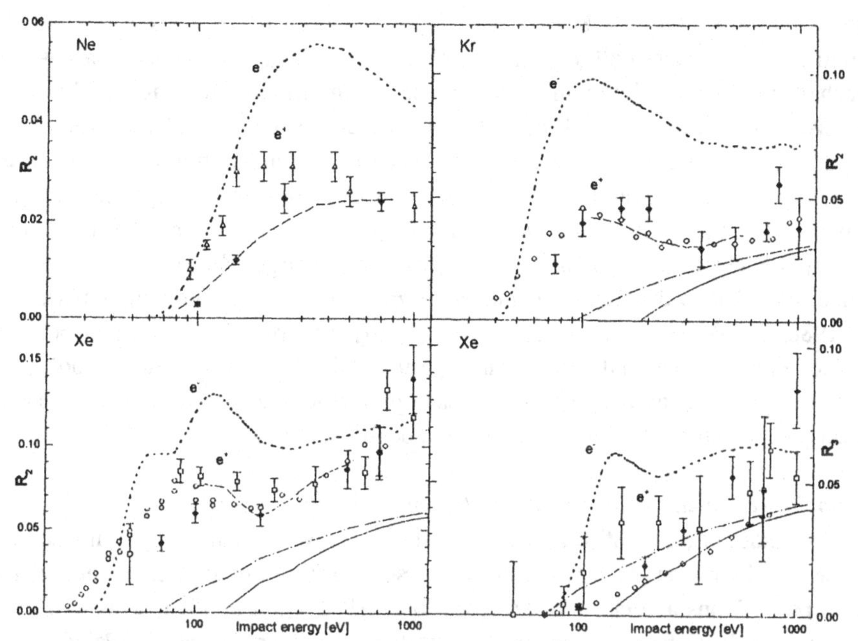


Fig. 25: The cross-section ratios for $e^-$ and $e^+$ $R_2 = Q_i^{2+}/Q_i^+$ for Ne, Kr and Xe and $R_3 = Q_i^{3+}/Q_i^+$ for Xe (Kara et al 1998). The dashed lines are predictions (see text). The dash-dot land solid lines are the contributions from M and N shell ionization (Helms et al 1995).

to even powers of the projectile charge q, interference among the scattering amplitudes should introduce a $q^3$ term. This charge dependence was initially confirmed by Andersen et al (1987) who found that the value of $R_2 = Q_i^{2+}/Q_i^+$ for $e^-$ and $p^-$ projectiles incident on helium merged at velocities of the order of MeV/amu whilst, in the same range, $R_2(p^+) \sim 0.5\, R_2(p^-)$. Additional confirmation was provided by corresponding measurements with positrons incident on helium, neon and argon (Charlton et al 1988, 1989).

Since these first investigations, further studies of double and even triple ionization with positrons have ensued with the most recent being those of Kara et al (1997, 1998). Their values of $R_2$ for neon, krypton and xenon as well as $R_3 = Q_i^{3+}/Q_i^+$ for xenon are compared to previous determinations (Charlton et al 1989, Helms et al 1995, Kruse et al 1991) and corresponding $e^-$ measurements (Krishnakumar and Srivastava 1988) in Fig. 25. In general the results of Kara et al are smaller than earlier data at lower energies, possibly due to better discrimination against ions accompanied by Ps formation. In all cases, $R_2(e^-) > R_2(e^+)$ as expected from the quantum interference term although, in the case of xenon, they appear to merge at high energies. Helms et al (1995) suggested that this is due to the dominance of inner-shell ionization followed by Auger decay, so that as one mechanism becomes dominant, interference is diminished. By adapting the semi-empirical formula of Lotz (1967, 1968) for inner and outer shell ionization by $e^-$ impact, they estimated the contribution to inner shell ionization to $R_2$ and $R_3$ for $e^+$. As shown in Fig. 25, their results broadly reproduce the energy dependence of the experimental data at high energies, confirming their suggestion on the importance of M and N shell ionization and Auger decay for the larger-Z targets in this energy range. Included in the figure are the values of $R_2$ ($e^+$) which, following Charlton et al (1988, 1989), were computed from $R_2(e^-)R_2(p^+)/R_2(p^-)$. The agreement between the computed values and the recent experimental data is surprisingly good as it implies that all phenomena specific to each projectile (e.g. exchange and trajectory effects) are equally important in single and double ionization.

*Dissociative ionization, dissociative attachment, positronium compounds*

Studies of dissociative ionization and dissociative attachment by $e^+$ impact are scarce. The high energy behaviour of the cross-sections for non-dissociative ionization of $N_2$ by positrons and electrons has been found (Poulsen et al 1994) to resemble the charge dependence found in the double ionization of lighter inert atoms (Paludan et al 1997a; Kara et al 1997). Ford and Reading (1990) have explained that the $q^3$ interference term should indeed be important in close collisions such as ionization-excitation processes of which dissociative ionization is an example. As the complexity of the target is increased, Poulsen et al (1994) found that the differences between $e^-$ and $e^+$ disappear and attributed this to the increasing number of channels available for ion creation. Xu et al (1994, 1995) have found similarities in the ionization/fragmentation patterns for positrons and photons. This work, in which the energies for ionization/fragmentation are provided by annihilation, will be discussed section 3.8.

Attachment may result in the production of new molecular species in which positronium substitutes, for example, H as in PsOH, PsH, PsF. By measuring the threshold for the dissociative attachment reaction

$$e^+ + CH_4 \rightarrow PsH + CH_3^+ \qquad (8)$$

Schrader et al (1992) deduced a binding energy of approximately 1 eV for the Ps compound. The status concerning calculations of the binding energies of these compounds, as well as current efforts towards experimental investigations in this area, have recently been reviewed (Schrader et al 1996, Schrader 1998).

### *3.7.2 Differential studies*

Major advances in the understanding of few-body physics involving correlated particles have been achieved through the measurement of electron impact differential ionization cross-sections and detailed comparisons with ever more sophisticated theories (see e.g. Lahmam-Bennani, 1991 and references therein). Part of the motivation to attempt similar work for positron impact is, again through comparisons with data for electrons, other simple projectiles and theory, to gain a greater understanding of collision dynamics, though with the new feature of strong electron-positron correlation. In particular this can lead to the phenomenon of electron capture to the continuum (ECC) in which the positron and electron emerge from the collision unbound but with similar velocities. It is expected that this effect will produce a cusp-like feature in the energy spectra of the ejected particles at half of the residual energy. This is similar to cusps found in the ejected electron energy spectra following the impact of heavy positively charged particles (e.g. Rødbro and Andersen 1979, Briggs 1989 and references therein). By contrast, negatively charged particles show the opposite, so-called anti-cusp behaviour, since the particles are unlikely to emerge velocity-matched (see e.g. Guang-yan et al 1992, Golden et al 1996, Briggs 1989 and comments below).

Successive layers of differential information are available in ionizing collisions. The simplest quantity is the single differential cross-section, $dQ_i^+/d\Omega$ where all positrons scattered (or electrons ejected) into a particular solid angle are measured, irrespective of their kinetic energy and the fate of the undetected particle. Collision parameters can be specified further by measuring the double differential cross-section, $d^2Q_i^+/dEd\Omega$, where the E and $\Omega$ refer to the energy and angle of the scattered positron or the ejected electron. Except where spin is expected to play an important role, the collision kinematics can be completely specified by determining the triple differential cross-section, $d^3Q_i/d\Omega_1 d\Omega_2 dE_1$ where the solid angles refer to the positron and electron (or two electrons for electron impact).

Measurements of $dQ_i^+/d\Omega$ for $e^+$-Ar collisions have been reported by Finch et al (1996) at a fixed angle of 60° over the energy range 40-150eV, with particular emphasis on the 50-60eV range where structures in $dQ_{el}/d\Omega$ were previously reported by Dou et al (1992). However, $dQ_i^+/d\Omega$ was found to be approximately constant between 40 and 60eV, before falling monotonically at higher energies. A more comprehensive series of measurements of $dQ_i^{z+}/d\Omega$ for both z=1 and 2 for argon and krypton targets has been reported by Falke et al (1997) at energies between 75 and 120eV.

The first reported study of $d^2Q_i^+/dEd\Omega$ for positron impact ionization was that of Moxom et al (1992) who, using a magnetically confined beam, conducted a search for ECC in $e^+$-Ar collisions. In this experiment, electrons ejected over a restricted angular

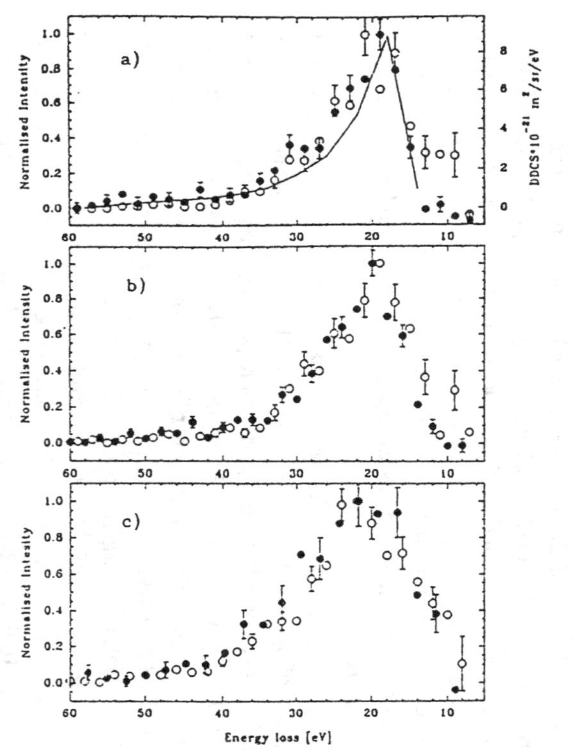


Fig. 26: Energy distributions for positrons (filled symbols) and electrons (hollow symbols) at impact energies of a) 100, b) 150 and c) 250eV at 0° (Kövér et al, 1993). The full curve represents the results of a CTMC calculations (Sparrow and Olson 1994).

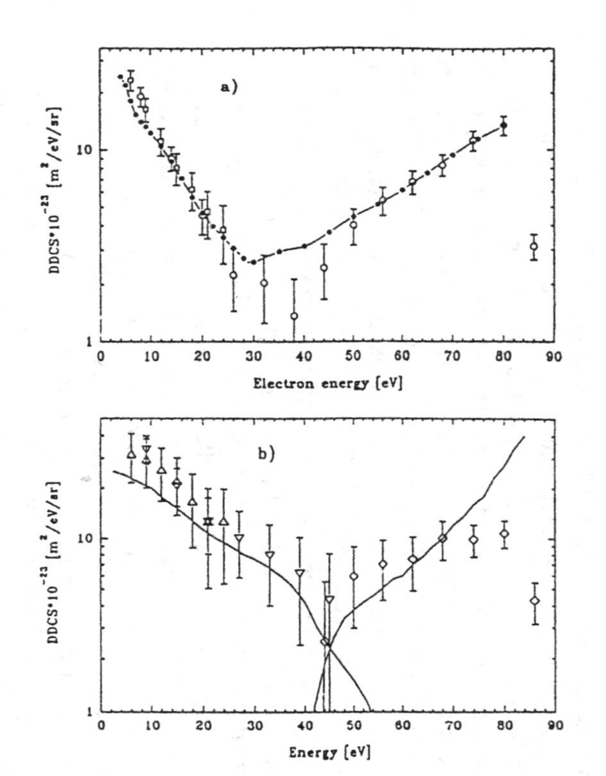


Fig. 27: Double differential cross-sections at 30° for a) $e^-$ and b) $e^+$ impact on Ar at 100eV (Kövér et al 1994). The filled points: DuBois and Rudd (1978). In b), triangles: ejected $e^-$; diamonds: scattered $e^+$ and full curve: CTMC calculation (Sparrow and Olson 1994).

range around 0° following positron impact at energies of 50, 100 and 150eV were energy analysed to search for evidence of an ECC cusp. It was found that the lower electron energies are most favoured, with the number ejected dropping rapidly to close to zero at around a half of the impact energy. Close inspection of the energy spectra revealed small bumps at 40 and 60eV for incident energies of 100 and 150eV respectively. The ECC peak would be expected at around 42 and 67eV at these impact energies. Whether or not these features are associated with this process, the experiment found that, in contrast, for instance, to proton impact, ECC makes a small contribution to $d^2Q_i^+/dEd\Omega$ for positrons. This was not in accord with quantum mechanical calculations at the time (see e.g. Bandyopadhyay et al 1994 and references therein) which seem to overestimate the ECC contribution, though in line with a Classical Trajectory Monte Carlo (CTMC) approach of Schultz and Reinhold (1990). The latter did not find a sharp cusp in $d^2Q_i^+/dEd\Omega$ at the ECC energy, but rather a broad ridge-like feature which they attributed to the fact that the light positron can scatter over a large angular range.

The next generation of experiments employed electrostatic techniques to allow the energy spectra to be determined directly. The measured distributions for positrons and electrons at impact energies of 100, 150 and 250eV (Kövér et al 1993) which have scattered close to 0° are shown in Fig. 26. In the case of $e^+$, no structure which could be attributed to ECC was found, in accord with Moxom et al (1992).

Double differential cross-sections for $e^+$-Ar scattering at angles other than 0° have been reported by Kövér et al (1994) and Schmitt et al (1994) at 100eV impact energy and Kövér et al (1997) at 60eV. The results of Kövér et al (1994) at 30° are presented in Fig. 27. The shape of the data for ejected electrons for $e^-$ impact (Fig. 27a) is in good accord with that found by DuBois and Rudd (1978) and both exhibit the same features, namely a preponderance of electrons at both low and high energies. The former are usually attributed to the liberated $e^-$ and the latter to the scattered projectile, though of course due to exchange this demarcation cannot be unambiguous. As expected, the probability of getting two $e^-$ emitted of roughly the same energy is low. There are no theoretical data for comparison. The corresponding $e^+$ impact data can be found in Fig. 27b for both scattered $e^+$ and ejected $e^-$. The latter, which can now be unambiguously identified, again lie mainly at low energies with the scattered positrons in the correspondingly higher energy range. Again there is no evidence of a substantial contribution to the cross-section from ECC. The solid line is the CTMC calculation of Sparrow and Olson (1994) which is in good accord for the ejected $e^-$ spectra, but higher than experiment for the scattered $e^+$, particularly in the low energy loss region. Similar conclusions have been forthcoming from the study of Kövér et al (1997), who tentatively linked this to close $e^+$-atom collisions. Clearly much further work remains to be done in the study of the differential behaviour of positron impact ionization cross-sections.

### 3.8 Annihilation

The annihilation of positrons and positronium in gases has been studied for almost fifty years (for a review see Charlton 1985). During most of this time, techniques consisted of i) recording the time intervals between the emission of the fast $e^+$ (usually detected via a nuclear de-excitation photon emitted within a few ps) and the annihilation of $e^+$ or positronium during or after thermalization in the gas and/or ii) studying the energy spectra of the annihilation quanta.

More recently, the lifetime of positrons with well defined energies have been measured under conditions of binary $e^+$-molecule encounters by measuring positron survival probabilities in a Penning trap (Surko et al 1988). These studies have been performed as a function of confinement time, target species and temperature (Iwata et al 1995 and references therein). Under similar conditions, the energy spectra of the annihilation $\gamma$-rays have also been investigated (Tang et al 1992, Iwata et al 1997). Furthermore, Xu et al (1994, 1995), by using a mass spectroscopy technique (Hulett et al 1991) have observed and studied extensive molecular fragmentation below the threshold for Ps formation (so-called 'sub-Ps-ionization' region). The ion yields and fragments (which by energy conservation must be produced by annihilation) have been found to depend on bond type, structure, molecular size and incident positron energy. As summarized by Hulett et al (1996), sub-Ps-ionization yields have generally been found to *increase* with a) decreasing positron energy and b) increasing size of the molecule but to *decrease* with increasing number of bonds. The structure of the molecule also appears to play an important role. The physical mechanism responsible for these phenomena has not yet been fully understood but Crawford (1994) has pointed out that if annihilation occurs with an electron lying below the highest occupied molecular orbital, the ionized molecule is left in an excited state from which molecular fragmentation might result.

The Doppler broadening of the 511keV annihilation radiation of thermalized positrons is mainly due to the momentum distribution of the target electrons. By monitoring the lineshape of the annihilation quanta, Iwata et al (1997) have concluded that positrons annihilate predominantly with valence electrons and that the majority of positrons annihilate with equal probability with any valence electron. In the case of the heavier inert atoms, they have found that a few percent of positrons annihilate with inner shell electrons.

In the study of positron annihilation from gaseous targets, it is customary to define an effective number of electrons $Z_{eff}$ in each gas molecule in terms of the annihilation rate $\Gamma$ through

$$\Gamma = Q_{2\gamma}\rho_e v = \pi r_o^2 c\rho Z_{eff} \tag{9}$$

where $Q_{2\gamma}=(\pi r_o^2 c/v)$ is the non-relativistic Dirac cross-section for annihilation into 2 $\gamma$-rays from a free electron, $r_o$ is the classical electron radius, $c$ is the speed of light and $v$ the relative speed of the annihilating pair. The free electron density, $\rho_e$, in a molecular gaseous medium is taken equal to $\rho Z_{eff}$ with $\rho$ being the target number density. In this way, $Z_{eff}/Z$ has been found to vary, for example, between $\sim 10^4$ for the heavier alkanes and typically 1-8 for the noble gases (Paul and St. Pierre 1963; Heyland et al 1982; Iwata et al 1995). The physical significance of such a variation had been interpreted as arising from the formation of long-lived positron-molecule bound states or resonances connected with vibrational excitations of the molecule (e.g.. Dzuba et al 1996). However, recent experiments with protonated and deuterated alkanes have shown that vibrational excitations are not dominant factors in determining $Z_{eff}$ (Iwata et al 1997).

An alternative interpretation has linked the large values of $Z_{eff}$ to the probability of virtual positronium formation and pick off annihilation from this state (Laricchia and Wilkin 1997, 1998; Laricchia 1997). By expressing the lifetime of a virtual positronium atom through the Uncertainty Principle as $\Delta t \sim \hbar / |E-E_{Ps}|$ where E is the positron incident energy, the model predicts a quasi-resonant enhancement in $Z_{eff}$ as $|E-E_{Ps}| \rightarrow 0$. The model has been evaluated using a semi-empirical approach and, as shown in Fig. 28, a variation for $Z_{eff}$ of almost six orders of magnitude has been found for a variety of targets (Laricchia and Wilkin 1997, 1998) in broad accord with experiment (e.g. Iwata et al 1995). The general features predicted by the same model for the annihilation cross-section $Q_{ann}=(\pi r_o^2 c/v)Z_{eff}$ for He, H, Xe, $H_2$ and $C_{14}H_{10}$ as a function of the incident positron energy are shown in Fig. 29 (Laricchia 1997, Laricchia and Wilkin 1998). The figure illustrates that, whilst the direct annihilation probability is indeed expected to decrease as the positron incident energy is increased due to the reduced interaction time, the probability of annihilation via the occurrence of a virtual process increases as the relevant threshold for the process is approached, reaching a maximum (equal to the collision cross-section) at the threshold itself. The enhancement of the annihilation probability at the Ps formation threshold has now been confirmed by elaborate variational calculations (Humberston and Van Reeth 1998, Van Reeth and Humberston 1998) and awaits experimental investigation.

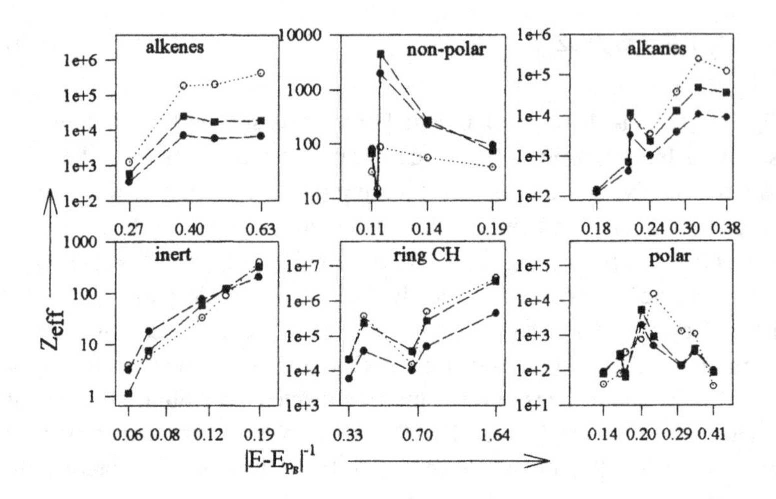


Fig. 28: Variation of $Z_{eff}$ with positronium formation threshold ($E_{Ps}$) calculated by Laricchia and Wilkin (1997, 1998) and compared to experiment - hollow symbols - (e.g. Iwata et al 1995). E is the positron incident energy.

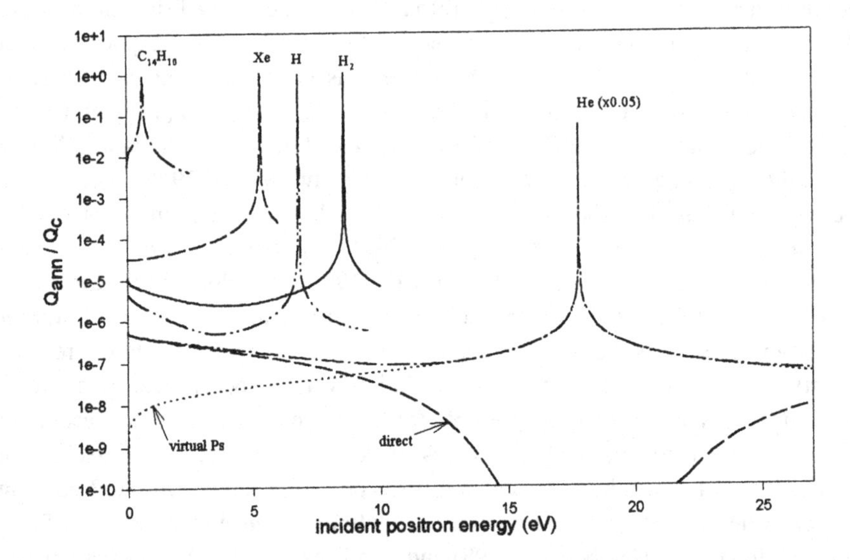


Fig. 29: Variation of $Q_{ann}/Q_c$ with incident positron energy for He, H, Xe, $H_2$ and $C_{14}H_{10}$ (Laricchia and Wilkin 1998). In the case of He, the contributions arising from virtual Ps formation and direct annihilation are shown explicitly.

# 4 Interactions of positronium with atoms and molecules

## *4.1 Introduction*

As mentioned in section 1, due to the coincidence of charge and mass in positronium, the mean static interaction and the adiabatic polarization are zero in the scattering of positronium from an atom or molecule. As a consequence, effects arising from the exchange interaction are expected to be more prominent than in $e^-$-atom scattering (e.g. Mott and Massey 1965, Drachman 1987). Unlike other (invariably much heavier) neutrals, recoil effects are expected to be as important for positronium as for $e^\pm$ (Manson and Ritchie 1985). Until very recently, information on Ps scattering was restricted to indirect methods and low energies. For example, the scattering of positronium from a range of atoms and molecules permeating the voids in silica aerogel has been studied using a one-dimensional Angular Correlation of the Annihilation Radiation (ACAR) technique (Nagashima et al 1992, 1995, 1998). In this work, the Ps atoms are ejected from the silica grains into free space with an average kinetic energy of ~0.8 eV and lose energy via collisions with other silica grains and/or gas atoms/molecules. A simple elastic scattering model has been used to obtain from the data a value of the momentum transfer cross-section. Values of the elastic scattering cross-sections for positronium-noble gas scattering averaged over 0-6.8eV have also been indirectly deduced by Coleman et al (1994) from a two-dimensional ACAR study.

Theoretical activity in the area of energetic Ps collisions has remained sporadic since the pioneering study of Massey and Mohr (1954). For a compilation of earlier calculations, see Charlton and Laricchia (1991). New impetus, however, has recently been imparted by developments in the production of collimated monoenergetic Ps atoms which are enabling direct determinations of total cross-sections from simple atomic and molecular systems.

Aside from the intrinsic interest of studying the atomic and molecular interactions of a neutral light particle such as positronium, a knowledge of Ps scattering cross-sections are important also for understanding $e^+$ slowing down in certain media (e.g. Laricchia and Jacobsen 1985), in studies of Ps diffraction from surfaces (Canter 1983), in modelling interactions of positronium injected into a plasma (Surko and Murphy 1990) and precision studies of positronium properties (e.g. Al-Ramadhan and Gidley 1994).

### *4.2 Positronium beams*

Collimated, energetic positronium may be produced by passing a positron beam through a neutralizing medium. The angular and energy distribution of the ensuing Ps beam depend on the medium and on the characteristics of the incident $e^+$ beam itself. Several methods have been investigated for producing fast positronium. These have employed either a gas or a solid as the neutralizer, the latter in the form of a film (transmission geometry) or a thick target (reflection geometry). Various methods (together with their relative merits with respect to efficiency of production, energy and angular distributions, as well as quantum state population) have been recently appraised by Laricchia (1995c). Table 3 provides a summary.

Methods employing a solid target are more easily suited to studies requiring ultra-high-vacuum (UHV) conditions. They are based on the formation of positronium by positrons which have not thermalized, either because of their shallow implantation, due to their low energies (Howell et al 1986) and/or geometry (Gidley et al 1987), or because they are sufficiently energetic to be transmitted through a thin target (Mills and Crane 1985). As a result, the ensuing positronium is generally characterized by wide energy distributions and may include excited-states. Efficiencies for the production of (n=2) positronium of the order of $10^{-3}$ have been measured for clean surfaces (Schoepf et al 1992) and in the range $10^{-4}$-$10^{-2}$ in non-UHV conditions (Canter et al 1975; Steiger and Conti 1992; Day 1993). Depending on experimental conditions, therefore, the contribution of Ps* to a beam might be non-negligible, especially for S-states whose lifetimes scale approximately as $n^3$.

Neutralization of a $e^+$ beam through a gaseous target has been demonstrated to produce energy tunable beams of considerably smaller energy spreads. The energy resolution of the Ps beam depends on the $e^+$ energy spread, the relative production of positronium in its various quantum states and its production simultaneous to other scattering processes.

Brown (1985) first observed energy-tunable positronium by identifying wings on either side of the 511keV positron annihilation line as being due to Doppler-shifted $\gamma$-rays from the in-flight annihilation of p-Ps formed by a positron beam multiply traversing a cell containing helium gas. At UCL, the overall efficiency for formation, transport and detection of collimated o-Ps was measured (Laricchia et al 1987). The positronium was formed in a 2cm-long gas cell and detected downstream by a channel-electron-multiplier-array (CEMA) in coincidence with one of the $\gamma$-rays from the annihilation of the positron. A time-of-flight method (Laricchia et al 1988b) was subsequently developed to characterize the Ps beam with respect to incident energy and, to some extent, quantum states (Zafar et al 1991; Laricchia et al 1992). In this work, a

Table 3. Summary for fast positronium production. $n$ indicates the principal quantum number of Ps, $\varepsilon_s$ its survival probability to the detector and $\varepsilon_d$ the ratio of the Ps to $e^+$ detection efficiencies (adapted from Laricchia 1995c).

| Method | % efficiency/($e^+$.sr) |
|---|---|
| Carbon foil, all $n$ (Mills and Crane 1985) | $0.5/\varepsilon_d$ |
| Glancing angle, all $n$ (Gidley et al 1987) | $1.5/(\varepsilon_d\varepsilon_s)$ |
| Backscattering, all $n$ (Howell et al 1986) | 5 |
| Backscattering, ($n$=2) | |
| non-UHV (Steiger and Conti 1992; Day 1993) | 0.3 |
| UHV (Schoepf et al 1992) | 0.03 |
| He, ($n$=1) (Zafar et al 1991; Laricchia et al 1992) | $1/\varepsilon_d$ |
| Ar, ($n$=1) (Zafar et al 1991; Laricchia et al 1992) | $3/\varepsilon_d$ |
| $H_2$, all $n$ (Tang and Surko 1993) | $8/\varepsilon_d$ |
| He, ($n$=1, high pressure) (Garner et al 1996) | $\leq 6.5/\varepsilon_d$ |
| Ar, ($n$=1, high pressure) (Garner et al 1996) | $\leq 4/\varepsilon_d$ |
| $H_2$, ($n$=1, high pressure) (Garner et al 1996) | $\leq 11/\varepsilon_d$ |

quasi-monoenergetic positron beam ($\Delta E \sim 3$eV) was accelerated and magnetically guided to a remoderator where it was "tagged" by detecting the secondary electrons thus liberated. These provided one of the timing signals in a delayed-coincidence sequence which was terminated by the detection of the o-Ps atoms (Laricchia et al 1988b). A similar method was later used to measure $Q_t$ for Ps scattering from atomic and molecular targets (Zafar et al 1996; Garner et al 1996).

A recent version of the system (comprising the tagger, a double cell arrangement and $\gamma$-ray detectors) is schematically illustrated in Fig. 30 (Garner et al 1996). The first cell is used for o-Ps creation whilst in the second cell a target gas may be introduced for Ps scattering studies. A retarder situated between the gas cells serves to prevent charged particles from entering the second cell. Both positrons and positronium are detected by a CEMA and, in close proximity to this, a NaI or CsI scintillator is used to detect the annihilation photons produced at the CEMA. A standard coincidence circuit is established between the signals from the two detectors to improve signal/background levels.

At sufficiently low neutralizer gas pressure, the Ps beam production efficiency $\varepsilon_{Ps}$ is approximately equal to $F_{Ps}$ as given by Eq. (5). At higher pressures, however, Ps scattering from the neutralizer may be significant and sets an upper limit on $\varepsilon_{Ps}$ according to

$$\varepsilon_{Ps} \propto \{1 - \exp(-\rho l_+ Q_{t+})\} F_{Ps} \exp(-\rho l_o Q_{to}) \qquad (10)$$

where the first term represents the fraction of scattered positrons, $F_{Ps}$ (as given by Eq. 5) the probability of forming positronium within an angle θ' and the last term the transmission probability of positronium through a target of number density $\rho$ and length $l$ (the subscripts + and o denote quantities associated with positrons and positronium respectively).

Fig. 31 shows the results of Garner et al (1996) for the variation in $\varepsilon_{Ps}$ with gas pressure for helium, argon and molecular hydrogen at several values of the Ps kinetic energies. Here it can be seen that, at the lower energies, molecular hydrogen possesses the highest $\varepsilon_{Ps}$ by up to a factor of 2, whilst at the highest energy investigated, argon provides the highest $\varepsilon_{Ps}$, by almost a factor of 2. From the pressure range where a linear dependence was observed, values were extracted to compare with theoretical predictions, as was shown in Fig. 18.

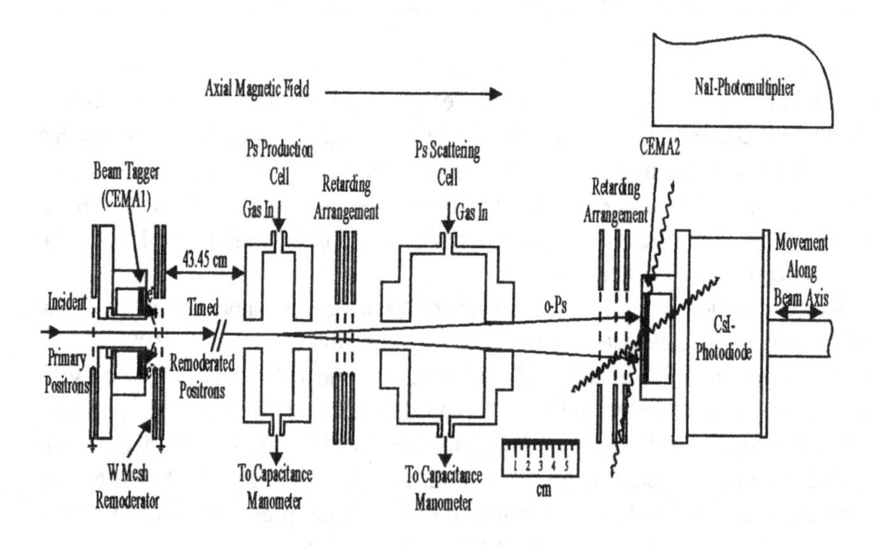


Fig. 30: Positronium beamline illustrating the tagger and double cell arrangement for positronium production and scattering (Garner et al 1996).

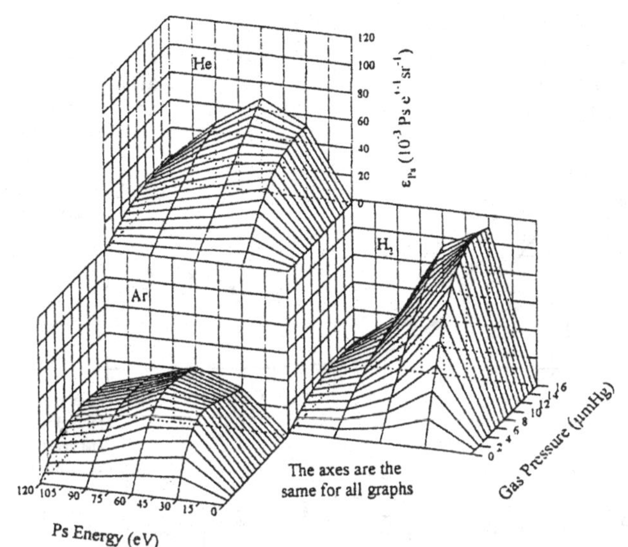


Fig. 31: Collimated Ps production efficiency versus gas pressure for helium, argon and molecular hydrogen at several values of the Ps kinetic energies (Garner et al 1996).

### *4.3 Total cross-sections*

Indirect estimates of Ps total cross-sections were deduced through Eq. (10) by Zafar et al (1991) using a single cell-technique by measuring the Ps beam intensity as a function of the pressure of the neutralizing gas. More recently, using the double cell arrangement shown in Fig. 30 and an attenuation method, the first direct measurements of the total cross-sections have been obtained (Zafar et al 1996; Garner et al 1996, 1998). By varying the angular acceptance of the detector, Garner et al (1998) have also been able to correct for the effect of forward scattering.

Results of the total cross-sections for Ps scattering from helium, argon and molecular hydrogen are shown in Fig. 32. In all cases, the experimental cross-sections display a rapid rise up to broad maximum before decreasing gradually at higher energies. More specifically, the direct measurements of Garner et al (1996) performed with an angular acceptance $\theta$=1.5° are, in the case of He, approximately 50% higher than the earlier indirect estimates of Zafar et al (1991). Also included in the Fig. 32 are the indirect estimates from the ACAR studies of the momentum and elastic scattering cross-sections of Nagashima et al (1995) and Coleman et al (1994), respectively. The calculations of Peach (1993), which implicitly include electron exchange but treat the target as frozen, agree qualitatively with the direct

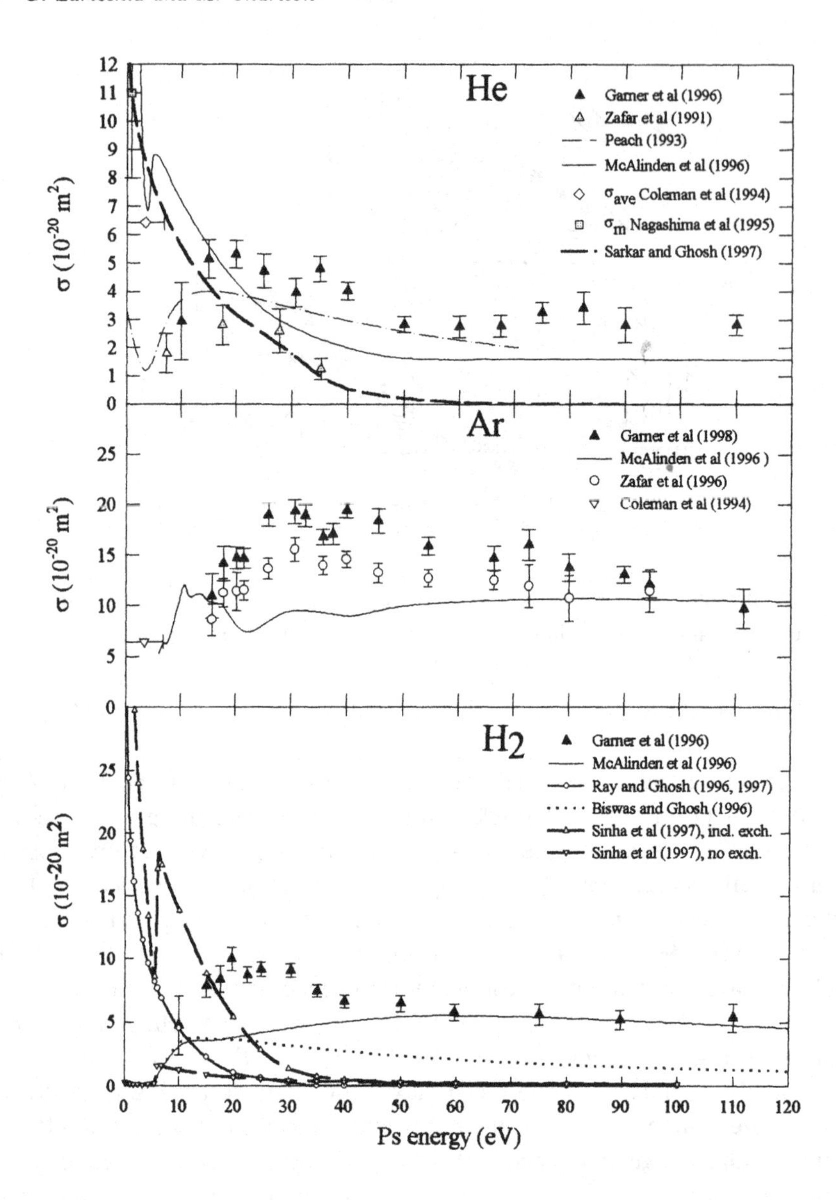


Fig. 32 Measurements of the total cross-sections for Ps scattering from He, Ar and $H_2$ compared with other available experimental data and theories.

experimental determinations over the whole energy range. Disagreements are found between experiment and the calculations of McAlinden et al (1996) which employ a coupled pseudostate approximation for collisions in which no change in the internal energy of the target occurs (so called target elastic) and the FBA otherwise (target inelastic). Exchange is not included in this theory. A static-exchange model has been employed by Sarkar and Ghosh (1997) in the calculation of the elastic scattering cross-section. By comparison with the first Born results, the importance of electron exchange has been found to persist up to 100eV or so. The model excludes all long range interactions which would tend to reduce the effect of the repulsive nature of the exchange interaction especially at the lower energies.

In the case of argon, the results of Garner et al (1998) are about 25% higher than those of Zafar et al (1996) due to an improved angular discrimination. The calculations of McAlinden et al (1996) are similar in approach to those for helium and agree with experiment only at the highest energies. Also included in the figure is the estimate for the average elastic cross-section of Coleman et al (1994).

For molecular hydrogen, the measurements of Garner et al (1996) performed with the same angular acceptance as for helium, are shown in Fig. 32 together with the calculations of Biswas and Ghosh (1996) for target elastic cross-section calculated by using the FBA. The theory neglects target inelastic processes, elastic scattering and even parity state transitions of the positronium-excitation channel. For comparison the results of Ps - H scattering of McAlinden et al (1996), Ray and Ghosh (1996, 1997) and Sinha et al (1997) are also shown multiplied by 2. The FBA calculations of McAlinden et al include target inelastic processes and, as for helium and argon, agree best, as expected, at the highest energies. The calculations of Ray and Ghosh (1996, 1997) use a static exchange model (in which the ground states of both atoms are retained) whilst those of Sinha et al (1997) employ a three-state Ps model close-coupling approximation (comprising elastic scattering and excitation to 2s and 2p states) with and without exchange. The results of the latter are shown in the figure where the importance of exchange below 50eV or so is evident.

A comparison of the total cross-sections for positronium, positron and electron has been made by Garner et al (1998). Here it was noted that $Q_t(Ps)$ exceeds that for $e^+$ at all energies investigated and that this was also true with respect to $Q_t(e^-)$, except at the lowest energies. Additionally it was noted that at energies where cross-sections $Q_t(e^+) \sim Q_t(e^-)$, the Ps cross-sections from helium, argon and molecular hydrogen, are approximately equal to the sum of $Q_t(e^+)$ and $Q_t(e^-)$. Much further experimental and theoretical work is anticipated.

## 5 Antihydrogen

### 5.1 Introduction

In January 1996, antihydrogen ($\overline{H}$) was observed at CERN's Low Energy Antiproton Ring (LEAR) for the first time (Baur et al 1996). The experiment exploited pair production by an antiproton in the vicinity of an atomic nucleus, Z, followed by capture of the positron by the antiproton. The reaction can be summarised as,

$$\overline{p}+Z \rightarrow (\overline{p}+e^{+}+e^{-}+Z) \rightarrow \overline{H}+Z+e^{-} \tag{12}$$

Similar processes had been studied for some time for ordinary matter (particularly pair production with subsequent electron capture by a highly charged ion) such that the cross-section for the antihydrogen reaction could be estimated readily (Munger et al 1994). The cross-section was known to increase, though rather weakly, with kinetic energy but to depend upon the second power of the charge on the atomic nucleus. Accordingly the collision target chosen was xenon and the antiproton kinetic energy set to 1.2GeV, close to the maximum at which LEAR could operate. Even so, the cross-section was only around $10^{-36}m^2$, approximately sixteen orders of magnitude lower than typical cross-sections discussed elsewhere in this chapter.

To compensate for this Baur et al (1996) used the stored, circulating antiproton beam at LEAR. The beam was made to cross the xenon gas target in the middle of one of the straight sections of the ring and any neutral object, such as antihydrogen, could be separated at the first corner of the ring as the antiprotons traversed the first bending magnet. Forty nanoseconds after creation, the antihydrogen atoms, travelling at around 0.9c, struck the first solid objects in their path, an array of three silicon detectors. The positron was stripped from the antiproton in either of the first two, whereupon it stopped and (predominantly) annihilated at rest. The two 511keV gamma rays were registered in coincidence using a segmented scintillator detector surrounding the silicon counters to signal the presence of a positron. The kinetic energy which the positron received upon break-up (~0.66MeV) of the antihydrogen was deposited in the silicon counters in addition to some of the 1.2GeV retained by the antiproton. This particle traversed all three silicon counters, an array of thin scintillators designed to measure its time of flight along a given path length and drift chambers to measure its position both before and after a bending magnet.

Signals from each of these detectors were used to isolate the antihydrogen from numerous other particles which could trigger some, but not all of the detectors. The result was that after around fifteen hours of beam on the xenon target, distributed over several days of running, 11-2±1 antihydrogen atoms were observed. This number is consistent with expectations based upon the cross-section, beam and target parameters and the expected detection efficiencies ( 2 here refers to events which could still possibly be due to some kind of background, and it has an associated statistical error of ±1). A follow-up experiment, which has a similar methodology and arrangement, is underway at Fermilab such that a total of around one hundred antihydrogen atoms has now been observed.

### *5.2 Prospects for the production of low energy antihydrogen*

A major goal of current antimatter research is the production of antihydrogen at low energies. It is hoped that this will allow precision comparisons of hydrogen and antihydrogen to be undertaken as tests of the CPT theorem and perhaps even the Weak Equivalence Principle for antimatter (see e.g. Charlton et al 1994; Eades 1995; Charlton and Holzscheiter 1998). In order to achieve these aims it is necessary to combine low energy positrons (or positronium atoms) and antiprotons and methods of doing this have been proposed and are being developed. One of the most successful techniques for accumulating low energy positrons in vacuum is the buffer gas cooling scheme developed by Surko and co-workers. We summarise this here, though note that extensive discussions of this accumulator can be found in the literature (Murphy and Surko 1992; Greaves et al 1994).

The cylindrically symmetric electrode structure of the so-called Penning-Malmberg trap and the variation of the electrical potential and gas pressure along the trap are shown in Fig. 33. The beam enters the electrode arrangement in region I where it encounters $N_2$ gas at a pressure of around $10^{-1}$ Pa . The gas is introduced at the centre of the first electrode, and differential pumping gives the required pressure gradients. The positrons are confined radially by an axial magnetic field of typically 0.1-0.2T. They pass through regions I, II and III and are then reflected by the electrical potential in the latter region. During the transit there is a reasonable chance, around 30%, of kinetic energy loss by electronic excitation of the $N_2$. Positrons which undergo this process are then trapped and eventually lose further energy by exciting vibrational and rotational transitions of the molecule. Thus, they are cooled to room temperature after approximately one second, whereupon they reside in region III at a pressure of $10^{-4}$ Pa. In this manner, positrons can be continuously trapped and this instrument has accumulated over $10^8$ positrons in times of the order of 100s. The positron lifetime in the trap is limited largely by annihilation on the buffer gas and is around 60s in region III, though this increases to 30 minutes if the $N_2$ gas is pumped out. Surko et al (1997) have also described how use of this system can be modified to allow its application for antihydrogen formation where stringent vacuum requirements must be met in order to avoid antiproton, or antihydrogen, annihilation. This involves the transfer of positrons to a separate storage region held at ultra-high vacuum, by temporarily opening and closing a gate valve which isolates the trap shown in Fig. 33 and the storage stage.

The production and manipulation of antiprotons is routine at some high energy accelerator laboratories. The Low Energy Antiproton Ring (LEAR) at CERN (although it closed in December 1996) has been a unique facility for the storage of antiprotons at low energies (≥6MeV). It is by extracting bursts of antiprotons from LEAR that advances in their deceleration and subsequent trapping and cooling have been possible. This machine will be replaced by the Antiproton Decelerator (AD) in 1999.

The first successful capture of antiprotons in a Penning trap was achieved by Gabrielse et al (1986) with similar work reported later by Holzscheiter et al (1996). In a typical trapping sequence, antiprotons in a 200ns burst are moderated by their passage through matter as they enter the vacuum chamber of the Penning trap. Radial confinement of the antiprotons is ensured by the large magnetic field provided by a superconducting solenoid. The electrostatic elements of the trap are an arrangement of cylindrical tubes located along the axis of the magnetic field to form a so-called open endcaps Penning trap (Gabrielse et al 1989) and appropriate potentials applied to the inner elements of this array form a harmonic well. Superimposed on this is a catching trap comprising an endcap, to which a (negative) dc potential of several kV is applied,

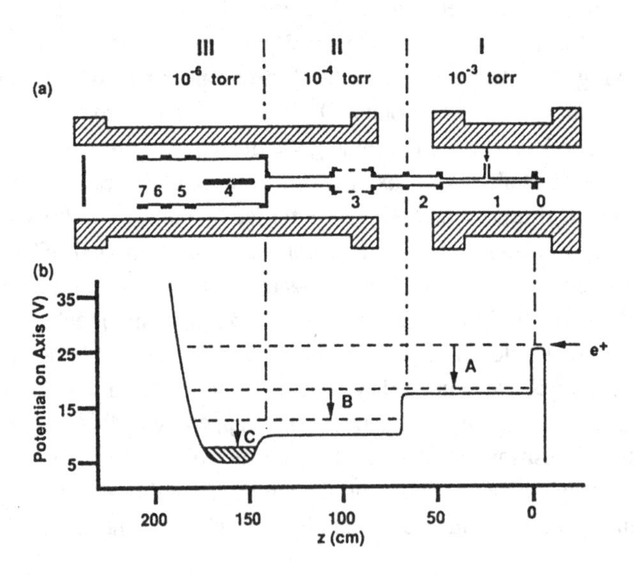


Fig. 33 Schematic diagram of a positron trap (Murphy and Surko 1992).

and an aluminium degrader foil at the trap entrance. Thus, some of the antiprotons which leave the degrader foil are repelled by the endcap voltage and return towards the entrance where the sudden application of a large negative potential to the foil completes the trapping procedure. Overall capture efficiencies (the number captured relative to that ejected by LEAR) can be as high as 0.5% for a 15kV trapping potential.

In the absence of a cooling process, the antiprotons travel back and forth in the trap maintaining, to a high degree, their initial kinetic energy. Fortunately, electrons can also be introduced into the trap in large numbers where they rapidly cool to the ambient temperature (close to 4.2K) and settle in the small axial harmonic well mentioned earlier. As the antiprotons pass to and fro, they cool by dissipating their kinetic energy to the trapped electrons. Within a time period of around tens or hundreds of seconds they reach thermal equilibrium with the electrons and are also trapped in the well.

Confinement times of hours were typical in the work reported by Holzscheiter et al (1996) and Feng et al (1997), though this was solely limited by the vacuum conditions in their trap. Under the extreme vacuum achieved in the fully cryogenic system of Gabrielse et al (1990) a cloud of antiprotons was held for a period of 3.4 months without noticeable loss.

Much discussion has been devoted to the possible mechanisms by which recombination might be achieved; here we simply list the reactions which are considered as likely candidates, together with recent appropriate references. The reactions are: radiative recombination, perhaps with laser stimulation (Wolf 1993, Müller and Wolf 1997), ternary recombination in dense positron plasmas (Gabrielse et al 1988, Pajek and Schuch 1997) and antiproton-positronium collisions (Charlton 1996, 1997), and can be summarised as

$$e^+ + \overline{p} \rightarrow \overline{H} + h\nu$$
$$e^+ + \overline{p} + mh\nu \rightarrow \overline{H} + (m+1)h\nu$$
$$e^+ + e^+ + \overline{p} \rightarrow \overline{H} + e^+$$
$$Ps + \overline{p} \rightarrow \overline{H} + e^-$$

Work on the matter counterparts of these reactions has been going on for some time (see Charlton et al 1994 and Charlton and Holzscheiter 1998 for overviews), with much of the more recent work concentrating on implementation in a trap-type environment. At the time of writing, though, none of the mechanisms has been demonstrated under such conditions. The reaction of protons and positronium to form hydrogen has recently been observed (Merrison et al 1997) in a crossed-beam experiment.

## 6 Conclusions and outlook

The developments of monoenergetic positron and positronium beams have been crucial for the controlled study of atomic and molecular physics phenomena involving these exotic particles. In this way, considerable advances have been made

Table 4. Outlook on experimentation in $e^+$ and Ps collisions (from Laricchia 1996).

*Positron scattering:*

1. as E, E' →0 (e.g. bound states, correlation effects);
2. resonance searches;
3. differential studies in Ps formation and ionization (interference effects and collision dynamics);
4. electronic excitations;
5. vibrational excitations, dissociation, fragmentation, positronic compounds;
6. spin polarized $e^+$ from paramagnetic molecules.

*Positronium scattering:*

1. as E →0 (e.g. bound states);
2. partial cross-sections;
2. spin polarized Ps (exchange and quenching).

not only in the understanding of their interactions with atomic and molecular matter but also in the treatment of unsolved problems in few-bodies collision physics in general. In recent years, particularly fruitful have been the studies of Ps formation and positron impact ionization. Stimulated by recent experimental progress, a new area of considerable activity has emerged concerning positronium collisions. With the on-going improvements in beam intensity, brightness and energy resolution, it is hoped that investigations of the type listed in Table 4 (Laricchia 1996) and detailed experimentation with atomic antimatter will become amenable to experimentation in the not-too-distant future.

## 7 Acknowledgements

It is a pleasure to thank all our co-workers and collaborators at UCL and elsewhere. We are also grateful to Andrew Garner for proof reading the script and to the Engineering and Physical Science Research Council, The Royal Society and NATO for their continuing support of positron research at UCL.

## 8 References

Alekseev A Sov Phys JETP 34 (1958) 826
Alekseev A Sov Phys JETP 36 (1959) 1312
Allison SK Rev Mod Phys 30 (1958) 1137
Al-Ramadhan AH and Gidley DW Phys Rev Letts 72 (1994) 1632
Andersen LH, Hvelplund P, Knudsen H, Møller SP, Sorenson AH, Elsener K, Rensfelt KG and Uggerhøj E Phys Rev A 36 (1987) 3612
Arbic BK, Hatamian S, Skalsey M, Van House J and Zhang W Phys Rev A 37 (1988) 3189
Asai S, Orito S and Shinohara N Phys Letts B 357 (1995) 475
Ashley P, Moxom J and Laricchia G Phys Rev Letts 77 (1996) 1250
Bandyopadhyay A, Roy K, Mandal P and Sil NC J Phys B 27 (1994) 4337
Bartschat K, McEachran RP and Stauffer AD J Phys B 21 (1988) 2789
Baur G, Boero G, Brauksiepe S, Buzzo A, Eyrich W, Geyer R, Grzonka D, Hauffe J, Kilian K, LoVetere M, Macri M, Moosburger M, Nellen R, Oelert W, Passagio S, Pozzo A, Röhrich K, Sachs K, Schepers G, Sefzick T, Simon RS, Stratmann R, Stinzing F. and Wolke M. Phys Letts B 368 (1996) 251
Berko S and Pendleton HN Ann Rev Nucl Part Sci 30 (1980) 543
Bhatia AK and Drachman RJ Phys Rev A28 (1983) 2523
Biswas PK and Ghosh AS Phys Lett A 223 (1996) 173
Biswas PK, Mukherjee T and Ghosh AS J Phys B 24 (1991) 2601
Bodwin GT and Yennie DR Phys Rep 48 (1978) 267
Bransden BH, Joachain CH and McCann JF J Phys B25 (1992) 4965
Briggs JS Comments At Mol Phys 23 (1989) 155
Brown BL in "Positron Annihilation" ed. PC Jain, RM Singru and KP Gopinathan (Singapore, World Scientific) (1985) 328
Campeanu RI, Fromme D, Kruse G, McEachran RP, Parcell LA, Raith W, Sinapius G and Stauffer AD J Phys B20 (1987) 3557
Canter KF in "Positron Scattering in Gases" ed. JW Humberston and MRC McDowell (New York, Plenum) (1983) 219
Canter KF, Coleman P G, Griffith TC and Heyland GR J Phys B. 5 (1972) L167
Canter KF, Mills AP Jr and Berko S, Phys Rev Letts 34 (1975) 177
Caswell WE and LePage GP Phys Rev A 18 (1978) 810
Chang TB, Hsaiowei T and Yaoquing L in "Positron Annihilation" ed. PC Jain, RM Singru and KP Gopinathan (Singapore, World Scientific) (1985) 212
Charlton M Rep Prog Phys 48 (1985) 737
Charlton M and Holzscheiter MH Rep Prog Phys (1998) in press
Charlton M and Laricchia G Comm At Mol Phys 26 (1991) 253
Charlton M and Laricchia G J Phys B 23 (1990) 1047
Charlton M Can J Phys 74 (1996) 483

Charlton M Hyp Int 109 (1997) 269
Charlton M, Clark G, Griffith TC and Heyland GR J Phys B. 16 (1983) L465
Charlton M., Anderson LH, Brun-Nielsen L, Deutch BI, Hvelplund P, Jacobsen FM, Knudsen H, Laricchia G, Poulsen MR and Pedersen JO J Phys B 21 (1988) L545
Charlton M, Brun-Nielsen L, Deutch BI, Hvelplund P, Jacobsen FM, Knudsen H, Laricchia G and Poulsen MR J Phys B 22 (1989) 2779
Charlton M, Eades J, Hörvath D, Hughes RJ and Zimmermann C Phys Reports 241 (1994) 65
Chu S, Mills AP Jr and Hall JL Phys Rev Letts 52 (1984) 1689
Coleman PG and McNutt JD Phys Rev Letts 42 (1979) 1130
Coleman PG, Johnston KA. Cox AMG, Goodyear A and Charlton M J Phys B 25 (1992) L585
Coleman PG, Rayner S, Jacobsen F M, Charlton M and West RN J Phys B 27 (1994) 981
Costello DG, Groce DE, Herring DF and McGowan JW Can Phys 50 (1972) 23
Crawford OH Phys Rev A 49 (1994) R3147
Crothers DSF J Phys B 19 (1986) 463
Crowe DM, Guo XQ, Lubell MS, Slevin J and Eminyan M J Phys B. 23 (1990) L325
Cvejanovic S and Read FH J Phys B7 (1974) 1841
Day DJ (1993) PhD Thesis University of London
Deb NC, McGuire JH and Sil NC Phys Rev A36 (1987) 1082
Deutsch M and Dulit E Phys Rev 84 (1951) 601
Donahue JB, Gram PAM, Hynes MV, Hamm RW, Frost CA, Bryant HC, Butterfield KB, Clark DA and Smith WW Phys Rev Lett 48 (1982) 1538
Dou L, Kauppila WE, Kwan CK, Przybyla D, Smith SJ and Stein TS Phys Rev Letts 68 (1992) 2913
Drachman RJ in "Atomic Physics with Positrons" ed. JW Humberston and EAG Armour (Plenum; New York) (1987) 203
Drachman RJ ed Proceedings of NASA Workshop on "Positron Annihilation in Gases and Galaxies" (NASA 3058) (1990)
DuBois RD and Rudd ME Phys Rev A 17 (1978) 843
Dzuba VA, Flambaum VV, Gribakin GF and King WA J Phys B. 29 (1996) 3151
Eades J Comments on Atomic and Molecular Physics 31 (1995) 51
Egan PO, Hughes VW and Yam MH Phys Rev A 15 (1977) 251
Falke T, Raith W, Weber M and Wesskamp U J Phys B 28 (1995a) L505
Falke T, Raith W and Weber M Phys Rev Letts 75 (1995b) 3418
Falke T, Brandt T, Kuhl O, Raith W and Weber MJ Phys B 30 (1997) 3247
Fee MS, Mills AP Jr, Chu S, Shaw ED, Danzmann K, Chichester RJ and Zuckerman DM Phys Rev Letts 70 (1993a) 1397

Fee MS, Chu S, Mills AP Jr, Chichester RJ, Zuckerman DM, Shaw ED and Danzmann K Phys Rev A 48 (1993b) 192
Fell RN Phys Rev Letts 68 (1992) 25
Feng X, Holzscheiter MH, Charlton M, Hangst J, King NSP, Lewis RA, Rochet J and Yamazaki Y Hyp Int 109 (1997) 145
Finch RM, Kövér A, Charlton M and Laricchia G J Phys B 29 (1996) L667
Floeder K, Horner P, Raith W, Schwab A, Sinapius G and Spicher G Phys Rev Letts 60 (1988) 2363
Fojón OA, Gayet R, Hanssen J and Rivarola RD Physica Scipta 51 (1995) 204
Ford AL and Reading JF J Phys B 23 (1990) 2567
Fornari L, Diana LM and Coleman PG Phys Rev Letts 51 (1983) 2276
Friedman JR, Guo XQ, Lubell MS and Frankel MR Phys Rev A46 (1992) 652
Frolov AM and Yeremin AYu J Phys B 22 (1989) 1263
Fromme D, Kruse G, Raith W and Sinapius G Phys Rev Letts 57 (1986) 3031
Gabrielse G, Fei X, Helmerson K, Rolston SL, Tjoekler R, Trainor TA, Kalinowsky H, Haas J and Kells W Phys Rev Letts 57 (1986) 2504
Gabrielse G, Haarsma L and Rolston SL Int J Mass Spectrom Ion Processes 88 (1989) 319
Gabrielse G, Rolston SL, Haarsma L and Kells W Hyp Int 44 (1988) 287
Gabrielse G, Fei X, Orozco LA, Tjoekler R, Haas J, Kalinowsky H, Trainor TA and Kells W Phys Rev Letts 65 (1990) 1317
Garner AJ, Laricchia G and Ozen A J Phys B 29 (1996) 5961
Garner AJ, Ozen A and Laricchia G Nucl Instr Meth B (1998) in press
Gidley DW, Mayer R, Frieze WE and Lynn KG Phys Rev Letts 58 (1987) 595
Gninenko SN, Klubhakov YuM, Poblaguev AA and Postoev VE Phys Letts B 237 (1990) 287
Golden DE, Xu Z, Bernhard J and Mueller DW J Phys B 29 (1996) 3741
Greaves RG, Tinkle MD and Surko CM Phys Plasmas 1 (1994) 1439
Griffith TC and Heyland GR Phys Rep 39C (1978) 169
Guang-yan P, Hvelplund P, Knudsen H, Yamazaki Y, Brauner M and Briggs JS Phys Rev A 47 (1992) 1531
Hagena D, Ley R, Weil D, Werth G, Arnold W, Rükert M and Schneider H Hyp Int 76 (1993) 297
Hatamian S, Conti RS and Rich A Phys Rev Letts 58 (1987) 1833
Helms S, Brinkmann U, Deiwiks J, Hippler R, Schneider H, Segers D and Paridaens J J Phys B 28 (1995) 1095
Heyland GR, Charlton M, Clark G, Griffith TC and Wright GL Can J Phys 60 (1982) 503
Hippler R, Klar H, Saaed K, McGregor I, Duncan AJ and Kleinpoppen H J Phys B. 16 (1983) L617
Hirose T Nucl Inst Meth B (1998) in press

Ho YK J Phys B. 16 (1983) 1530
Ho YK Phys Lett A 102 (1984) 348
Ho YK in Proceedings of NASA Workshop on "Annihilation in Gases and Galaxies" (ed RJ Drachman) NASA publication 3058 (1990) pp243-256
Hofmann A, Falke T, Raith W, Weber M, Becker D and Lynn KG J Phys B30 (1997) 3297
Holzscheiter MH, Feng X, Goldman T, King NSP, Lewis RA, Nieto MM and Smith GA Phys Lett A 214 (1996) 279
Howell RH, Alvarez RA and Stanek M Appl. Phys. Lett. 40 (1982) 751
Howell RH, Rosenberg IJ and Fluss MJ Phys Rev B 34 (1986) 3069
Hulett LD Jr, Lewis TA, Alsmiller RG, Peelle R, Pendyala S, Dale JM and Rosseel TM Nucl Instr Meth B24/25 (1987) 905
Hulett LD, Donohue DL, Lewis TA Rev Sci Instr 62 (1991) 2131
Hulett LD, Xu J, McLuckey SA, Lewis TA and Schrader DM Can J Phys 74 (1996) 411
Humberston JW and Van Reeth P Nucl Instr Meth B (1998) in press
Ihra W, Macek JH, Mota-Furtado F and O'Mahony PF (1997) Phys Rev Letts 78 4027
Iwata K, Greaves RG, Murphy TJ, Tinkle MD and Surko CM Phys Rev A 51 (1995) 73
Iwata K, Greaves RG and Surko CM Phys Rev A 55 (1997) 3586
Jacobsen FM, Charlton M, Chevallier J, Deutch BI, Laricchia G and Poulsen MR J Appl Phys 67 (1990) 575
Jacobsen FM, Frandsen NP, Knudsen H, Mikkelsen U and Schrader DM Phys B 28 (1995a) 4691
Jacobsen FM, Frandsen NP, Knudsen H and Mikkelsen U J Phys B 28 (1995b) 4675
Jones GO, Charlton M, Slevin J, Laricchia G, Kövér Á, Poulsen MR and Chormaic SN J Phys B 26 (1993) L483
Kahali S and Ghosh AS Phys Letts 239 (1995) 344
Kara V, Paludan K, Moxom J, Ashley P and Laricchia G J Phys B 30 (1997) 3933
Kara V, Paludan K, Moxom J, Ashley P and Laricchia G Nucl Instr Meth B (1998) in press
Katayama Y, Sueoka O and Mori S J Phys B 20 (1987) 1645
Kauppila WE and Stein TS Adv At Mol Opt Phys 26 (1990) 1
Kauppila WE, Kwan CK, Przybyla D, Smith SJ and Stein TS Can J Phys 74 (1996) 474 and references therein
Kernoghan AA, Robinson DJR, McAlinden MT and Walters HRJ J Phys B. 29 (1996) 2089
Khan P and Ghosh AS Phys Rev A 28 (1983) 2181
Khriplovich IB and Yelkhovsky AS , Phys Lett B 246 (1990) 520
Khriplovich IB, Milstein I and Yelkhovsky AS, Phys Lett B282 (1992) 237

Klar H J Phys B 14 (1981) 4165
Knudsen H and Reading JF Phys Rep 212 (1992) 107
Knudsen H, Brun-Nielsen L, Charlton M and Poulsen MR J Phys B 23 (1990) 3955
Knudsen H, Mikkelsen U, Paludan K, Kirsebom K, Moller SP, Uggerhøj E, Slevin J and Charlton M Phys Rev Letts 74 (1995) 4627
Kövér Á, Laricchia G and Charlton M J Phys B 26 (1993) L575
Kövér Á, Laricchia G and Charlton M J Phys B 27 (1994) 2409
Kövér Á, Finch RM Charlton M and Laricchia G J Phys B 30 (1997) L507
Krishnakumar E and Srivastava SK J Phys B 21 (1988) 1055
Kruse G, Quermann A, Raith W, Sinapius G, Weber M J Phys B 24 (1991) L33
Lahmam-Bennani A J Phys B 24 (1991) 2401
Laing EP and Dermer CD Optics Comm 65 (1988) 419
Laricchia G (1995a) in "The Physics of Electronic and Atomic Collisions" ed. LJ Dube, JBA Mitchell, JW McConkey and CE Brion (New York, AIP) p385
Laricchia G Nucl Instr Meth B 99 (1995b) 363
Laricchia G in "Positron Spectroscopy of Solids" Proceedings of the International School of Physics <<Enrico Fermi>> vol. 125 A Dupasquier and AP Mills, eds (IOS; Amsterdam) (1995c) p401
Laricchia G Hyp Int 100 (1996) 71
Laricchia G Mat Sc Forum 255 (1997) 228
Laricchia G and Jacobsen FM in "Positron (electron)-gas scattering" ed. WE Kauppila,S Stein and JM Wadehra (World Scientific, Singapore) (1985) 329
Laricchia G and Moxom J Phys Lett A174 (1993) 255
Laricchia G and Wilkin C Phys Rev Letts 79 (1997) 2241
Laricchia G and Wilkin C Nucl Instr Meth B (1998) in press
Laricchia G, Charlton M, Clark G and Griffith TC Phys Lett A109 (1985) 97
Laricchia G, Charlton M, Davies SA, Beling CD and Griffith TC J Phys B 20 (1987) L99
Laricchia G, Charlton M and Griffith TC J Phys B 21 (1988a) L227
Laricchia G, Davies SA, Charlton M and Griffith TC J Phys E 21 (1988b) 886
Laricchia G, Zafar N, Charlton M and Griffith TC Hyp Int 73 (1992) 133
Laricchia G , Moxom J and Charlton M Phys Rev Letts 70 (1993) 3229
Ley R Hyperfine Interactions 109 (1997) 181
Lotz W Astrophys J Suppl. 14 (1967) 207
Lotz W Z Phys. 216 (1968) 241
Lynn KG and Jacobsen FM Hyperfine Interactions 89 (1994) 19
Mandal P, Guha S and Sil NC J Phys B12 (1979) 2913
Manson JR and Ritchie RH Phys Rev Letts 54 (1985) 785
Massey HSW and Mohr CBO Proc Roy Soc 67 (1954) 695
Massey HSW Can Journ Phys 60 (1982) 461
McAlinden MT and Walters HRJ Hyperfine Interactions 73 (1992) 65

McAlinden MT and Walters HRJ Hyperfine Interactions 84 (1994) 407
McAlinden MT, MacDonald FRGS and Walters HRJ Can J Phys 74 (1996) 434
McEachran RP, Ryman AG and Stauffer AD J Phys B 12 (1979) 1031
McGuire JH Phys Rev Letts 49 (1982) 1153
Merrison JP, Bluhme H, Chevallier J, Deutch BI, Hvelplund P, Jørgensen LV, Knudsen H, Poulsen MR and Charlton M, Phys Rev Letts 78 (1997) 2728
Meyerhof WE and Laricchia G J Phys B 30 (1997) 2221
Meyerhof WE, Laricchia G, Moxom J, Watts MST and Humberston JW Can J Phys 74 (1996) 427
Mills AP Jr Phys Rev Letts 46 (1981) 717
Mills AP Jr Phys Rev Letts 50 (1983) 671
Mills AP and Crane WS Phys Rev A31 (1985) 593
Mills AP Jr and Bearman GH Phys Rev Letts 34 (1975) 246
Mills AP Jr and Zuckermann DM Phys Rev Letts 64 (1990) 2637
Mills AP, Voris SS and Andrew TS J Appl Phys 76 (1994) 2556
Mitroy J J Phys B 29 (1996) L263
Mitsui T, Fujimoto R, Ishisaki Y, Ueda Y, Yamazaki Y, Asai S and Orito S Phys Rev Letts 70 (1993) 2265
Mori S and Sueoka O J Phys B 27 (1994) 4349
Mott NF and Massey HSW "The Theory of Atomic Collisions" (Clarendon; Oxford) 1965
Moxom J, Laricchia G, Charlton M, Jones GO and Kövér Á J Phys B 25 (1992) L613
Moxom J, Laricchia G and Charlton M J Phys B 26 (1993) L367
Moxom J, Laricchia G, Charlton M, Kövér Á and Meyerhof WE Phys Rev A 50 (1994) 3129
Moxom J, Ashley P and Laricchia G Can Jour Phys 74 (1996) 367
Müller A and Wolf A Hyperfine Interactions 109 (1997) 233
Munger CT, Brodsky SJ and Schmidt I Phys Rev D 49 (1994) 3228
Murphy TJ and Surko CM Phys Rev A 46 (1992) 5696
Nagashima Y, Hyodo T and Fujiwara K Mat Sc Forum 105-110 (1992) 1671
Nagashima Y, Kakimoto M, Hyodo T, Fujiwara K, Ichimura A, Chang T, Deng J, Akahnae T, Chiba T, Suzuki K, McKee BTA and Stewart AT Phys Rev A 52 (1995) 258
Nagashima Y, Hyodo T, Fujiwara K and Ichimura A J Phys B (1998) 329
Nico JS, Gidley DW, Skalsey M and Zitzewitz PW Materials Science Forum 105-110 (1992) 401
Ore A and Powell JL Phys Rev 75 (1949) 1696
Orito S, Yoshimura K, Haga T, Minowa M and Tsuchiaki M Phys Rev Letts 63 (1989) 597
Overton N, Mills RJ and Coleman PG J Phys B26 (1993) 3951
Pajek M and Schuch R Hyperfine Interactions 108 (1997) 185

Palathingal JC, Asokakumar P, Lynn KG and Wu XY Phys Rev Letts 67 (1991) 3491
Palathingal JC, Asokakumar P, Lynn KG, Wu XY Phys Rev A51 (1995) 2122
Paludan K, Laricchia G, Ashley P, Kara V, Moxom J, Bluhme H, Knudsen H, Mikkelsen U, Møller SP, Uggerhøj E and Morenzoni EJ Phys B 30 (1997a) L581
Paludan K, Bluhme H, Knudsen H, Mikkelsen U, Møller SP, Uggerhøj E and Morenzoni EJ Phys B 30 (1997b) 3951
Paul DAL and St Pierre L Phys Rev Letts 11 (1963) 493
Peach G (1993) Private Communication to G. Laricchia (1995c)
Peterkop R J Phys B 4 (1971) 513
Petlentz P and Smith Jr VH Phys Rev A36 (1988) 5125
Pirenne J Arch Sci Phys Nat 28 (1946) 233
Poulsen MR, Frandsen NP and Knudsen H Hyperfine Interactions 89 (1994) 73
Przybyla DA, Kauppila WE, Kwan CK, Smith SJ and Stein TS Phys Rev A55 (1997) 4244
Rau ARP Phys Rev A4 (1971) 207
Ray H and Ghosh AS J Phys B 29 (1996) 5505
Ray H and Ghosh AS J Phys B 30 (1997) 3745
Rich A Rev Mod Phys 53 (1981) 127
Rødbro M and Andersen FD J Phys B 12 (1979) 2883
Rost JM Phys Rev Lett 72 (1994) 1998
Rost JM and Heller EJ Phys Rev A 49 (1994) R4289
Sarkar NK and Ghosh AS J Phys B 30 (1997) 4591
Schmitt A, Cerny U, Möller H, Raith W and Weber M Phys Rev A 49 (1994) R5
Schoepf DC, Berko S, Canter KF and Sferlazzo P Phys Rev A45 1407 (1992)
Schrader DM Nucl Instr Meth B (1998) in press
Schrader DM, Jacobsen FM, Frandsen NP and Mikkelsen U Phys Rev Letts 69 (1992) 57
Schrader DM, Hulett LD, Xu J, Laricchia G, Lewis TA and Horsky TN Can J Phys 74 (1996) 497
Schultz DR and Olson RE Phys Rev A38 (1988) 1861
Schultz DR and Reinhold CO J Phys B 23 (1990) L9
Schultz DR, Olson RE and Reinhold CO J Phys B 24 (1991) 521
Sinha PK, Chaudhury P and Ghosh AS J Phys B 30 (1997) 4643
Sil NC and Roy K Phys Rev A 54 (1996) 1360
Skalsey M and Conti RS Phys Rev D51 (1995) 6292
Smith SJ, Hyder GM, Kauppila WE, Kwan CK and Stein TS Phys Rev Letts 64 (1990) 1227
Sparrow RA and Olson RE J Phys B 27 (1994) 2647
Sperber W, Becker D, Lynn KG, Raith W, Schwab A, Sinapius G, Spicher G and Weber M Phys Rev Letts 68 (1992) 1019

Steiger TD and Conti RS Phys Rev A45 (1992) 2744
Stein TS, Jiang J, Kauppila WE, Kwan CK, Li H, Surdutovich A and Zhou S Can J Phys 74 (1996) 313
Surdutovich A, Jiang J, Kauppila WE, Kwan CK, SteinTS, Zhou S Phys Rev A 53 (1996) 2861
Surko CM and Murphy TJ Phys Fluids B2 (1990) 1372
Surko CM, Passner A, Leventhal M and Wysocki F Phys Rev Letts 61 (1988) 1831
Surko CM, Greaves RG and Charlton M Hyperfine Interactions 109 (1997) 181
Tang S and Surko CM Phys Rev A47 (1993) 743
Tang S, Tinkle MD,Greaves RG and and Surko CM Phys Rev Letts 68 (1992) 3793
Temkin A Phys Rev A 30 (1984) 2737
Tripathi S, Sinha C and Sil NC, Phys Rev A39 (1988) 2924
Van Reeth P 1997 private communication
Van Reeth P and Humberston JW J Phys B 30 (1997) L95
Van Reeth P and Humberston JW J Phys B. 31 (1998) L231
Varracchio Ficocelli E J Phys B. 22 (1990) L779
Walters HRJ, Kernoghan AA and McAlinden MT in "The Physics of Electronic and Atomic Collisions" ed LJ Dube, JBA Mitchell, JW McConkey and CE Brion (New York; AIP) (1995) 397
Wannier GH Phys Rev 90 (1953) 817
Watts MST and Humberston JW J Phys B 25 (1992) L491
Weber M, Hofman A, Raith W, Sperber W, Jacobsen FM and Lynn KG Hyperfine Interactions 89 (1994) 221
Wetmore AE and Olson RE Phys Rev A 34 (1986) 2822
Wheeler JA Ann NY Acad Sci 48 (1946) 219
Wolf A Hyperfine Interactions 76 (1993) 189
Wolfenstein LN and Ravenhall DG, Phys Rev 88 (1952) 279
Xie R, Petkov M, Becker D, Canter KF, Jacobsen FM, Lynn KG, Mills R and Roellig L Nucl Instr Meth B 93 (1994) 98
Xu J, Hulett LD, Lewis TA, Donohue DL, McLuckey SA and Crawford OH Phys Rev A 49 (1994) R3151
Xu J, Hulett LD, Lewis TA and McLuckey SA Phys Rev A 52 (1995) 2088
Yang CN Phys Rev 77 (1949) 242
Yang J, Chiba M, Hamatsu R, Hirose T, Matsumoto T and Yu J Phys Rev A 54 (1996) 1952
Zafar N, Chevallier J, Jacobsen FM, Charlton M, Laricchia G Appl Phys A 47 (1988) 409
Zafar N, Chevallier J, Laricchia G, Charlton M J Phys D 22 (1989) 868
Zafar N, Laricchia G, Charlton M and Griffith TC J Phys B 24 (1991) 4461
Zafar N, Laricchia G, Charlton M and Garner A Phys Rev Letts 4 (1996) 1595
Zecca A, Karwasz GP, Brusa RS Nuovo Cimento 19 (1996) 1

Zhou S, Kauppila WE, Kwan CK and Stein TS Phys Rev Letts 73 (1994) 236
Zhou S, Li H, Kauppila WE, Kwan CK and Stein TS Phys Rev A 55 (1997) 361
Ziock KP, Dermer CD, Howell RH, Magnotta F and Jones KM J Phys B 23 (1990a) 329
Ziock KP, Howell RH, Magnotta F, Failor RA and Jones KM Phys Rev Letts 64 (1990b) 2366

CHAPTER 4

# THE FATE OF SLOW POSITRONS IN CONDENSED MATTER

R.M. NIEMINEN
*Laboratory of Physics*
*Helsinki University of Technology*
*P.O. Box 1100*
*021015 HUT, Finland*
*E-mail: rniemine@csc.fi*
*http://www.hut.fi/Units/Physics*

This Chapter summarizes the basic processes for positrons entering condensed matter. The energy loss mechanisms and slowing-down of energetic positrons are discussed, and the key features of the ensuing implantation distributions are presented. This includes a discussion of the elementary excitations and scattering processes in the medium. The ballistic, diffusive and quantum regimes of positron motion are then described. The Chapter also discusses the physics of positron-ion interactions and electron-positron correlation. Their effects on the thermalised positron wave functions are presented, with the view of quantitative predictions of positron energetics (affinities, positron-lattice coupling *etc.*) and annihilation characteristics (annihilation rates, momentum distributions). The main features of positron trapping at and detrapping from bulk and surface defects are summarized. Finally, the different scenarios for positrons in the vicinity of surfaces and interfaces are discussed.

## 1 Introduction

Collimated, monoenergetic beams of positrons, with energies ranging from eV's to GeV's, are enormously useful over vast areas of physics and applications. This Chapter summarizes the basic physical properties of low-energy (up to tens on keV) positrons as they slow down in and interact with condensed matter. The ultimate fate of positrons in condensed matter is, of course, annihilation in the nanosecond timescale. However, during that time a rich variety of physical phenomena are explored, and they are reflected in the temporal, spatial and energetic distributions of the annihilation radiation.

Initially swift positrons *lose rapidly their energy* in a medium. Energy is lost to electron excitations, including target atom ionisations and collective plasmon-like processes. Phonon excitations take over at lower energies, which eventually lead to positron thermalisation.

Near and at thermal equilibrium, which is typically reached in a few picoseconds at room temperature, the positron motion continues as a *quantum diffusion* process, during which the positron momentum distribution is preserved. During diffusion, a positron scatters quasielastically off host atoms and their collective modes. Mainly during this stage positrons can get *trapped* at attractive potential wells due a *quantum-mechanical capture process*. The attractive wells are usually due to regions of low atomic density, such as open-volume lattice defects. In addition, the long-range Coulomb tails of charged impurities can act as traps at low temperatures, as can the image-potential wells outside surfaces.

The thermal stage ends when the positron *annihilates* with an electron. The concomitant radiation conveys information which can be used to reconstruct host electronic properties near the site of annihilation.

Many of the positrons implanted in the primary beam may reach the entrance surface prior to their annihilation, either before or after thermalisation. Several scenarios are then possible. Firstly, even thermalised positrons may be *spontaneously ejected* from the surface, as the positron work function can be negative. Secondly, positrons may escape by capturing a surface electron and emerging as *positronium* (Ps), as typical electron work functions are lower than the positronium binding energy. Thirdly, positrons may get trapped at image-potential induced *surface states* at the solid-vacuum interface. The surface state can be unstable towards thermal desorption. Finally, positrons as wave-mechanical particles may be *reflected* back to the interior from the potential step at the surface.

The versatility of scenarios implies that positrons can be used to extract different kinds of physical information of the surface and near-surface region. However, as charged particles positrons are strong probes which tend to disturb their electronic environment. Electron-positron correlation effects have to be considered when models and theories for positrons as surface probes are constructed.

Several recent reviews and monographs deal with positrons as condensed-matter and (near-)surface probes (Schultz and Lynn 1988, Ishii 1992, Puska and Nieminen 1994, Dupasquier and Mills 1995, Nieminen 1995a, 1995b). These should be consulted for more extensive bibliographies and details. This article summarizes the main concepts and provides an update of the more recent theoretical and computational developments.

## 2 Elementary excitations and scattering

Kilovolt electrons and positrons lose their energy in inelastic processes involving electronic and ionic degrees of freedom: radiative losses can usually be neglected. The electronic mechanisms include core ionisation and excitation, as well as single-particle (electron-hole) and collective (plasmon-like) excitations of the more loosely bound valence electrons. At low energies of a few eV and below, energy exchange with ionic degrees of freedom (phonon scattering) becomes operative. The ionic scattering mechanisms also include the scattering off collective density fluctuations in disordered amorphous and liquid systems.

### *2.1 Inelastic scattering*

#### ***2.1.1 Electronic excitations***

At high energies, the Möller and Bhabha cross sections (Rohrlich and Carlson 1954) describe electron and positron inelastic scattering off electrons, respectively. In this energy range, electrons lose their energy faster than positrons, but a crossover takes place at around 100 keV, where also the widely used Bethe-Bloch description for the stopping power becomes valid (Schultz *et al.* 1990). For electrons, the upper limit for energy transfer is half the particle energy due to the indistinguishability from the target electrons; additional differences between electrons and positrons arise from the exchange symmetry. Positrons are thus expected to lose their energy a little faster. The Bethe-Bloch theory ceases to be valid at energies corresponding to typical core excitations, although parametrisations do exist also in this energy region.

Models for inelastic scattering in the kilovolt energy region can be divided into two classes. One is based on the dielectric response of the medium (Quinn 1963), which naturally emphasises the collective and delocalized nature of the excitations. The other approach is based on scattering off individual atoms in the medium (Gryzinski 1965, Salvat *et al.* 1985), and uses localised atomic concepts such as generalised oscillator strengths. It is argued (Nieminen 1988) that both are needed in a quantitative theory of slowing down and implantation phenomena of electrons and positrons.

In linear response theory, the doubly-differential inelastic scattering cross section is proportional to the imaginary part of the inverse of the dielectric function of the medium. A well-studied model system is the electron gas. Following the pioneering work of Lindhard, numerous constructions for the energy- and momentum-dependent dielectric response function for solids have been presented (Ritchie *et al.* 1975, Ashley *et al.* 1979). The formalism includes both electron-hole and plasmon excitations in a natural way, with various schemes to account for electron-electron correlations in terms of so-called local-field corrections. The generalisation to describe the electron gas response to positrons is straightforward (Oliva 1980, Nieminen and Oliva 1980).

For real solids, the calculation of the dielectric function based on the electronic band structure is possible but technically cumbersome. Penn (1987) has presented a construction where experimentally available optical data can be used instead. Alternatively, the dielectric function can be approximately obtained by using electron-gas data and a local-density approximation. The inelastic scattering cross sections, valid at high energies (Gryzinski 1965, Salvat *et al.* 1985) and augmented with the dielectric response at low energies give a robust description for metals and semiconductors (Jensen and Walker 1993).

In insulators, the presence of the energy gap between the valence and conduction bands prohibits ionization and electron-hole pair creation processes when the positron has slowed down to energies equal to and smaller than the gap. Thereafter, processes which lead to the formation of excitons or positronium are possible until the positron kinetic energy is a few eV below the gap. Finally, only phonon scattering is possible (see below). As the relevant phonon energies are much smaller than, say, the band gaps of rare-gas solids, positrons in insulators can remain "hot" for relatively long periods of time.

Fig. 1. presents schematic data for the stopping power of electrons and positrons in the energy range relevant for positron-beam applications. The stopping power $S(E)$ at kinetic energy E is defined as

$$S(E) = -\frac{dE}{dx} \qquad (1)$$

*i.e.* the energy lost per distance travelled.

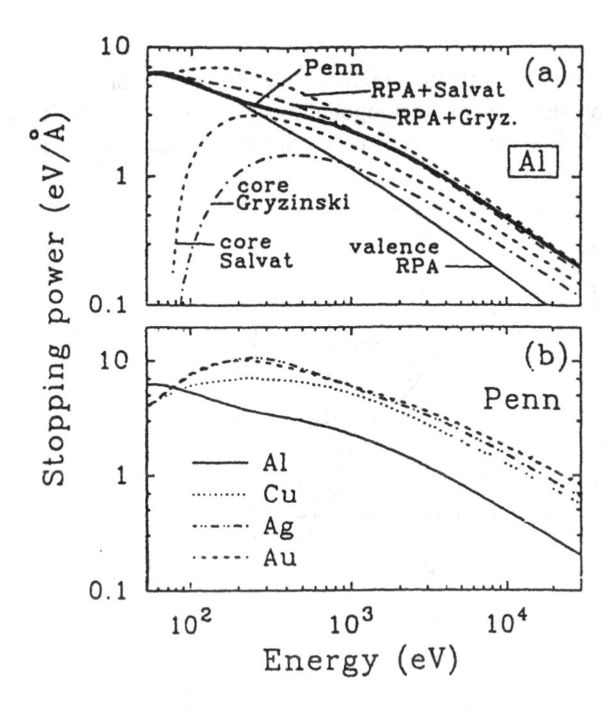


Fig. 1. Positron stopping powers. Panel (a) shows results for Al using different models for inelastic scattering. Panel (b) shows the Penn model results for different solids. From Jensen and Walker (1993).

### *2.1.2 Phonon scattering*

Scattering off density fluctuations and phonons only becomes important at the energies of eV and below, as the positrons approach thermalisation. The most important scattering mode is that due to longitudinal acoustic vibrations. Useful expressions for the cross sections, based on the Debye model and the deformation-potential approximation, have been presented by Nieminen and Oliva (1980) and Jensen and Walker (1990).

Fig. 2. shows the calculated mean free path of positrons as a function of kinetic energy. The electronic excitations dominate, and the temperature dependence is due to the onset of phonon scattering at low energies.

### *2.2 Elastic scattering off stationary atoms*

At energies above 100 keV, the relativistic Mott cross section describes elastic scattering of both electrons and positrons off the nuclei. At lower energies, the cross section should be obtained from the true crystalline potential. For electrons, this potential includes the exchange contribution as the scattering electron state must be orthogonal against the core states. Thus a good approximation for the electron potential in the core region is given by

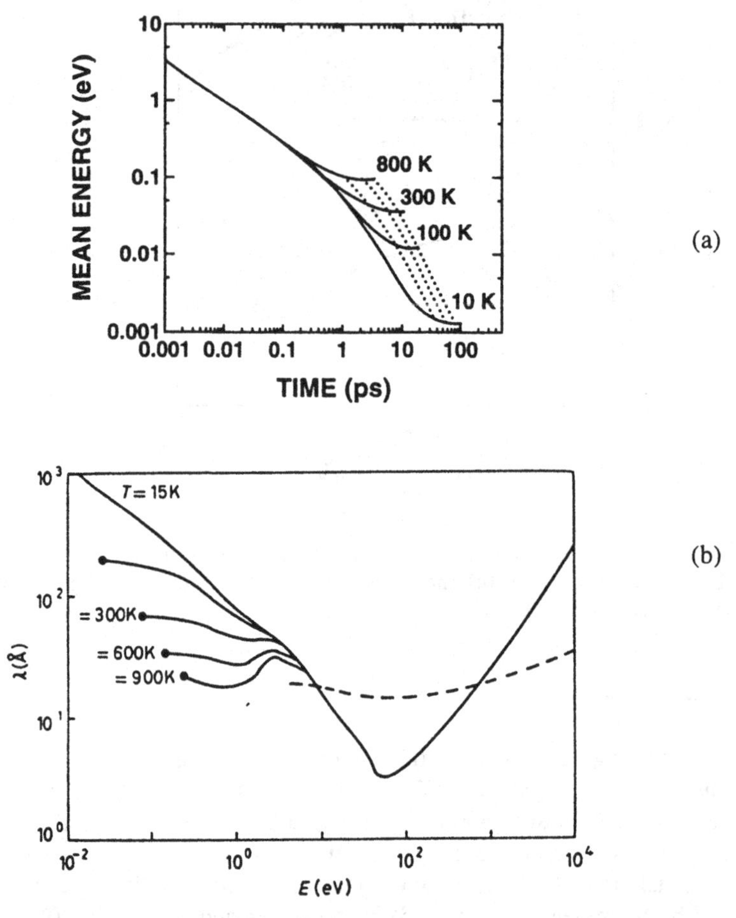


Fig. 2. (a) Positron mean energy as a function of time after implantation in Al at different temperatures (Jensen and Walker 1990).
(b) Positron mean free path in Al as a function of kinetic energy at different temperatures (Nieminen and Oliva 1980).

the self-consistent Hartree-Fock potential. Positrons see only the electrostatic potential plus a relatively weak correlation contribution (see Section 5 below).

The differential cross section for elastic scattering off a single atom can be calculated from the standard partial-wave formula for a spherically symmetric potential. At kilovolt energies, the positron cross section is somewhat less forward peaked than that of the electrons (Valkealahti and Nieminen 1984).

In the energy region below 1 keV, the wave nature of positrons eventually initiates multiple scattering, and in ordered media positrons ultimately become Bloch states. This feature can be utilised in low-energy positron diffraction (LEPD), which reaches its maximum surface sensitivity at around 100 eV, when the inelastic mean free path reaches its minimum. Calculations of LEPD intensity curves can utilise the same methods (Pendry 1974) as used for low-energy electron diffraction (LEED).

Apart from LEPD, the positron excited-state band structure is of interest when considering slowing-down in insulators. The existence of the bandgap prohibits low-energy electronic excitations, and phonons are much less effective in consuming the energy in the few-eV region. It follows that positrons hitting an insulator surface at these energies will show enhanced Bragg scattering at energies which correspond to the forbidden energy band gaps of positrons in the material. The width of the Bragg peak gives the value of the energy gap at the corresponding Brillouin-zone (BZ) boundary (Gullikson *et al.* 1988), and the position of the peak can be used to determine the positron work function. Puska and Nieminen (1992) have calculated the positron band structures in rare-gas solids and extracted the work functions and band gaps. The values for the gap at L point of the BZ are between 1.6 eV (Ne) and 1.1 eV (Xe), in excellent agreement with experiment.

In metallic materials, the positrons keep on relaxing fast towards the ground state. The crossover from single-atom to multiple-atom elastic scattering takes place at an energy of tens of eV, but the exact onset of the quasielastic regime has little quantitative importance for the implantation profiles discussed below.

Positrons emanating from β-decay in radioactive nuclei have a high degeree of spin polarisation. The polarisation largely retained during slowing down, as the magnetic interactions are weak compared to electronic ones. This enables the preparation of spin-polarised positron beams. Their application to studies of magnetism is discussed elsewhere in this volume.

## 3 Ballistic motion, energy loss and implantation

### *3.1 Implantation profiles*

An important question for positron beam studies of solids and surfaces is the quantitative treatment of the implantation and transport stages preceding eventual annihilation or positron/positronium re-emission. In the early stages of thermalisation, the positron motion is ballistic between collisions, and it is convenient to use trajectory calculations based on the collision cross sections and Monte Carlo (MC) simulation techniques (Valkealahti and Nieminen 1983, 1984, Jensen and Walker 1993, Ritley *et al.* 1993, Ghosh and Aers 1995). The MC methods provide accurate statistical distributions of the slowing-down positrons in arbitrary geometry: several computer codes are available, with various degrees of sophistication in the cross sections they use for the elastic and inelastic processes. The MC methods also provide detailed information of the energy- and angle-dependent backscattering at surfaces, as well as the spectrum of secondary electrons. As the motion ceases to be ballistic at low energies and the cross section data becomes less accurate, the MC simulations are usually terminated at a cutoff energy of the order of 10-50 eV. For initial energies of keV and above, the relative changes in the implantation distribution below the cutoff energy are small. Fig. 3 displays a typical implantation distribution of positrons impinging on semi-infinite aluminium at the energy of 3 keV. Fig. 4 displays vertical implantation profiles at various energies, fitted either to the popular Makhov form (Valkealahti and Nieminen 1983, 1984)

$$P(x,E) = -\frac{d}{dx}\left[\exp( -( \frac{x}{x_o} )^m\right] \tag{2}$$

or the more sophisticated form proposed by Baker *et al.* (1991)

$$P(x,E) = -\frac{d}{dx}\left[\exp(-(\frac{x}{x_o^*}(1+\frac{x}{x_o^*})^2)^{m^*}\right] \tag{3}$$

Above, $x_o$, $m$, $x_o^*$ and $m^*$ are material-dependent parameters. The mean vertical implantation range $\langle x\rangle$ scales with the initial energy as

$$\langle x\rangle = AE^n \tag{4}$$

where $A$ and $n$ are also material parameters. Comparison with experiment shows in general good agreement, also for the backscattering spectra (Coleman *et al.* 1992, Massoumi *et al.* 1993). Tabulations of the implantation parameters are available in the literature (Puska and Nieminen 1994, Jensen and Walker 1993, Ritley *et al.* 1993).

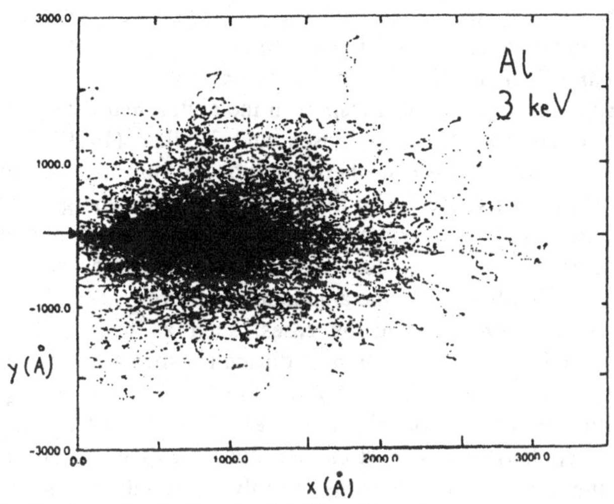


Fig.3. Distribution of endpoints for 3 keV positrons implanted in Al from Monte Carlo simulations (Weber and Nieminen 1998).

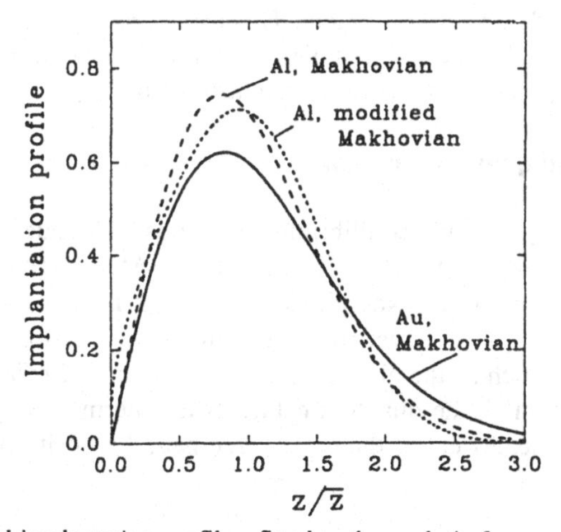


Fig.4. Vertical implantation profiles, fitted to the analytic forms of Eqs. (2) and (3). As the distance is in units of the mean implantation depth, the curves are valid at different incident energies (Jensen and Walker 1993).

### *3.2 Epithermal transport*

Below the cutoff energy of 10-50 eV but prior to full thermalisation, the positrons move as epithermal particles. The description of transport during this stage has received considerable attention. At low implantation energies a substantial part of the positrons can escape from the surface before thermalisation, and attempts have been made to use the epithermal part of the positron spectrum also in quantitative surface studies.

The proper description of transport in the epithermal energy range is based on the Boltzmann equation. Jensen and Walker (1990) used this approach in their study of positron thermalisation in bulk matter, including both electron and phonon scattering in the evaluation of the energy loss rates. For isotropic scattering, the Boltzmann equation can be solved numerically. They confirm earlier results for fast thermalisation towards a Maxwell-Boltzmann velocity distribution for thc annihilating positron.

For spatially nonhomogeneous cases, such as near surfaces, the solution of the full Boltzmann equation is difficult, especially if directional scattering is included. The method can be simplified using the spatially averaged two-flux approximation (Huttunen *et al.* 1989, Kong and Lynn 1990). While the epithermal transport can be modelled qualitatively in terms of elastic and inelastic mean free paths, a truly quantitative description is problematic. Britton (1991) has formulated the problem in terms of generalized diffusion equation, where the epithermal component is included in terms of an effective diffusion coefficient. However, one can conclude (see Section 7 below) that the use the epithermal positrons in quantitatively in surface or near-surface studies is far from straightforward

## 4 Diffusive and quantum motion

After reaching thermal equilibrium, positrons in condensed matter continue to scatter quasielastically off phonons. At the relevant long wavelengths, the dominating acoustic-phonon scattering is mostly isotropic, and thus the motion is a homogeneous random walk, possibly under an external driving force, such as the electric field near the surface of an insulator or semiconductor. Although the quantum nature of the thermalised positron is manifest, over length scales larger than mean free path the positron motion is best described as diffusive.

### *4.1 Positron diffusion*

A convenient way to describe the transport of thermalised positrons is through the continuity equation

$$\frac{\partial n(x,t)}{\partial t} + \nabla \cdot \vec{J}(x,t) = -\lambda(x) n(x,t) \tag{5}$$

Here $n(x,t)$ is the time-dependent positron density, $J(x,t)$ is the positron current density and $\lambda(x)$ stands for the total depletion rate of diffusion positrons. It is the sum of the annihilation rate $\lambda_{ann}$ and the spatially varying trapping rate $\kappa(x)$:

$$\lambda(x) = \lambda_{ann} + \kappa(x) \tag{6}$$

The current and number densities are tied together by Fick's law, which reads in the presence of an electric field $E(x)$

$$\vec{J}(x,t) = -D_+ \nabla n(x,t) + \eta_+ \vec{E}(x) \tag{7}$$

Above, $D_+$ is the positron diffusivity and $\eta_+$ the mobility. Thermalised positrons are isotropic random walkers, and the Nernst-Einstein relation holds:

$$D_+ = \frac{\eta_+}{e} k_B T \tag{8}$$

The mobility is typically limited by phonon scattering, and the standard cross sections can then be utilised to provide values for the relaxation times, mobility and thus the positron diffusivity. For longitudinal acoustic phonons, the diffusion constant scales with temperature as $D_+(\mathrm{LA}) \sim T^{-1/2}$. In polar materials, optical phonons also play a role at high temperatures, with $D_+(\mathrm{LO}) \sim T^{-3/2}$. At low temperatures, impurity scattering can become important, with leading contributions as $D_+ \sim T^{3/2}$ (neutral impurities).

The diffusion equation for thermalised positrons reads

$$\left(D_+ \nabla^2 - \eta_+ \vec{E}(x) \cdot \nabla - \eta_+ \nabla \cdot \vec{E}(x) - \lambda(x)\right) n(x,t) = \frac{\partial n(x,t)}{\partial t} \tag{9}$$

This can be solved by standard numerical techniques. In several applications, it is customary to integrate Eq. (9) over the lateral dimensions and consider the depth information only. If the epithermal particles are neglected, the initial condition for the diffusion problem is

$$n(x,t=0) = P(x,E) \tag{10}$$

Boundary conditions must be specified at surfaces and interfaces. Deep into the solid, the positron density vanishes. The ejection current at the surface is often assumed to be directly proportional to the density of positrons at the surface. Thus the boundary conditions read

$$-D_{+}\frac{\partial n(x,t)}{\partial t}\bigg|_{x=0} + \eta_{+}\vec{E}(0) = -\upsilon n(0,t)$$

$$\lim_{x \to \infty} n(x,t) = 0 \tag{11}$$

where $\nu$ is the transition rate from the surface to the vacuum or to the surface state, to be discussed more later. The diffusion model provides the framework for defect-profile analysis (see Section 7) as well as studies of the positron mobility in electric fields (Mäkinen *et al.* 1990, Shan *et al.* 1996).

### *4.2 Quantum reflection*

The ground-state wavefunction of a thermalised positron in condensed matter is that of a thermal de Broglie wave, modulated by the atomic-scale wiggles due to the repulsion from the positive ionic cores. The "thermal-wave" nature of delocalized positrons in condensed matter is important for low-temperature phenomena. For example, simple arguments show that the propagating wave should be reflected from any discontinuities in the average potential, ranging over distances large compared to the thermal wavelength. Thus a thermal positron propagating towards a surface, either external or that of a large internal void, will have an increasing elastic reflection coefficient as temperature is lowered. Eventually, any potential step at a surface or at a void should become totally reflecting when temperature goes to zero, provided that inelastic or strong-coupling effect do not distort the behavior.

Positron quantum reflection was first observed and explained for large voids in metals (Nieminen *et al.* 1979), and later predicted for positrons approaching surfaces (Nieminen and Oliva 1980). Experiments have confirmed this prediction (Britton *et al.* 1989, Jacobsen and Lynn 1996).

The question how inelastic processes (such as electron-hole and phonon excitations) alter the picture is somewhat controversial and unsettled, as the result depends on how strongly the positron is coupled to the surface and its excitation modes. For positrons at metal surfaces, the coupling seems to be relatively weak (Walker *et al.* 1992) and hence the reflection is expected to be enhanced at low temperatures.

## 5 Electron-positron correlation and annihilation

Electron-positron correlation manifests itself in the pileup of electrons around a positrons. It has important physical consequences. Firstly, it affects the ground-state energetics of positrons. The correlation energy is an important part of the positron total energy, both inside materials and at surfaces. Secondly, the annihilation characteristics depend on the details of the positron screening cloud. Both the annihilation rate, which is proportional to the contact electron density at the positron, and the momentum distribution of the annihilation quanta are affected.

As is the case for electronic structure, a reasonable starting point for quantitative calculations of positron wavefunctions, energies and annihilation characteristics is the (two-component) density-functional theory and the local-density approximation (LDA) for electron-positron correlation effects (Boronski and Nieminen 1986). This is based on the variational minimisation of the total energy functional $E[n_-(r),n_+(r)]$ with respect to the electron and positron densities, $n_-(r)$ and $n_+(r)$, respectively. The densities can be constructed in terms of single-particle electron wavefunctions $\psi_-(r)$ and the positron wavefunction $\psi_+(r)$. For localised positron states, the positron-induced deformations of the surrounding electronic structure can be substantial, even beyond the short-range pileup screening, and the variational equations have to be solved self-consistently for both densities (Gilgien *et al.* 1994, Puska *et al.* 1995). In many cases, however, the "conventional" scheme where the effect of the positron on the average electron density is neglected, is rendered satisfactorily accurate due to the cancellation effects between short-range and long-range screening effects (see *e.g.* Korhonen *et al.* 1996).

The positron ground-state energetics is governed by the eigenvalue

$$E_+ = \frac{\hbar^2}{2m_+}\int d\bar{r}\left|\nabla\psi_+(\bar{r})\right|^2 + \int d\bar{r}\left|\psi_+(\bar{r})\right|^2 V(\bar{r}) \qquad (12)$$

where $m_+$ is the positron mass. The positron potential is

$$V(\vec{r}) = \Phi(\vec{r}) + \frac{\delta E_{xc}[n_+]}{\delta n_+(\vec{r})} + \frac{\delta E_{corr}[n_+, n_-]}{\delta n_+(\vec{r})} , \qquad (13)$$

where $\Phi(\mathbf{r})$ is the electrostatic potential

$$\Phi(\vec{r}) = \int d\vec{r}' \frac{n_{nucl}(\vec{r}') - n_-(\vec{r}') + n_+(\vec{r}')}{|\vec{r} - \vec{r}'|} \qquad (14)$$

arising from the nuclei, the electrons, and the positron itself. $E_{xc}[n]$ is the one-component exchange-correlation energy functional and $E_{corr}[n_+, n_-]$ the two-component correlation energy functional. The "conventional scheme" corresponds to taking the limit $n_+ \rightarrow 0$, and invoking the LDA for both electrons and the positron.

The total annihilation rate reads

$$\lambda_{ann} = \pi r_o^2 c \int d\vec{r} n_+(\vec{r}) n_-(\vec{r}) \gamma(n_+, n_-) \qquad (15)$$

The pair momentum distribution $\rho(\mathbf{p})$ can be written as

$$\rho(\vec{p}) = \pi r_o^2 c \sum_i \left| \int d\vec{r} e^{-i\vec{p}\cdot\vec{r}} \psi_+(\vec{r}) \psi_-(\vec{r}) \sqrt{\gamma_i(\vec{r})} \right|^2 \qquad (16)$$

where the summation is over the participating electron states. In Eqs. (15) and (16), $\gamma$ stands for the density- and state-dependent enhancement factor, which has received wide attention in literature. The presently popular choices for $\gamma$ are based on the LDA, using results for the homogeneous electron-positron gas mixture. Fig. 5. Presents the enhancement factor for the contact density of electrons around a positron, for different values of the mean density ratio $x=n_+/n_-$ and the electron density parameter $r_s$.

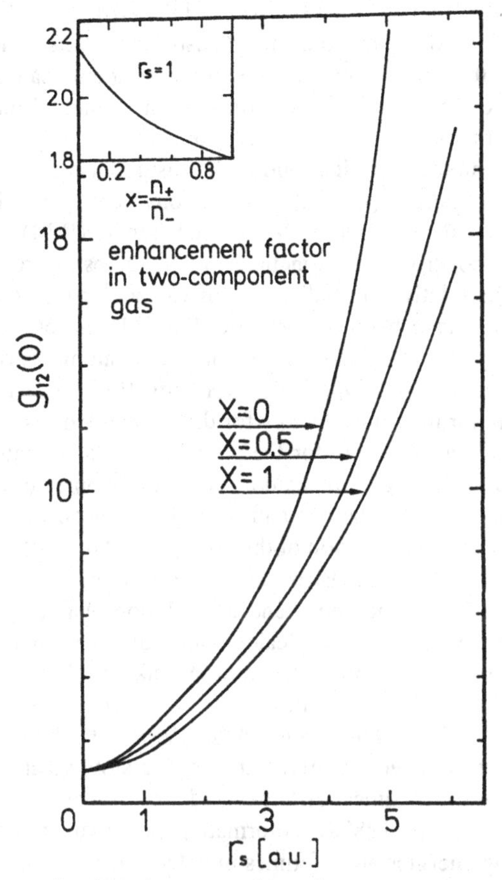


Fig.5. Contact density enhancement factor in a two-component electron-positron gas. $r_s$ is the electron gas density parameter ($1/n_- = 4\pi r_s^3/3$). From Nieminen *et al.* (1985)

As the LDA is based on the electron-gas data, it tends to overestimate the screening effects. In real solids, the core electrons are less polarisable. Phenomenological models have been devised to describe core electron contributions, with improved quantitative accuracy especially in insulators (Puska *et al.* 1989b). Barbiellini *et al.* (1995, 1996) have developed a scheme which accounts for the weaker response of the tightly bound core electrons in terms of the gradient of the total electron density. The generalised-gradient approximation (GGA) for positron-electron correlations reads

$$\gamma_{GGA} = 1 + (\gamma_{LDA} - 1)e^{-\alpha s} \quad , \tag{17}$$

Where the relative density gradient is gauged by $s = |\nabla n_-|^2/(n_- q_{TF}(n_-))^2$, $q_{TF}$ is the local Thomas-Fermi wavevector, and the constant $\alpha \approx 0.22$. The GGA scheme, reminiscent of the one popular in electronic structure calculations, leads to systematic improvement in lifetime estimates for elemental materials, where the LDA tends to predict a little too short lifetimes.

The shapes of momentum distributions, measured via one- or two-dimensional angular correlation (ACAR) distributions, are generally less affected by many-body effects: the independent-particle model (IPM) based on accurate electron and positron wavefunctions often gives a reasonable description. However, the relative normalisations of valence and core parts of so-calculated distributions have been problematic. Recently, an efficient and unambiguous first-principles scheme for calculating momentum distributions has been presented ( Alatalo *et al.* 1996, Barbiellini *et al.* 1997). It is based on the use of the IPM, with normalisation factors for different electrons based on the GGA or LDA enhancement factors for the partial annihilation rates. This method is very accurate in the high-momentum region in Doppler lineshapes, arising from the sum of core annihilations and the valence electron Umklapp events. It also provides a good description of the normal ACAR spectra.

The equations briefly summarised above constitute the "*standard model*" for calculations of the energetics and annihilation characteristics of positrons in condensed matter. Extensive calculations of positron densities, ground-state energetics and annihilation rates in both bulk and defected solids have been reported in literature (for summaries, see *e.g.* Jensen 1989, Puska 1991, Puska and Nieminen 1994, Nieminen 1995a, 1995b, Barbiellini *et al.* 1996, 1997, Korhonen *et al.* 1996). Useful numbers for slow-positron beam studies include positron bulk lifetimes, affinities and work functions in solids (Boev *et al.* 1987, Puska *et al.* 1989a), deformation potential constants for phonon coupling, binding energies and lifetimes in defects and at surfaces, and the shape parameters for momentum distributions.

## 6 Trapping and detrapping

Positron trapping is a quantum-mechanical transition between a delocalized propagating state and a localized one, at an open-volume defect, around a negatively charged impurity or at a surface. The capture rate is proportional to the density of defects or surface area, and the proportionality constant is called the trapping coefficient ν. It can be estimated from the Fermi Golden Rule (Hodges 1970) and depends on the particular energy release mechanism whereby the binding energy to the trap is dissipated.

In metals, energy release through electron-hole excitation is the dominating channel. Phonon excitations can play a role if the binding energy to the trap is weak, *i.e.* not much larger than typical phonon (Debye) energy.

For spatially small defects such as vacancies, positron trapping is transition-limited. In the semiclassical picture, this means that positron diffusion to the defect does not limit the overall trapping rate: it is determined by the transition probability defined by the Golden Rule. Typical values for $\nu$ are of the order of $10^{-14}$ – $10^{-15}$ $s^{-1}$, *i.e.* about five orders of magnitude larger than normal annihilation rates. This makes it possible to detect defect concentrations of the order of $10^{-7}$ – $10^{-4}$.

The temperature dependence of the positron trapping rate (per unit concentration of defects) is, especially in the case of semiconductors, an important experimental issue as it can be used to make conclusions of the defect type. In metals, the trapping rate is temperature independent, except for very large defects (voids or surfaces), where the quantum reflection of the initial positron wave plays a role.

In semiconductors and insulators, the defects can be charged which has consequences for the trapping coefficient, as the initial positron wavefunction is modified by the Coulomb tail of the defect potential. Positively charged defects effectively repel positrons, and the trapping rate is expected to be very small. On the other hand, negatively charged defects enhance the initial positron density in the defect region and increase the trapping coefficient; they may also support a series of shallow Rydberg-like states at low temperatures. These can be precursors to the final deep state at the open volume defect. Moreover, even negatively charged impurities without any open volume can act as traps at low temperatures, where the Rydberg-state positrons are stabilized. In both cases, the trapping rate to negatively charged defects increases proportional to $T^{-1/2}$ as temperature is lowered. This can used as a fingerprint for the charge state of the defect. The calculation of positron trapping rates at defects in semiconductors, including the role of possible deep electron states, has been discussed in detail by Puska *et al.* (1990).

The trapping cross section can also exhibit resonances in the eV energy region for slowing down positrons. It seems that their overall effect on trapping-related phenomena is small, at least in metals (Jensen and Walker 1990) as the energy-loss rate is much larger than the resonant trapping rate.

Thermal fluctuations can lead to positron detrapping. In thermal equilibrium, the ratio between the detrapping (δ) and trapping rates (κ) for point defects in three dimensions is

$$\frac{\delta}{\kappa} = \frac{1}{c_v}\left(\frac{m_+ k_B T}{2\pi\hbar^2}\right)^{3/2} e^{-E_b/k_B T} \tag{18}$$

where $E_b$ is the binding energy to the trap and $c_v$ the defect concentration.

Optically excited escape ("laser detrapping") is in principle also possible, but the due to the short lifetime of positrons the effect is difficult to detect, as the relevant cross section is small.

## 7 Surface and near-surface processes

The different surface and near-surface processes for positrons are schematically summarized in Fig. 6. and their mean features are summarised below.

### *7.1 Trapping at below-surface defects*

A useful application of positron beam is the study of defects below the surface or at buried interfaces. The signal is observed either through the arrival and subsequent annihilation of positrons reaching back at the outer surface after initial implantation or through the direct annihilation signals of positrons being consumed below the surface. For the latter, momentum distribution measurements usually based on the Doppler-broadening technique are used, although lifetime measurements with time-structured beams are now also feasible.

In all cases, the analysis of the experiment is based on the following:

(i) the knowledge of an accurate implantation profile, either from simulation or experiment

(ii) the solution of the diffusion-annihilation equation in the presence of traps, using the implantation profile as the initial condition; only thermalised positrons should strictly speaking be considered

(iii) the correlation between the observed annihilation signal and calculated characteristics, based on models for the defect types and their distributions

(iv) the iterations of the steps (ii) and (iii) with different defect profiles to achieve agreement between experiment and modelling

Program packages with varying degrees of sophistication are available for the steps (i) and (ii). The applications of the technique to near-surface and interfacial defect profiling in mesoscopic semiconductor and metal structures is discussed in detail elsewhere ( Schultz and Lynn 1988, van Veen *et al.*, this volume).

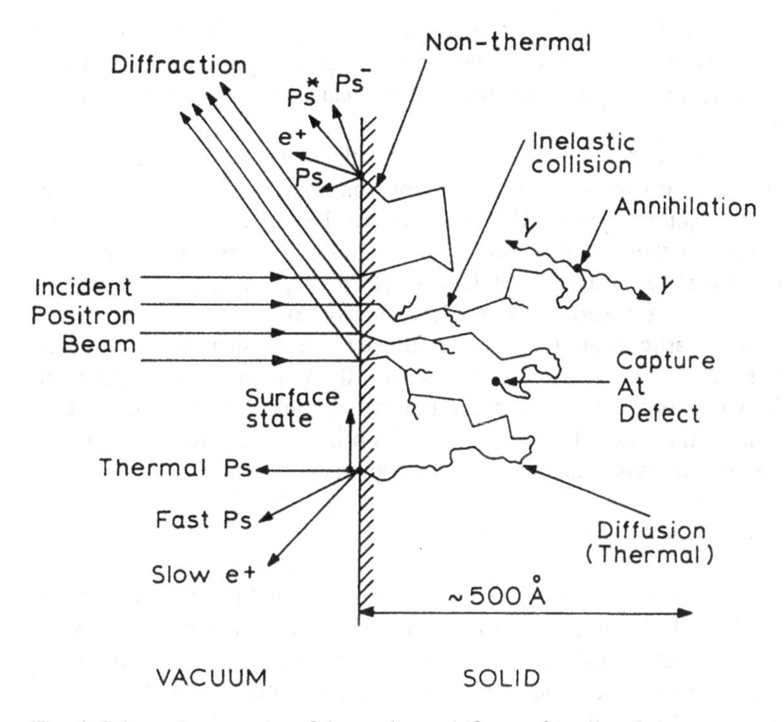


Fig. 6. Schematic summary of the positron surface and near-surface processes.

*7.2 Diffraction and reflection*

Low-energy positrons incident on a surface can be diffracted with a small but nonvanishing probability: typical total backscattering probability in the energy range of 100 eV is $10^{-3}$. The elastic scattering events resulting in the diffraction take place in the time scales of the order $10^{-16}$ s, before appreciable inelastic collisions take place. This is the regime of low-energy positron diffraction (LEPD), whose utility complementing LEED in surface structure determinations has been reviewed in detail by Duke (1995).

The construction of the multiple-scattering potential for dynamical LEPD calculations is less ambiguous than for electrons. No exchange term is present in the positron potential. The positrons are also strongly repelled from the ionic cores, and move mainly in the interstitial regions, where the electron density is relatively slowly varying. Thus a local-density approximation for the correlation potential is a good starting point, and can easily be improved by the GGA method.

Another useful observation is that the relative strength of the scattering of similar-energy electrons and positrons depends on the target atom as well as the incident energy. For heavy atoms the forward cross section is typically much higher for electrons than for positrons; for light atoms the situation is reversed. Therefore it is possible to exploit this difference in compound systems, *e.g.* surfaces with adsorbate overlayers.

The inelastic mean free path at LEPD energies is shorter for positrons than for electrons, by as much as 30 % for 50 eV particles impinging on metals. This causes a substantial reduction in the contribution of subsurface layers to the total diffracted intensity, and thus brings about a reduction in the computing time for dynamical intensity calculations.

*7.3 Re-emission phenomena*

Part of the implanted positrons survive annihilation and defect trapping and reach the entrance surface, either fully or partially thermalised. The epithermal fraction and its energy and momentum spectrum depend on the implantation energy in a complicated manner, while the thermalised positrons have an isotropic Maxwellian velocity distribution determined by the substrate temperature.

Once at the surface, a number of scenarios are possible, and these are discussed separately below. Good surveys of the relevant physics has been given by Mills (1995) and Weiss (1995) in the 1995 Varenna Summer School proceedings (Dupasquier and Mills 1995).

If the material has a negative work function $\phi_+ < 0$ for positrons, spontaneous ejection of free positrons is possible. Since the positronium binding energy 6.8 eV is comparable to typical electron affinities, a pickup process is also possible where the escaping positron takes along an electron and emerges as positronium. One can define the positronium work function (emission potential) as

$$\phi_{Ps} = \phi_+ + \phi_- - 6.8\,eV \qquad (19)$$

where $\phi_-$ is the electron work function.

Thirdly, the positron can become trapped at an image-potential-induced surface state (Hodges and Stott 1973, Nieminen and Hodges 1976), which may also be thermally destroyed to produce desorbed positronium.

Ignoring for a moment any strong-coupling or dynamical phenomena, one can simply view all these processes as transitions between well-defined initial states, where the energy difference is taken up by a set of elementary excitations, such as electron-hole pairs and phonons. The rate $\Lambda$ of such processes is given by the Golden Rule expression

$$\Lambda(E) = \frac{4\pi}{\hbar} \sum_{i,f} \left| M_{fi} \right|^2 \delta(E + E_i - E_f) \tag{20}$$

where $E$ is the initial kinetic energy of the positron, and $i$ and $f$ denote the initial and final states, respectively, of the system, including the ensemble of excitations. $M_{f\,i}$ is the matrix element of the (screened) electron-positron interaction potential between the initial and final states. The delta function conserves the total energies of the combined system (positron, electrons, possible excitations). The detailed form of the expression depends on the actual transition process under study as well as the nature of the solid surface, *i.e.* whether there is an electron band gap or not, whether there are electronic surface states, possibly pinning the Fermi level *etc.*

### *7.3.1 Positron re-emission*

Positron emission from a metal with a negative positron work function $\phi_+ < 0$ is basically an adiabatic process, since the maximum kinetic energy for the emitted positron is usually much smaller than the Fermi energy $\varepsilon_F$. The positron leaves the surface relatively slowly compared to typical conduction electron response times, and the electron system is left mainly in its ground state. This implies that slow positron escape from the surface mainly elastically, *i.e.* with maximum kinetic energy $|\phi_+|$, peaked around the surface normal with the width of the angular distribution proportional to $k_B T / |\phi_+|$.

Inelastic processes (electron-hole excitations) are possible and produce an inelastic tail to the energy spectrum of emitted positrons. The shape of this tail can be estimated from Golden Rule calculations (Neilson *et al.* 1986, 1988), which also predict the modifications in the angular distribution brought by inelastic electron-hole excitations. The matrix element effects produce a broadening of the angular spectrum of the emitted positrons. The overall influence of the inelastic broadening is, however, relatively weak. In particular, if one assumes a constant matrix element between initial and final states and a flat electron density of states near the Fermi level, the transition rate becomes

$$\Lambda(E) \propto const \cdot (\phi_+ - E)\sqrt{E} \tag{21}$$

Then the current distribution of emitted positrons having the kinetic energy larger than $E_z \leq |\phi_+|$ in the direction perpendicular to the surface is

$$J(E_z) \propto const \cdot |\phi_+|^{5/2} (\frac{4}{15} + 2x^{3/2} (\frac{1}{3} - \frac{x}{5})) \tag{22}$$

where $x = E_z / |\phi_+|$. This result is analogous to the classic Berglund-Spicer (1964) model for the photoemission current.

*7.3.2 Positronium re-emission*

The emission of positronium resulting from the capture of a Fermi level electron near the surface can also be reasoned using the Golden Rule expression. In this case, if inelastic effects are ignored, the total kinetic energy for the ejected positronium is given by $\phi_{Ps}$ , but unlike for positrons, the angular spectrum can be wide.

The angle-resolved energy spectrum of the positronium emitted from the solid carries information about the electron momentum distribution of the surface region, complementary to that provided by angle-resolved photoemission measurements. In principle, angle-resolved positronium emission spectroscopy offers a possibility to probe the electronic states within a few eV of the Fermi level, without the complication of the initially empty electron states involved in photoemission. However, the matrix element effects conveyed by the electron-positron interaction can be significantly stronger in this case (Walker and Nieminen 1986). The interaction leads to the formation of energetic positronium and affect the shape of its momentum distribution. The effect is stronger for the momentum distribution of the self-annihilation of the positronium, as measured by 2D ACAR, than for the center-of-mass energy distribution measured by time-of-flight spectroscopy. There is some evidence that the time-of-flight and ACAR spectra reflect the shape of the the projected electronic band structure of the metal surface (Mills 1995), but the interpretation is not straightforward.

In insulators, positronium can exist inside the material, and usually has a negative affinity for them. In this case the Ps is emitted monoenergetically, provided that it has thermalized before diffusing to the surface.

### *7.3.3 Branching ratios*

The Golden Rule picture based on first-order perturbation theory also provides a first estimate of the relative probabilities of the various positron surface processes. The absolute values of the rates can be estimated as well, but tend to be sensitive to values of the model parameters such as the actual form of the surface potential, the screening lengths *etc.* (Neilson *et al.* 1986). The capture to the surface state usually has the highest probability (if a stable surface state exists), while the positron and positronium re-emission probabilities are roughly equal.

Assuming free-electron like dispersion relations near the surface and smooth matrix elements coupling the initial and final states, one can make simple estimates of the branching ratios between positron and positronium emission rates. This is basically a density-of-states argument, which leads to the positronium/free-positron branching ratio

$$\frac{\Lambda_{Ps}}{\Lambda_{e^+}} \propto const \cdot \frac{\left|\phi_{Ps}\right|^{3/2}}{\left|\phi_+\right|^{1/2}} \tag{23}$$

and the slow positron yield $y$ as a function of $\phi_+$:

$$y = \frac{\sqrt{|\phi_+|}}{\sqrt{|\phi_+|+b}} \tag{24}$$

where $b$ is a constant.

An alternative picture of the branching ratio between positron and positronium emission can be based on resonant-charge transfer model, originally derived for ion neutralisation at surfaces. It is based on the notion that as thermal positrons leave the surface, their increasing velocity, proportional to $|\phi_+|^{1/2}$ , leads to a diminishing probability for positronium formation, as the electron system has less time to respond and "hop on" to the escaping positron. Using time-dependent nonequilibrium perturbation theory, the following expression can be derived (Nieminen and Oliva 1980) for the slow-positron yield

$$y = const \cdot \exp(-\sqrt{\frac{\phi_0}{\phi_+}}) \tag{25}$$

where $\phi_0$ is a constant characteristic of the metal in question. Similar reasoning has been used with quantitative success in ion neutralisation spectroscopy (Yu and Lang 1983). As for positron spectroscopy, the data cannot resolve between the static and dynamic models, but agree with both Eqs. (24) and (25).

### *7.3.4 Epithermal re-emission*

Depending on the nature of the substrate and the implantation energy, a substantial fraction of the positrons may reach the surface before full thermalisation. This is the case in particular in insulating solids where a band gap puts a lower limit to the energy transfer to electronic excitations. For epithermal emission, the positron emission spectrum is expected to be broad. The re-emission process can be modelled using the Boltzmann equation appraoch, but quantitative information on the physical properties of the surface is difficult to extract. An interesting application of epithermal emission is the determination of positron thermalisation lengths and work functions for surface where $\phi_+ > 0$ (Knights and Coleman 1996).

### *7.4 Glancing-angle positronium formation*

Positronium can also be formed in glancing-angle scattering of a positrom beam from the external surface (Gidley *et al.* 1987). Again, the Golden Rule expression (Eq.(20)) provides a good starting point to discuss the phenomenon. The positron scatters from the surface potential and captures an electron. If any matrix-element effect are ignored, the component of the momentum parallel to the surface is conserved. After the positronium has formed, a hole is left in the substrate , very much like in photoemission. Thus information could possibly be extracted of the occupied electronic states, with a probe that has very high surface sensitivity.

Earlier, a similar experiment utilizing spin-polarised positrons has been used measure the positronium formation probability as a function of the spin direction of the incoming beam (Gidley *et al.* 1982). This is sensitive to the spin polarisation of electrons on the surface, and provides a way to probe surface magnetism.

### *7.5 Capture to the surface state*

The density-of states argument leads to the capture rate scaling as

$$\Lambda_{ss} \propto (E_b - \phi_{Ps})^2 \tag{26}$$

where $E_b$ is the binding energy to the surface trap. More accurately, the Golden Rule expression can be evaluated for models of surface traps. Walker *et al.* (1992) extended the calculations of Neilson *et al.* (1986) to evaluate the capture rate into the surface state as a function of the positron kinetic energy, approaching the surface either from inside (epithermal and thermal positrons) and outside. As pointed out earlier (Nieminen and Oliva 1980), this approach leads to a vanishing trapping probability as temperature approaches zero, when the surface reflectivity becomes unity. In general, the trapping rate depends on the (epithermal) kinetic energy. The dependence is due the variation of the matrix element with positron energy and the phase space available to to the trapped positron and the excited electron-hole pair.

Surface resonances in the scattering cross section can also be present, as first pointed out by Jennings (1988) in the context of LEPD, where they would be seen in the form of interference features below the threshold of the specularly reflected beam (Horsky *et al.* 1989). However, these are mainly an elastic phenomenon where the positron is temporarily trapped in an intermediate state, with the wavefunction perpendicular to the surface bound in the image well but with the total energy positive. Since true surface trapping is an inelastic event, resonance phenomena are likely to be unimportant for capture.

### *7.6 Desorption from the surface*

At elevated temperatures, thermal emission of positrons and positronium may become important, analogous to thermionic emission of electrons and thermal desorption of atoms, respectively. Thermally emitted positrons are relevant for positive-workfunction surfaces only, and their emission rate $\Gamma$ obeys the classical version of the Richardson-Dushman-type equation

$$\Gamma = n(0)\sqrt{\frac{k_B T}{2m_+}}\exp(-\frac{\phi_+}{k_B T}) \tag{27}$$

where *n(0)* is the steady-state density of diffusing positrons at the surface, solved from Eq. (9). Except for very low workfunction values, the rate in Eq. (27) is expected to be small and probably overwhelmed by the other surafce processes, in particular spontaneous positronium emission and surface state capture.

A more important process, for which there is also ample experimental evidence, is the thermal desorption of positronium where the positron originates from the surface state and the electron comes from the Fermi level. The activation energy $E_a$ for thermal emission is

$$E_a = E_b + \phi_- - 6.8\,eV \tag{28}$$

where $E_b$ denotes the surface state binding energy with respect to vacuum.

The fraction of positrons $f_{th}$ that will be thermally desorbed as positronium, having originally been trapped to the surface state is

$$f_{th} = A(T)\frac{e^{-Ea/k_BT}}{\lambda_s + A(T)e^{-Ea/k_BT}} \tag{29}$$

where $\lambda_s$ is the annihilation rate in the surface state and *A(T)* a prefactor depending through detailed-balance arguments on the velocity-dependent sticking coefficient of positronium atoms impinging on the surface. The determination of this in the low-energy limit is of interest to the so-called quantum sticking problem (Martin *et al.* 1993).

### *7.7 Positron-induced Auger emission*

The positron potential and wavefunction on an adsorbate-covered surface are depicted in Fig. 7. The wavefunction shows remarkable localisation in the surface region.

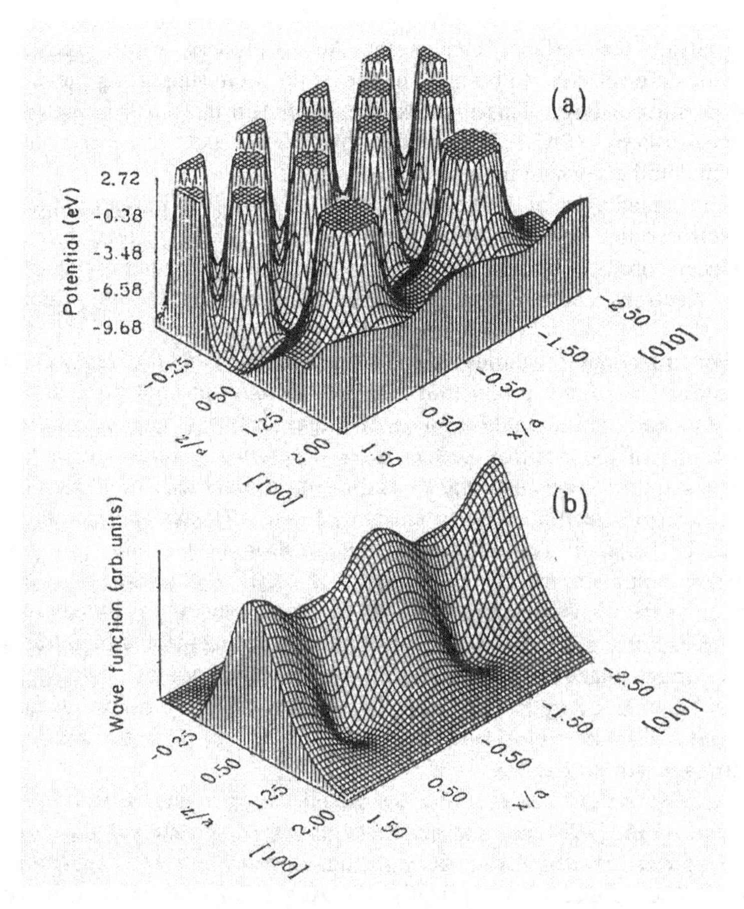


Fig. 7. (a) Positron effective potential on a Cs p(2x2)/Ni(100) surface. (b) Positron surface state wavefunction . The length unit a is the Ni lattice constant of 3.52 Å. (Nieminen and Jensen 1988).

The positron surface state is naturally unstable against annihilation with a nearby surface electron. If this electron is a deep core state, a localised hole is created. The hole will quickly be filled by an electron from either one of the outer atomic shells or from the valence bands. The liberated energy can be transferred to another electron, which is then ejected and may eventually be detected outside the surface. This is the Auger process, whose spectral characteristics are sensitive to both the nature of the atom supporting the core hole and its surroundings. This process is exploited in the positron-induced Auger spectroscopy (PAES), pioneered by Weiss and coworkers and discussed in detail elsewhere in this volume.

The advantages of PAES over electron-, ion-, or photon-induced Auger spectroscopies are its extreme surface sensitivity (the positron surface state produces core hole only in the uppermost atomic layers) and the lack of secondary electrons and the lack of surface damage due to the primary radiation.

For quantitative studies, the interpretation of PAES requires a reliable estimation of the production rates of various relevant core holes, which have to lead to observable amounts of Auger electrons. This must based on calculations of the positron surface state properties, a topic which has attracted substantial interest over the years. In these calculations, the "standard model" of positron-electron states in condensed matter (Puska and Nieminen 1994) can be applied. An additional complication is that the positron wavefunction amplitude in the core region is small and sensitive to the construction of the image-potential well. As the image potential is by nature a non-local phenomenon, LDA-based theories have to be augmented to mimic a physically correct image tail. The argument can be turned around to note that PAES can provide stringent tests for the quality of the positron surface wavefunction. This is useful since lifetime and ACAR data for positron surface states is still very scarce.

Jensen, Weiss, Fazleev and coworkers (Jensen and Weiss 1990, Fazleev *et al.* 1995, 1998) have carried out extensive computational studies of the PAES process by calculating the annihilation rates of surafce positrons with specified core electrons. Typically, the Auger emission is associated with 1 – 5 % of the annihilations at the surface. The Auger cross sections are sensitive to any adsorbate coverage and atomic geometries, which enables in principle both chemical analysis and (through computation) structural studies of surfaces. A downside is the sensitivity of the positrons to surface defects and possibly preferential annihilation around specfic atoms, which may render large parts of the surface invisible to PAES.

## 8 Concluding remarks

This article has briefly summarised the main physical concepts of positrons striking a solid target in the form of a monoenergetic beam. There is a rich variety of physical phenomena involved, and many of them can also be detected experimentally. The proper interpretation of the experiments requires a quantitative understanding of the positron probe, which usually gives the experimental fingerprint in the indirect form of annihilation radiation.

Computational methods for analysing positron-beam experiments have matured and are reported in recent literature, which should be consulted for detailed information. Some of the main results have been highlighted here, with the emphasis on situations where positrons can provide useful information of material properties near surfaces and interfaces.

## 9 References

Alatalo, M., Barbiellini, B., Hakala, M., Kauppinen, H., Korhonen, T., Puska, M.J., Saarinen, K., Hautojärvi, P. and Nieminen, R.M., 1996, Phys.Rev. B **54**, 2397-2409.

Ashley, J.C., Tung, C.J., and Ritchie, R.H., 1979, Surf.Sci. **81**, 409-426.

Baker, J.A., Chilton, N.B., Jensen, K.O., Walker, A.B., and Coleman, P.G., 1991, J. Phys.: Condens. Matter **3**, 4109-4120.

Barbiellini, B., Puska, M.J., Torsti, T. and Nieminen, R.M., 1995, Phys.Rev. B **51**, 7341-7345.

Barbiellini, B., Puska, M.J., Korhonen, T., Harju, A., Torsti, T., and Nieminen, R.M., 1996, Phys.Rev. B **53**, 16201-16220.

Barbiellini, B., Hakala, M., Puska, M.J., Nieminen, R.M. and Manuel, A.A., 1997, Phys.Rev. B **56**, 7136-7142.

Berglund, C.N. and Spicer, W.E., 1964, Phys.Rev. **A136**, 1044-1056.

Boev, O.V., Puska, M.J., and Nieminen, R.M., 1987, Phys.Rev. B **36**, 7786-7795.

Boronski, E. and Nieminen, R.M., 1986, Phys.Rev. B **34**, 3820-3831.

Britton, D.T., 1990, J.Phys.: Condens.Matter **3**, 681-692.

Britton, D.T., Huttunen, P.A., Mäkinen, J., Soininen, E. and Vehanen, A., 1989, Phys.Rev.Lett. **62**, 2413-2417.

Coleman, P.G., Albrecht, L., Jensen, K.O., and Walker, A.B. 1992, J.Phys: Condens.Matter **4**, 10311-10322.

Duke, C.B. 1995, in *Positron Spectroscopy of Solids*, eds. Dupasquier, A. and A.P. Mills, Jr. (IOS Press: Amsterdam), 317-357.

Dupasquier, A. and Mills Jr., A.P. , 1995 (ed): *Positron Spectroscopy of Solids* (IOS Press: Amsterdam).

Fazleev, N.G., Fry, J.L., Kuttler, K.H., Koymen, A.R. and Weiss, A.H., 1995, Phys.Rev. B **52**, 5351-5363.

Fazleev, N.G., Fry, J-L., and Weiss, A.H. 1998, Phys.Rev. B **57**, 12506-12519.

Ghosh, V.J. and Aers, G.C., 1995, Phys.Rev. B **51**, 45-59.

Gidley, D.W., Koymen, A.R., and Capehart, T.W., 1982, Phys.Rev.Lett. **49**, 1779-1782.
Gidley, D.W., Amyer, R., Frieze, W.E., and Lynn, K.G., 1987, Phys.Rev.Lett. **58**, 595-598.
Gilgien, L., Galli, G., Gygi, C., and Car, R., 1994, Phys.Rev.Lett. **72**, 3214-3217.
Gryzinski, M., 1965, Phys. Rev. **A138**, 305-340.
Hodges, C.H., 1970, Phys.Rev. Lett. **25**, 284-287.
Hodges, C.H. and Stott, M.J., 1973, Solid State Commun. **12**, 1153-6.
Horsky, T.N., Brandes, G.R., Canter K.F., Lippel, P.H., and Mills, A.P. Jr., 1989, Phys.Rev. B **40**, 7898-7903.
Huttunen, P.A., Vehanen, A., and Nieminen, R.M., 1989, Phys. Rev. B **40**, 11923-11926.
Ishii, A., 1992 (ed): *Positrons at Metallic Surfaces* (Trans Tech Publications: Switzerland).
Jacobsen, F.M. and Lynn, K.G. , 1996, Phys.Rev.Lett. **76**, 4262-4264.
Jennings, P. J., 1988, Surf.Sci. **198**, 180-191.
Jensen, K.O., 1989, J. Phys.: Condens.Matter **1**, 10595-10608.
Jensen, K.O. and Walker, A.B., 1990, J. Phys: Condens.Matter **2**, 9757-9857.
Jensen, K.O. and Walker, A.B., 1992, in Ishii, A., 1992 (ed): *Positrons at Metallic Surfaces* (Trans Tech Publications: Switzerland).
Jensen, K.O. and Walker, A.B., 1993, Surf.Sci. **292**, 83-97.
Jensen, K.O. and Weiss, A.H., 1990, Phys.Rev. B **41**, 3928-3948.
Knights, A.P. and Coleman, P.G., 1996, Surf.Sci. **367**, 238-244.
Korhonen, T., Puska, M.J., and Nieminen, R.M., 1996, Phys.Rev. B **54**, 15016-15024.
Martin, T.R., Bruinsma, R., and Platzman, P.M., 1993, Phys.Rep. **233**, 135-158.
Massoumi, G.R., Lennard, W.N., Schultz, P.J., Walker, A.B., and Jensen, K.O., 1993, Phys.Rev. B **47**, 11007-11018.
Mills, A.P., Jr, 1995, in *Positron Spectroscopy of Solids*, eds. Dupasquier, A. and A.P. Mills, Jr. (IOS Press: Amsterdam), 209-258.
Mäkinen, J., Corbel, C., Hautojärvi, P., Vehanen, A., and Mathiot, D., 1990, Phys.Rev. B **42**, 1750-1762.
Neilson, D., Nieminen, R.M., Szymanski, J., 1986, Phys. Rev. B **33**, 1567-1571.
Neilson, D., Nieminen, R.M.,. Szymanski, J., 1988, Phys. Rev. B **38**, 11131-11134.
Nieminen, R.M., 1988, Scanning Microscopy **2**, 1917-1926.
Nieminen, R.M., 1995a, in *Positron Spectroscopy of Solids*, eds. A. Dupasquier and A.P. Mills, Jr. (London: IOS Press) 443-490.

Nieminen, R.M., 1995b, Mat. Sci. Forum **175-178**, 279-286 .
Nieminen, R.M., Boronski, E., and Lantto, L.J., 1985, Phys. Rev. B **32**, 1377-1380.
Nieminen, R.M. and Hodges, C.H,, 1976, Phys.Rev. B **18**, 2565-2588.
Nieminen, R.M. and Jensen, K.O., 1988, Phys.Rev. B **38**, 5764-5767.
Nieminen, R.M. and Oliva, J., 1980, Phys.Rev. B **22**, 2226-2251.
Nieminen, R.M., Laakkonen, J., Hautojärvi, P., and Vehanen A., 1978, Phys.Rev. B **19**, 1937-50.
Oliva, J., 1980, Phys. Rev. B **21**, 4909-4916.
Pendry, J.R., 1974, *Low Energy Electron Diffraction* (New York: Academic Press).
Penn, D.R. , 1987, Phys.Rev. B **35**, 482-486.
Puska, M.J., 1991, J.Phys.: Condens.Matter **3**, 3455-3469.
Puska, M.J., Corbel, C. and Nieminen, R.M., 1990, Phys.Rev. B **41**, 9980-9993.
Puska, M.J., Lanki, P., and Nieminen, R.M., 1989a, J. Phys.: Condens. Matter **1**, 6081-6093.
Puska, M.J., Mäkinen, S., Manninen, M., and Nieminen, R.M., 1989b, Phys.Rev. B **39**, 7666-7678.
Puska, M.J. and Nieminen, R.M. , 1992, Phys.Rev. B **46**, 1278-1283.
Puska, M.J. and Nieminen, R.M. , 1994, Rev.Mod.Phys. **66**, 841-897.
Puska, M.J. , Seitsonen, A.P. and Nieminen, R.M. , 1995, Phys.Rev. B **52**, 10947-10961.
Quinn, J.J., 1963, Appl.Phys.Lett. **2**, 167-169.
Ritchie, R.H., Tung, C.J., Anderson, V.E., and Ashley, J.C., 1975, Rad.Research **64**, 181-204.
Ritley, K.A., Lynn, K.G., Ghosh, V.J., Welch, D.O. and McKeown, M., 1993, J. Appl.Phys. **74**, 3479-3496.
Rohrlich F. and Carlson R.C., 1954, Phys.Rev. **93**, 38-48.
Salvat F., Martinez U.D., Mayol, R., and Parellada, J., 1985, J. Phys. D **18**, 299-315.
Schultz, P.J. and Lynn, K.G., 1988, Rev.Mod.Phys. **60**, 701-780.
Schultz, P.J., Logan, L.R. , Lennard, W.N. and Massoumi, G.R. , 1990, Scanning Microscopy Supplement **4**, 223-237.
Shan, Y.Y., Asoka-Kumar, P., Lynn, K.G., Fung, S. and Beling, C.D., 1996, Phys.Rev. B **54**, 1982-1986.
Valkealahti, S. and Nieminen, R.M., 1984, Appl.Phys. A **35**, 51-59.
Valkealahti, S. and Nieminen, R.M., 1983, Appl.Phys. A **32**, 95-106.
Walker, A.B., Jensen, K.O., Szymanski, J., and Neilson, D., 1992, Phys.Rev. B **46**, 1687-1785.

Walker, A.B. and Nieminen, R.M., 1986, J. Phys. F: Metal Phys. **16**, L295-L303.

Weber, F. and Nieminen, R.M., 1998, unpublished.

Weiss, A.J., 1995, in *Positron Spectroscopy of Solids*, eds. Dupasquier, A. and A.P. Mills, Jr. (IOS Press: Amsterdam), 259-284.

Yu, M.L. and Lang, N.D., 1983, Phys.Rev.Lett. **50**, 127-130.

CHAPTER 5

# SURFACE SCIENCE WITH POSITRONS

A.H. WEISS
*Department of Physics*
*University of Texas at Arlington*
*Arlington, Texas 76019, USA*
*E-mail: weiss@uta.edu*

P. G. COLEMAN
*Department of Physics*
*University of Bath*
*Claverton Down, Bath BA2 7AY, UK*
*E-mail: p.g.coleman@bath.ac.uk*

Positron beam technology has developed in tandem with our understanding of positron-surface interactions to enable the positron to be used in novel and unique ways to study the physical and chemical properties of surfaces. Most notable among the new positron spectroscopies are Low-Energy Positron Diffraction (LEPD) and Positron Annihilation-induced Auger Electron Spectroscopy (PAES) and these, along with a number of other applications of positron beams to the investigation of surface phenomena, are described and discussed in this chapter.

## 1 Introduction

Low energy positron interactions with matter provide numerous surface selective signals (see, for instance, excellent reviews by Mills [1], Schultz and Lynn [2] and Niemenen [3]). These signals can be exploited in spectroscopies that provide unique information about the surface often unavailable by other means. A number of these are summarized in figure 1. Many positron spectroscopies have no direct analogs in electron spectroscopy, while others - such as Positron Annihilation induced Auger Spectroscopy, have signal outputs which are identical to their electron counterparts but arise from unique excitation mechanisms. This chapter will provide a review of

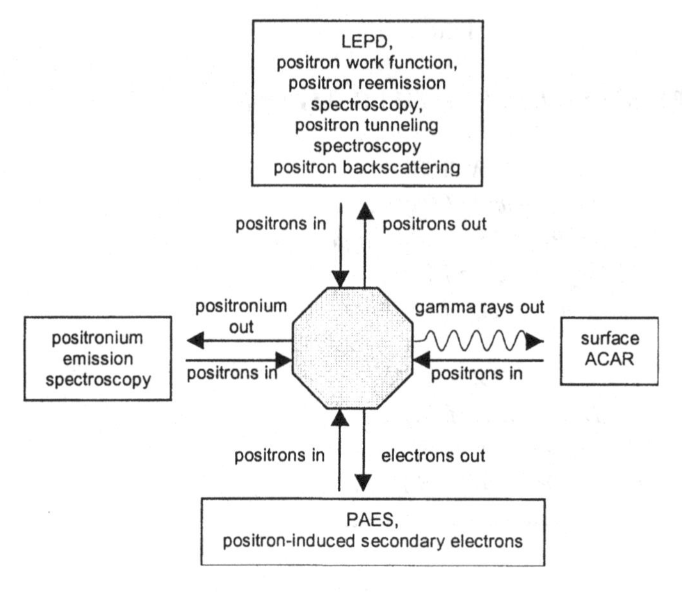


**Fig. 1** Positron surface spectroscopies

recent research aimed at the development and application of surface selective positron spectroscopies.

The surface selectivity of positron spectroscopies has a number of sources of origin. Low energy positrons have a large inelastic cross section and lose energy rapidly in most solids by (in order of decreasing energy loss) ionizing core electrons, generation of plasma oscillations, creating electron-hole pairs, and finally via phonon production. Positrons implanted at energies in the kilovolt range penetrate only on the order of $10^3$ Å in metals before they lose almost all of their energy and begin to diffuse.

In the case of Low Energy Positron Diffraction (LEPD), in which only the elastically scattered positrons are detected, the relatively large inelastic cross section results in a probe depth of only a few atomic layers below the surface. Positrons which penetrate deeper than a few Å have a low probability of elastically scattering out of the surface before undergoing an inelastic collision and dropping below the energy threshold of the detector. In general LEPD is more surface sensitive than Low Energy Electron Diffraction (LEED) due to the larger inelastic cross section of the positron.

Even higher surface specificity can be obtained in positron surface spectroscopies that exploit the fact that positrons implanted at low energies in metals and semiconductors have a high probability of diffusing to the surface and becoming trapped in a surface state before they annihilate. For example, in Positron Annihilation Induced Auger Electron Spectroscopy (PAES) and Surface Angular Correlation of Annihilation Radiation (Surface - ACAR) measurements, the signals originates as a result of the annihilation of positrons in the surface state. Consequently, PAES and Surface–ACAR typically probe only the top most atomic layer due to the short penetration depth of the positron surface state which is localized in an image correlation well just outside the surface.

Other spectroscopies, which involve measurements of low energy positron reemission or positronium emission from the surface, are highly surface selective because these emissions are strongly dependent on details of the interactions and potentials in the selvage region (e.g. those pertaining to the dipole layer and the image correlation well).

While it is not the purpose of this review to provide a complete history of positron surface spectroscopy, a brief (and necessarily incomplete) discussion of some early research using positrons as a surface probe will be attempted in order to provide context to current work (the reader is again referred to the reviews of Mills [1] and Schultz and Lynn [2] for additional details and references).

The study of positron surface interactions was inaugurated with the pioneering work of W.H. Cherry who in 1957 produced the first successful beam of low energy positrons and used it for the purpose of measuring positron stimulated secondary electron emission [4]. Cherry was able to produce a beam of 0.22 $e^+$/mCi-sec at 940eV with a conversion efficiency of about one slow positron per $10^8$ positrons from the source!! Cherry's work was not immediately followed up. In the late 60's papers by Madey [5] and Costello *et al* [6] on positron re-emission from solid surfaces rekindled interest in low energy positron interactions with surfaces. In the years that followed the search was on for for negative work function materials to provide a source of low energy positrons for gas scattering experiments, and for surfaces that would produce large amounts of low energy positronium *in vacuo* for lifetime measurements. The Brandeis discoveries of positronium formation at surfaces in the ground and excited states were made using high vacuum ($10^{-6}$ and $10^{-8}$ mbar) systems [7,8].

The first experiments done in ultra high vacuum (UHV) on a well-characterized surface were the series conducted by Mills and his collaborators; these included a study of positron work functions [9-11]. The results were in reasonably good agreement with previous theoretical calculations by Hodges and Stott [12] and Niemenen and Hodges [13]. The temperature dependence of positronium emission was modeled by Mills in terms of thermally-activated desorption of positronium trapped in a surface state [14]. The first surface experiments carried out with an electrostatic positron beam system resulted in the observation of Low Energy Positron Diffraction (LEPD) by Rosenberg *et al.* on a Cu(111) surface [15]. The first positron - secondary electron measurement conducted on a clean metal (Cu(100) was performed with this apparatus [16]. Subsequently Lynn and co-workers performed LEPD measurements at Brookhaven [17], as did Mills and Crane at Bell Labs [18]. Positronium diffraction – which is exceptionally surface-sensitive – was also pursued at Brookhaven [19].

More recent developments in the field have included Positron Annihilation induced Auger Electron Spectroscopy (PAES), initially demonstrated by Weiss and co-workers in 1987 [20], and the observation of positron tunneling and emission from pseudomorphically-grown films [42]

The rest of this chapter will be largely devoted to research carried out after the initial developmental stage of the field and will include major sections on positron re-emission studies, LEPD and PAES. In addition, we shall touch upon a number of other positron spectroscopies, including positron-induced secondary electron emission, positron backscattering and positronium emission which, while they have not been as widely applied as the first three, provide important insights into the interaction of positrons with surface and near-surface regions.

## 2 Positron re-emission studies

### *2.1 Introduction*

After implantation beneath a solid surface positrons can re-emerge after some combination of one or more inelastic and elastic collisions. Positrons that are re-emitted before they have lost a large fraction of their incident energy are characterized as part of the backscattered spectra. The subject of positron backscattering will be considered later in this chapter. Positrons that are reemitted after they have lost all but a small fraction of their energy but are still at energies well above kT with respect to the bottom of their band are said to be epithermal positrons. Note that the boundary between backscattering and epithermal emission is not clear-cut and becomes less so as the energy of the incident positrons decreases. Positrons that lose even more energy before reemission (or annihilation) and drop to an energy within ~ kT of the bottom of their band in the material are said to be thermalized.

In many metals and semiconductors, positrons implanted at energies ranging up to several keV have a relatively high probability of thermalizing and diffusing to the surface. In those cases where the surface has a negative positron work function, $\phi^+$, (i.e. the bottom of the positron band is above the local vacuum level just outside the surface) the positrons can be ejected into the vacuum with a net kinetic energy equal to the $-\phi^+$ (generally referred to as work function emission). The positron work function, like the electron work function, is highly sensitive to the physical and chemical structure of the surface. This sensitivity has been exploited in a number of surface science studies as will be discussed in the following section on work function positron emission. The presence of adsorbed or deposited overlayers of a different material can result in a more complicated behaviour of the positrons as it exits the surface as will be discussed in following sections on positron tunnelling, and reemission positron energy loss spectroscopy.

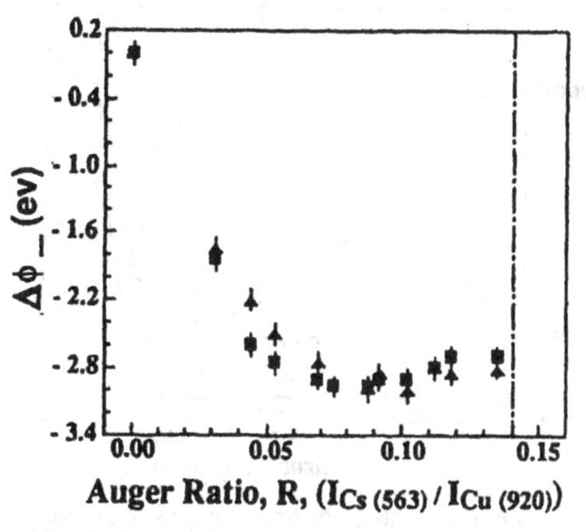


**Fig. 2** Change in electron work function as a function of Cs coverage on Cu, expressed as the Auger peak ratio (0.14 corresponds to monolayer coverage) [21].

## 2.2 *Work function positron emission*

The positron work function is surface sensitive through the surface dipole layer, which is caused by the spilling of electrons out from the surface. The electron work function can be written as the sum of the surface dipole $D$ and the bulk chemical potential $\mu^-$:

$$\phi^- = D - \mu^- \tag{1}$$

where the dipole contribution comes from the potential jump associated with crossing a dipole layer. When a positron is injected into a crystal there is typically only one positron and on the order of $10^{23}$ electrons.

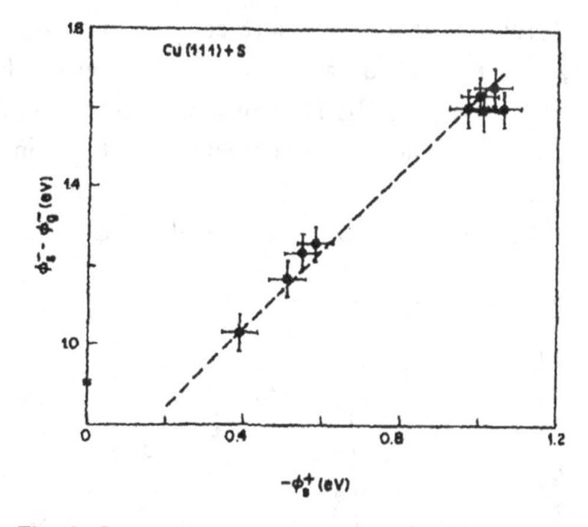


**Fig. 3** Correlation between change in electron work function (vertical scale) and positron work function (horizontal scale) for a S-covered Cu surface [2].

As a consequence, the charge density of the system is dominated by the electron contributions so that the magnitude of the surface dipole is not affected by the positron. However, the positron's positive charge means that $D$ is repulsive, as opposed to attractive for electrons, so that the positron work function can be written as

$$\phi^+ = -D - \mu^+ \tag{2}$$

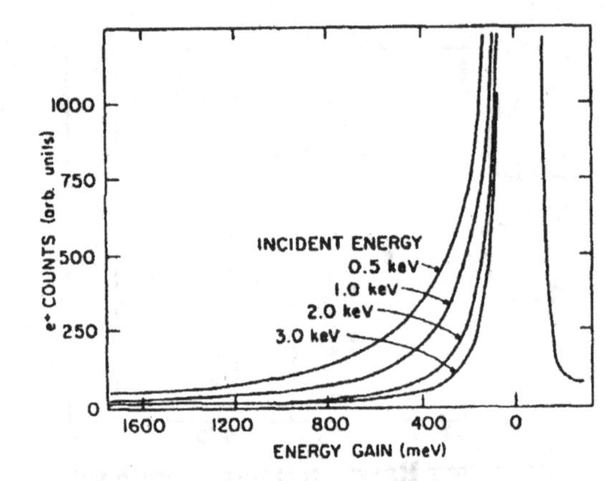


**Fig.4** Re-emitted positron energy spectra for positrons of four different energies incident on Ni(100), showing the increasing magnitude of the epithermal component with decreasing incident energy [ 3].

where $\mu^+$ is the chemical potential of the positron referenced to the mean electrostatic potential in the bulk. If (i) the positron loses energy and drops to an a level close to its ground state, (ii) it does so deep enough below the surface so that eq. (2) is appropriate, and (iii) it exits into the vacuum without losing energy, then the positron will leave the surface with a kinetic energy close to $-\phi^+$.

The surface dipole can be strongly affected by the presence of foreign atoms on the surface. As an example, a covering of less than an atomic layer of Cs on Cu can cause a reduction of almost 3 eV in the size of the dipole contribution (fig.2)[21]. This can be understood in terms of the Cs atoms donating electrons to the substrate metal result-ing in a so called "counter-dipole" layer which serves to reduce the net surface dipole.

Murray *et al.* [22] observed that $\Delta\phi^+ = -\Delta\phi^-$ by simultaneously measuring the changes in the positron and electron work functions as a function of sulfur coverages on a Cu(111) sample (Figure 3). They found that the $\phi^+$ became more negative as $\phi^-$ became more positive. This provided strong evidence in support of the picture implied by equations (1) and (2) in which many positrons and electrons leave the bulk with an energy near $\mu$ and traverse the region of the surface dipole with out losing energy.

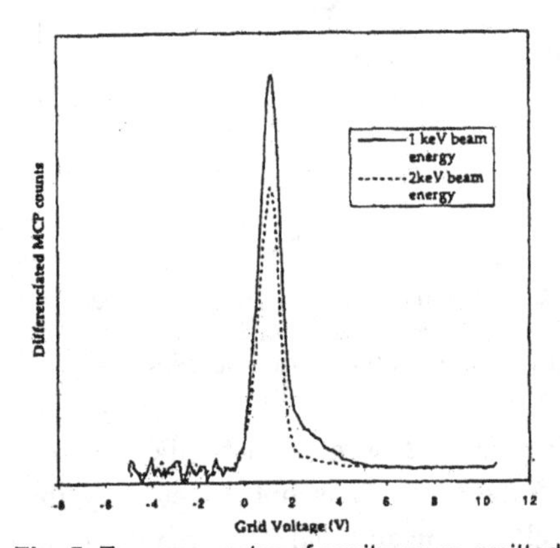


**Fig. 5** Energy spectra of positrons re-emitted from SiC for incident energies of 1keV (solid line) and 2 keV (broken line). The former exhibits a clear epithermal component [26].

A number of experiments have helped to elucidate dynamical features of emission (see Chapter 2). The physical picture of implantation followed by energy loss, random motion to the surface, and finally reemission suggest that the higher the energy of the incident positron the deeper the positron penetrates and the more likely it is to thermalize completely before re-emission. The utility of this picture is confirmed by the data shown in Figure 4, which were taken at four different incident beam energies [23]. The fraction of the spectra above the negative work function emission energy (*ie* the epithermal contribution) decreases significantly as the incident beam energy is raised. In addition, positrons are found to emerge largely in the forward direction (normal to the surface), consistent with the picture that they are "rolling down" a potential step in the direction normal to the surface due to surface dipole. In practice there is a spread in energy due to the initial thermal energy of the positron and a distribution of angles which includes contributions from (i) the initial (thermal) distribution of transverse momentum of the positrons in the bulk and (ii) elastic scattering of the positrons as they leave the surface.

Table 1 provides an updated version of the positron work function tables found in Mills [24] and Schultz and Lynn [25]. Notable additions include work function measurements for SiC [26], GaN and GaAs. These compound semiconductors have, unlike pure Si and Ge, shown themselves to be efficient emitters of work function positrons (Figure 5). References are given for the new entries in the work function table; those not listed here are to be found in refs. [24] and [25].

**Table I**

**Positron and Electron Work Functions**

Surfaces are polycrystalline unless otherwise identified.

Experimental errors shown in parentheses.

{note: positron affinities $A$ can be calculated from $A = -(\phi^{+}+\phi^{-})$}

| Z | Element | $\phi^{+}$ Exp.(eV) | $\phi^{+}$ Th (eV) | $\phi^{-}$ exp. (eV) |
|---|---|---|---|---|
| 3 | Li | | 4.90 [39] | 2.90 [39] |
| | Li (100) | | 3.44 [27] | 2.93 [27] |
| | Li (110) | | 3.51 [27] | 2.93 [27] |
| | Li (111) | | 3.53 [27] | 2.93 [27] |
| 4 | Be (0001) | > 0 | -1.3 [29] | 3.92 [29] |
| | | | -0.5 | 5.10(2) |
| 6 | C (0001) | 1.4(5) | | |
| | C (100) (diamond) | -3.03 [30] | | 3.7 [36] |
| 11 | Na | | 4.94 [39] | 2.75 [49] |
| | Na (100) | | 5.85 [27] | 2.70 [27] |
| | Na (110) | | 5.88 [27] | 2.70 [27] |
| | Na (111) | | 5.88 [27] | 2.70 [27] |

| Z | Element | $\phi^+$ Exp.(eV) | $\phi^+$ Th (eV) | $\phi^-$ exp. (eV) |
|---|---|---|---|---|
| 12 | Mg | | 3.4 [29] | 3.64 [29] |
| | | | 2.97 [39] | 3.66 [39] |
| 13 | Al | -0.2 [29] | 1.0 [29] | 4.25 [29] |
| | | | 0.15 [39] | 4.28 [39] |
| | Al (100) | -0.16(3) | -0.32, -0.5 | 4.41(3) |
| | | -0.19(5) | | |
| | Al (110) | | -0.19 | 4.28(3) |
| | Al (111) | 0.065(3) | -0.15 | 4.24(3) |
| 14 | Si (100) | | 2.04 | 4.91(5) |
| | Si (111) | ~0 | 2.21 | 4.74(5) |
| 19 | K | | 5.32 [39] | 2.30 [39] |
| | K (100) | | 4.78 [27] | 2.30 [27] |
| | K (110) | | 4.82 [27] | 2.30 [27] |
| | K(111) | | 4.83 [27] | 2.30 [27] |
| 21 | Sc | | 2.4 [29] | 3.5 [29] |
| 22 | Ti | > 0 [28] | -0.7 [29] | 4.3 [29] |
| | | | +0.1 [30] | |
| 23 | V | -0.6(2) [28] | -.04 | 4.3 [29] |
| 24 | Cr (100) | -1.76(5) | -1.9 [29] | 4.46(6) |
| | | | -1.76 | |
| 25 | Mn | | | |
| 26 | Fe | -1.2(2) [28] | -1.2 [29] | 4.4 [29] |
| | | | -0.7 [30] | |
| 27 | Co | | -1.5 [29] | 5.9 [29] |
| | | -0.8 [35] | | 5.0 [35] |
| 28 | Ni | -1.2(2) [28] | -0.1 [29] | 5.15 [29] |
| | | | -0.4 [30] | |
| | Ni (100) | -1.0, -1.3(1) | -0.77 | 5.22 |
| | Ni (110) | -1.4(1) | -0.59 | 5.04 |
| | Ni (111) | | -0.90 | 5.35 |
| 29 | Cu | | -0.08 [39] | 4.65 [39] |
| | Cu (100 | -0.3(2) [27] | -0.23 [27] | 4.59 [27] |
| | Cu (110) | -0.2(2) [27] | -0.14 [27] | 4.48 [27] |
| | Cu (111) | -0.4(2) [27] | -0.36 [27] | 4.85 [27] |
| 30 | Zn | | 0.9 [27] | 4.24 [29] |
| 31 | Ga | | 2.0 [27] | 3.96 [29] |
| 32 | Ge(100) | | 2.77 [33] | |
| | Ge (111) | | 1.98 | 4.8 |
| | | | 2.85 [33] | |
| 37 | Rb (100) | | 4.87 [27] | 2.26 [27] |
| | Rb (110) | | 4.82 [27] | 2.26 [27] |
| | Rb (111) | | 4.90 [27] | 2.26 [27] |
| 38 | Sr | | | |
| 39 | Y | | 2.7 [29] | 3.2 [29] |
| 40 | Zr | | 0.9 [29] | 4.1 [29] |
| 41 | Nb | < 0 | 0.6 [29] | 4.3 [29] |

| Z | Element | $\phi^+$ Exp.(eV) | $\phi^+$ Th (eV) | $\phi^-$ exp. (eV) |
|---|---|---|---|---|
| 42 | Mo | -2.2(2) [28] | -1.6 [29] | 4.6 [29] |
| | | | -2.0 [30] | |
| | Mo (100) | -1.7 | -2.65 | 4.53 |
| | Mo (110) | | -3.03 | 4.95 |
| | Mo (111) | <-3 [29] | -2.67 | 4.55 |
| 46 | Pd | | 0.8 [29] | 5.6 [29] |
| 47 | Ag (100) | 0.6(2) [37] | 0.72 | 4.64(2) |
| | Ag (110) | | 0.84 | 4.52(2) |
| | Ag (111) | | 0.62 | 4.74(2) |
| 48 | Cd | | 2.3 [29] | 4.1 [29] |
| 49 | In | | 3.2 [29] | 3.8 [29] |
| 50 | Sn | 0.7($2_s$) [38] | | |
| | Sn (100) | >0 | 2.7 | 4.4 |
| 55 | Cs (100) | | 4.89 [27] | 2.14 [27] |
| | Cs (110) | | 4.94 [27] | 2.14 [27] |
| | Cs (111) | | 4.93 [27] | 2.14 [27] |
| 73 | Ta | -1.2 | | |
| 74 | W | -3(1) | | |
| | | -2.75 [35] | | 5.25 [35] |
| | W (100) | -3.0 | | |
| | W (110) | -3.0(2) | | |
| | W (111) | -2.6(1) | -2.1 | 4.47(2) |
| 78 | Pt | -1.8(2) [28] | -0.2 [30] | |
| | Pt (100) | -1.9(1) | | |
| 79 | Au | 0.9($2_s$)[38] | 1.1 | 5.2 |
| 82 | Pb | 0.9($2_s$)[38]2.06 | | |
| | Pb (100) | | 1.55 | 4.01 |
| | GaAs | -0.6(1) [32] | | |
| | SiC(6H) | -2.1(1) [32] | | |
| | | -2.1(2) [31] | | 6.51(25)[31] |
| | SiC(4H) | -2.17(25)[31] | | 6.35(25)[31] |
| | SiC(3C) | -2.35(20)[31] | | 6.18(25)[31] |
| | CoSi | | -3.02 [34] | 4.87 [34] |
| | $CoSi_2$ | | -0.60 [34] | 4.55 [34] |
| | | | -0.46(5)[30] | 4.62(7)[30] |
| | NiSi | | -1.14 [34] | 4.56 [34] |
| | $NiSi_2$ | | -0.94 [34] | 4.64 [34] |
| | $Ni_2Si$ | | -2.06 [34] | 4.79 [34] |
| | NiSi | | -1.44 [34] | 4.93 [34] |
| | $NiSi_2$ | | -0.73 [34] | 4.54 [34] |

## 2.3 *Positron tunneling spectroscopy*

The presence of thin overlayers of dissimilar metals can dramatically effect the fraction of positrons reemitted from a metal surface. Gidley and co-workers have

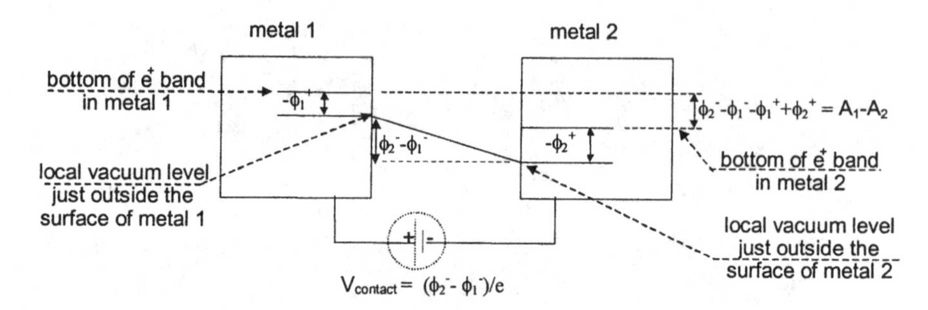


**Fig. 6** Positron energy levels in two dissimilar metals (see text for discussion)

made use of this fact to study pseudomorphic growth on metals and in experiments aimed at establishing contrast mechanisms in positron reemission microscopy [42, 43, and see Chapter 7]. The changes in reemission stem from a number of factors including overlayer-induced changes in the dipole layer at the surface, the different local lattice potential through which the positron must travel and the presence of internal electric fields resulting from dipole layers formed due to electron work function differences between layers. Gidley and co-workers showed that much of the behavior of the positrons may be understood by considering an averaged property of each metal termed the positron affinity, defined as $A = -(\phi^+ + \phi^-)$.

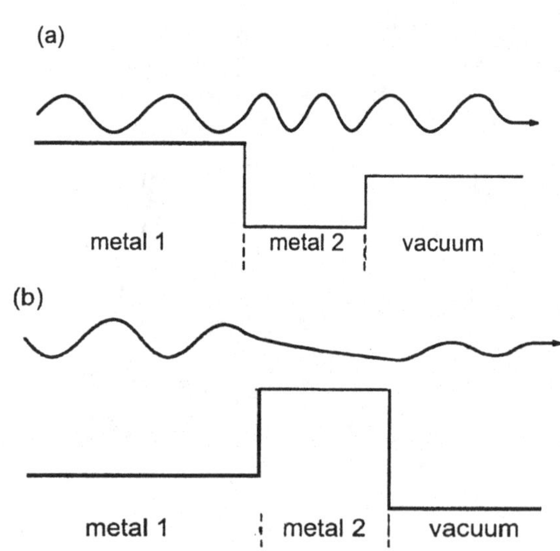


**Fig. 7** (a) Passage of positron through overlayer when $A_1>A_2$. (b) Tunneling of positron through overlayer barrier when $A_1<A_2$. See text for discussion.

To understand the utility of A consider two dissimilar metals connected by a wire (Figure 6). Electrons will move through the wire to the metal with the larger electron work function until, in equilibrium , an electric field is established so as to raise the electrostatic potential of the lower work function material so that the *electron* chemical potentials are equal (note that the charge rearrangement is determined by the electron work functions and not the positron work functions since there are ~$10^{23}$ electrons but, in

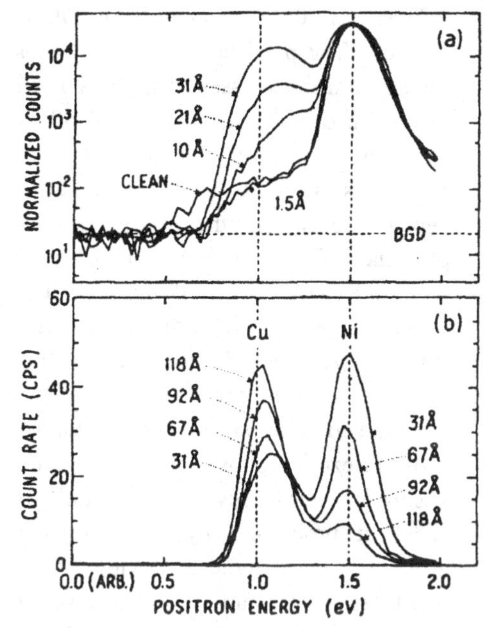


**Fig. 8** Reemitted positron energy spectra for various thicknesses of Cu on Ni (incident energy 4keV). (a) normalised to peak height, (b) to total intensity[44]

general, only one positron in the metal at one time). We can model this in terms of the equivalent circuit shown in fig. 6 in which this effect is modeled by a battery of voltage $V_{contact} = 1/e(\phi_2^- - \phi_1^-)$. For simplicity we take the example where both metals have a negative work function - however, the treatment applies in general. The ground state energy of the positron in the metal is at an energy $-\phi_1^+$ above the "local vacuum" level just outside the metal surface It may be readily seen from the diagram that the difference in the ground state energies of the positron in metal 1 and metal 2 is given by $A_1$-$A_2$. Thus a positron in metal 1 can spontaneously move to metal 2 but not vice-versa. Now imagine that the surfaces of metals 1 and 2 are brought into contact. The two surface dipoles and the potential step between metals 1 and 2 will merge into one net dipole layer at the interface. However, the relative positions of the positron ground state energies will be the same as in the separated case.

Gidley and Frieze exploited these ideas in using low energy positron reemission to study thin metal overlayers [44]. It is convenient to distinguish two cases: case 1 where $A_1 > A_2$, and case 2, where $A_1 < A_2$.

In case 1 (Figure 7(a)) there are traveling wave solutions to the Schrödinger equation in the overlayer for positrons with the ground state energy in the substrate. Thus ground state positrons from the substrate can traverse the overlayer and then the positron may enter the vacuum. In case 2 ($A_1 < A_2$) there is no traveling wave solution in the overlayer and the positron must tunnel through it to escape into the vacuum (Figure 7(b)).

This technique is illustrated in Fig. 8 as Cu was deposited on Ni(110). The value of the positron affinity for Ni is larger than that for Cu; $A_1$-$A_2$ (see Fig.6) is the difference in the peak energies in Fig. 8, ~ 0.5eV. The persistence of the Ni peak for Cu film thicknesses as high as 118Å is a result of the long mean free path of 0.5eV positrons in the overlayer; its mean energy is seen to be constant as it depends only on the bulk chemical potentials $\mu^+ + \mu^- = -(\phi^+ + \phi^-)$. The Cu peak in Fig. 8 only occurs at its 'expected' mean energy when the film thickness is at least 100 Å, when the majority of 'hot' positrons from the substrate have thermalized prior to reemission. Gidley and Frieze propose that reemitted positron spectroscopy (RPS) is thus an excellent probe of any processes that affect $\mu^+ + \mu^-$, including alloying of the overlayer film, and is insensitive to surface contamination. In addition, measurement of the elastic peak shift with temperature provides a means of measuring thermal volume expansion in thin films – a technique applied by Gidley [42] in the study of pseudomorphically-grown Ni films on Cu substrates. RPS of cobalt and nickel silicides showed no psedomorphic shifts, a result of incomplete positron thermalization [45]. The University of Western Ontario group have used RPS to study metal bilayers, including the pseudomorphic growth of Co on W(110) [46], alloying and defects in Pd films on Cu(110) [47] and the annealing properties of the resulting films [48].

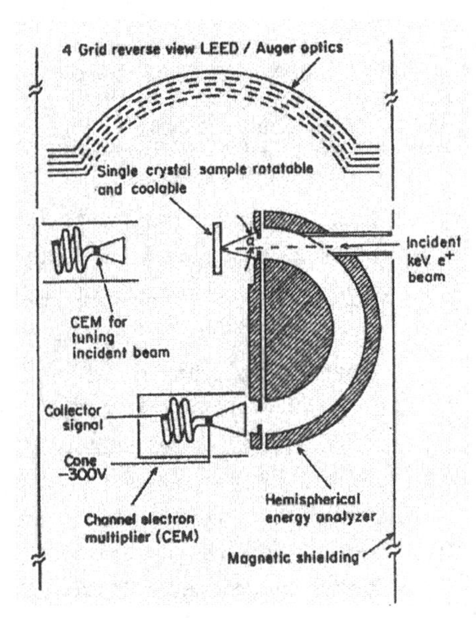


**Fig. 9** REPELS apparatus [23]

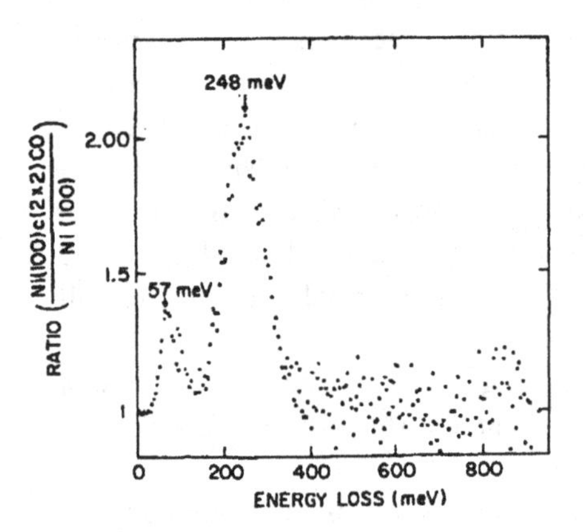


**Fig. 10** REPELS energy-loss peaks for positrons re-emitted from CO-covered Ni(100): 57meV – Ni-C stretch: 248meV – C-O stretch [49].

## *2.4 Reemitted Positron Energy Loss Spectroscopy (REPELS)*

Fischer et al. [23] developed a novel and ingenious method for measuring high-resolution, angle-resolved energy spectra for positron re-emitted from surfaces which involved the surface under study being placed at one port of a hemispherical energy analyser; re-emitted positrons of predetermined energy thus traveled to the detector at the other end of the hemispheres (see Figure 9). This apparatus was employed in 1983 to perform the only measurements to date of positron excitation of adsorbate vibrational levels [49]. The technique – Re-emitted Positron Energy Loss Spectroscopy (REPELS) – has features complementary to those of the standard electron equivalent, EELS. The most obvious difference between the two is that the source of the eV positrons in the former case is the substrate itself, allowing normal passage of the interacting particles. There is also the absence of the exchange interaction in the positron cases, and the probability that the mode of interaction will in general be different to that for electrons. Figure 10 shows the energy-loss peaks in the REPELS spectrum obtained for CO on Ni, clearly showing energy losses associated with the Ni-C and C-O stretches.

## 3 Low Energy Positron Diffraction (LEPD)

### *3.1 Introduction*

Low Energy Positron Diffraction (LEPD) like Low Energy Electron Diffraction (LEED) relies on measurements of the diffraction of elastically scattered particles from ordered surfaces to obtain information on the arrangement of atoms in the first

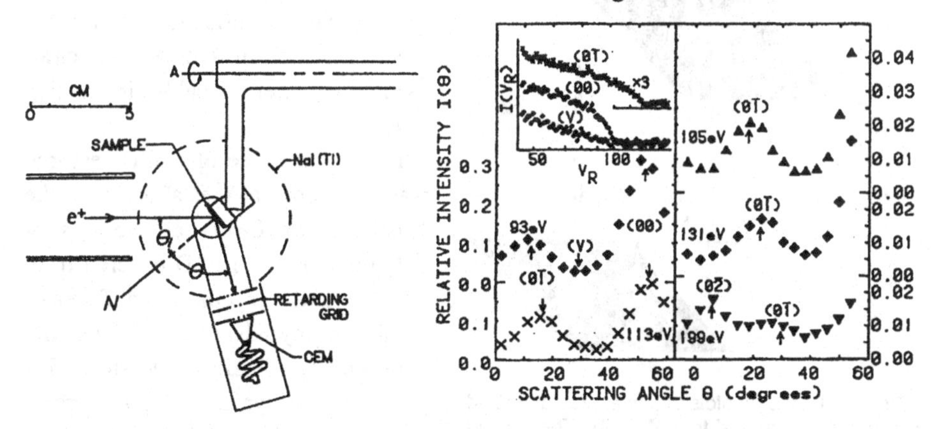


**Fig. 11** Left: Detail of apparatus used to obtain first LEPD data. Right - intensity of elastically-scattered e⁺ vs angle and energy. Arrows point tothe calculated diffraction angles. The inset shows the measured intensity vs detector retarding grid potential. [58]

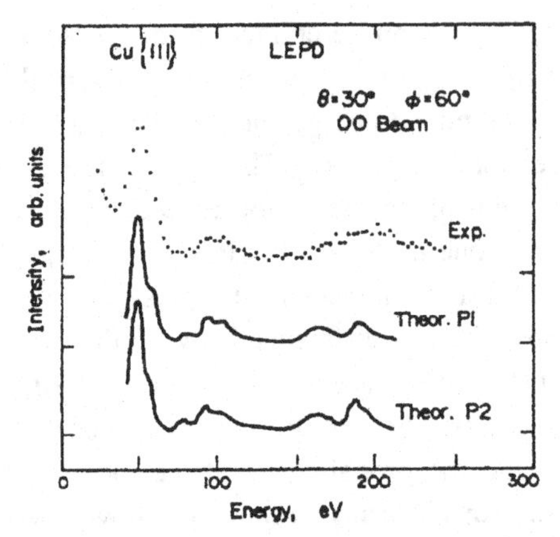


**Fig. 12** . Experimental and theoretical 00 spectra for LEPD from Cu{111} at θ = 30° and ϕ = 60° [60]. *P1* and *P2* refer to two different potentials used for the calculations in [60].

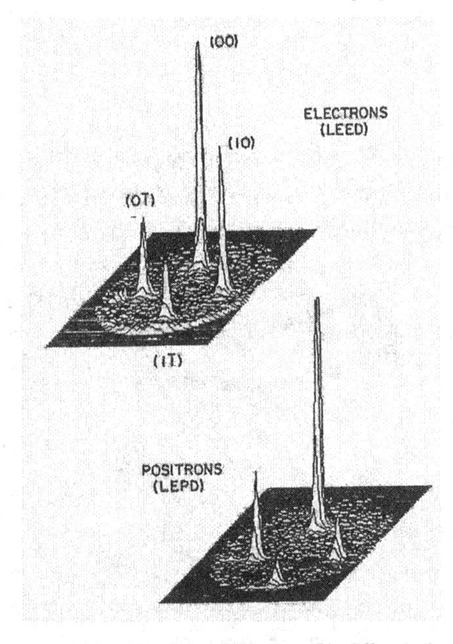


**Fig. 13** Isometric plots of low-energy diffracted intensities of $e^-$ and $e^+$ from W(110). The specular and three diff-racted spots are shown in each case. The 250eV beams were incident parallel to the [ $\bar{1}$10 ] bulk direction at ϕ = 20 ° to the surface plane. [63]

few atomic layers (surface structure). LEED has long been one of the work-horse techniques in identifying surface structures and has obvious advantages over LEPD owing to the experimental ease associated with the ready availability of electrons. However, a body of recent experimental [50-54] and theoretical work [54,55] has demonstrated that LEPD can have substantial advantages in identifying surface structures in systems where LEED has failed to give accurate or conclusive results (e.g. compound semiconductors and hydrogenated surfaces). It has been suggested that positrons are, in fact, an ideal surface structural probe due to the fact that, like electrons they are highly surface sensitive, but unlike electrons, they are repulsed from the atomic cores [56]. This core repulsion results in theoreti-cal simplifications that allow positrons to be treated in some cases as if they were weakly interacting [57].

The initial phase of LEPD research was aimed at establishing the feasibility of LEPD as well as at determining how LEPD differed from LEED and if these differences conferred any advantages. The first experimental LEPD measurements performed at typical LEED energies were carried out in 1979 at Brandeis University [58,59] on the (111) surface of Cu. These measurements demonstrated that LEPD could be performed using laboratory based positron beams (see Fig. 11), and showed significant differences between the energy dependence of LEED and LEPD intensities [59,60]. At roughly the same time as the Brandeis group reported their first LEPD studies, Mills and Platzman reported [61] measurements of first-order Bragg reflection from Al and Cu at ~10eV and Feder [62] reported the results of the first calculations of LEPD intensities in a (purely) theoretical study of the W(001) surface. Subsequently, Jona et al. [60] provided the first comparison between theory and experiment demonstrating that modified versions of theoretical models used for LEED could be used to calculate LEPD intensity versus energy (*I-V*) curves in good agreement with the Brandeis LEPD measurements on Cu (see Fig. 12]. Following their first work on Cu(111), the Brandeis group compared the results of measurements of the 00 and $\bar{1}0$ peaks of Cu(100) and Cu (111) surfaces [59]. Subsequent experimental work was carried out by Frieze et al. [63] who used a brightness enhanced beam and a position sensitive channel plate detector to acquire a two dimensional image of a LEPD diffraction pattern from W[110] (see Fig. 13). D.R. Cook et al. [64] measured positron reflectivity from a air cleaved LiF surface and Mills and Crane measured normal incidence elastic backscattering probability from air and vacuum cleaved 100 surfaces of NaF and LiF [65]. Mayer et al.[66] made measurements of LEPD at glancing incidence from Ni(110), Cu(100), and NaF(100). In another paper, Mayer et al. [67] reported on the results of a second LEPD and LEED study of Cu(100) in which a brightness enhanced beam was used to achieve better collimation, eliminating some of the uncertainties associated with the angular averaging necessary in previous work [59]. As in the work of Weiss et al. [59],

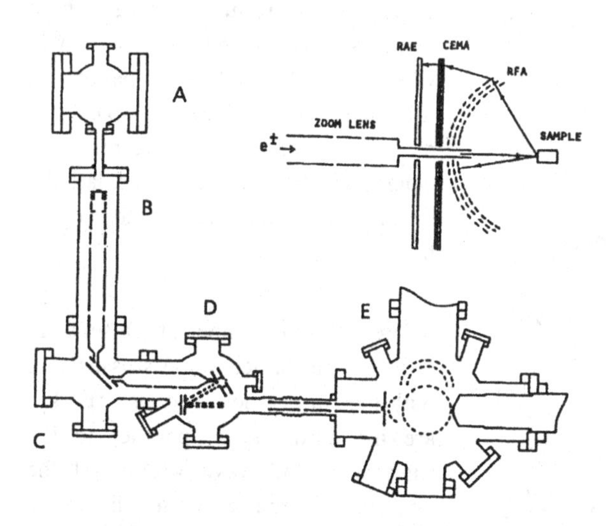


**Fig. 14.** Schematic diagram of the Brandeis slow positron apparatus. *B* - positron source, *C* - 90° cylindrical mirror, *D* - remoderation stages for brightness enhancement. *E* is the sample interaction region and diffractometer. Inset: detail showing Channel Electron Multiplier Array (CEMA) and spherical sector LEED-LEPD optics. [54]

Mayer et al. found that good agreement could be obtained between experimental LEPD *I-V* curves and *I-V* curves calculated using modified LEED codes in which the sign of the coulomb interaction was reversed and the exchange term was eliminated [67].

Following up on discoveries made in the initial demonstration experiments, the Brandeis group has exploited the advantages of LEPD in a number of studies of compound semiconductors for which definitive surface structures had not been previously established including: analysis of the $(10\bar{1}0)$ and $(11\bar{2}0)$ surfaces of CdSe [50,51,54], the (110) surfaces of GaAs [52,55] and InP [52] and CdTe [53]. The brightness enhanced LEPD system used in these studies is shown in Fig. 14.

### 3.2 *Analysis of surface structure*

In both LEPD and LEED, a monoenergetic beam of particles is scattered from the sample. If the surface is well ordered, the elastically scattered part-icles are diffracted into directions determined from constructive inter-ference of the scattered particle wave. Typically, the incident beam is well collimated and circular in cross section resulting in a set of cylindrical diffracted beams which, when imaged, result in a pattern of 'diffraction spots' (see fig. 13). The directions of these beams are determined from the 2-D translational symmetry of the surface according to the equation [68]:

$$\mathbf{k}_{\parallel out} = \mathbf{k}_{\parallel in} + \mathbf{G}_{SN} \tag{3}$$

where $\mathbf{G}_{SN}$ is a member of the set of 2-D reciprocal lattice vectors associated with the surface net., and $\mathbf{k}_{\parallel out}$ and $\mathbf{k}_{\parallel in}$ are the components, parallel to the surface, of the wave vectors of the outgoing electron and incoming electrons respectively (it is straightforward to show using eq. (3) that, for normal incidence of the primary beam, if the diffraction spots

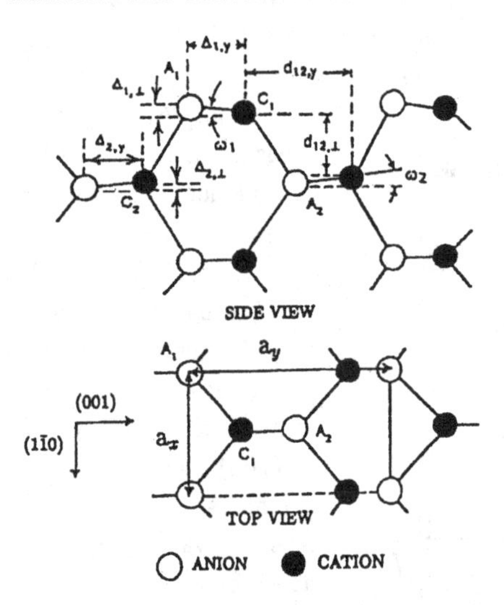


**Fig. 15.** Schematic indication of the atomic geometry and independent structural parameters of the zincblende (11) surface. The first- and second-layer shear angles are given as $\omega_1$ and $\omega_2$. A negative $\omega_1$ and a positive $\omega_2$ are depicted in the drawing. The anion (cation) in the i[th] layer is designated by $A_i(C_i)$. [52]

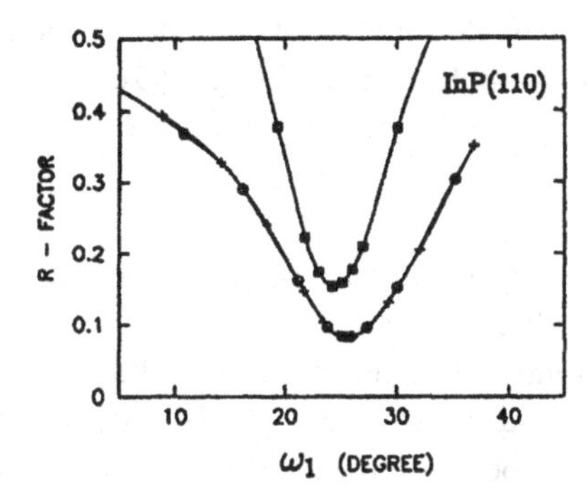


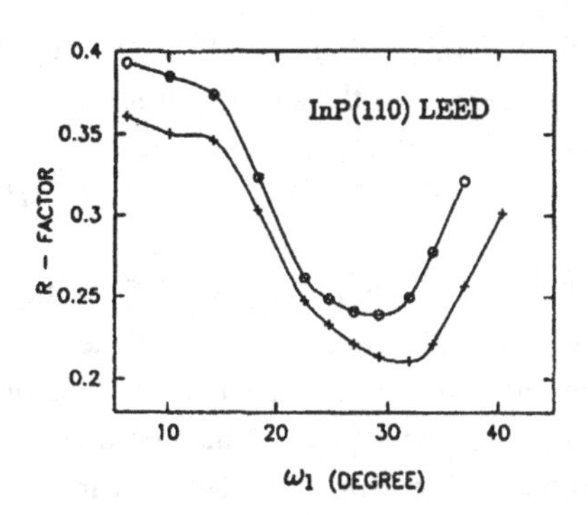


**Fig. 16.** *R* factor (a measure of the goodness of fit) as a function of $\omega_1$ for LEPD (left) and LEED (right). All structural and nonstructural parameters were initially determined by the simplex search procedure. Then all parameters except $\omega_1$ were kept fixed in the calculation of *R* as a function of $\omega_1$. The two curves in each panel refer to two different methods of calculating of calculating the *R* factor and different scattering models. [52]

are imaged on a spherical detecting surface centered on the sample then, the resulting spot pattern when projected onto a plane parallel to the surface is congruent to the 2-D reciprocal lattice of the surface net). Because of their relatively large cross sections for inelastic scattering, electrons or positrons that penetrate more than a few layers into the surface have a high probability of dropping below the 'elastic-energy' threshold set for the detector (typically an eV or so below the incident beam energy). As a result, both LEED and LEPD are highly surface specific.

While the 2-D translational symmetry of the surface may be determined directly from the diffracted spot patterns, a determination of the actual three dimension structure of the surface layers requires an analysis of the intensities of the diffracted beams as a function of energy (*I-V* profiles). In general, multiple scattering can not be neglected at the energies used to obtain surface structural information (~20 eV – ~400 eV). As a result, significant computation effort must be expended to calculate theoretical diffracted intensities and simple inversion of the diffracted data is precluded. Typically, structural information is obtained by constructing a parameterized model surface structure (Fig. 15) and optimizing the parameters to obtain the best agreement between measured and calculated values of *I-V* profiles ( Fig. 16). It is assumed that the true surface structure is best approximated by the model structure that gives optimum agreement between theory and experiment. The Brandeis group has demonstrated that LEPD can be used to achieve qualitatively and quantitatively better agreement between experimental and theoretical *I-V* profiles (compare Figs. 17 and 18) leading to significantly more reliable determinations of the surface structure than is possible using LEED [50-55].

### *3.3 Differences between LEED and LEPD*

Although both LEPD and LEED share many common features, a number of important differences have been identified [50-67,69-71]. The genesis of these differences lie in the facts that the Coulomb interaction for electrons and positrons are opposite in sign and that unlike a scattered electron, a scattered positron is distinguishable from the electrons in the solid being studied. Thus the electrons encounter an attractive nucleus and repulsive screening electrons; positrons encounter a repulsive nucleus and attractive screening electrons. As a result, at the scattering energies relevant to LEED and LEPD the electron spends significantly more time in the core region than does the positron. A summary of the differences between the low energy scattering processes of positrons and electrons and their significance in determining those cases where LEPD has advantages over LEED are enumerated below [71]:

*(i) Positron scattering is less sensitive to Z:*

Duke and Lessor [70] have shown that positron phase shifts for elastic scattering from the atomic core potentials in the energy range relevant for LEPD vary much more slowly with $Z$ than is the case for the corresponding electron scattering phase shifts. They attribute this behavior to the fact that for positrons both the Coulomb interaction and the "centrifugal barrier," (i.e. the term $[l(l+1)/2mr^2]$ in the partial wave decomposition of the Schrödinger Equation), are both repulsive. In the electron case the cancellation of the attractive Coulomb part and repulsive centrifugal term results in resonances which produce 180 degree jumps in the scattering phase shifts. The position of these resonances is strongly $Z$ dependent leading to significant differences in the phase shifts of even isoelectronic atoms [70]. The $Z$ independence of the positron scattering phase shifts leads to a maximal interference between waves scattered from different elements and hence to an enhanced sensitivity to structural parameters in multi component systems. The reduced $Z$ dependence of the positron phase shifts have been cited as a reason that, LEPD has been more successful than LEED in arriving at accurate surface structures for a number of compound semiconductor surfaces including, perhaps most notably, InP(110) [52,72] (see Figs. 15-18).

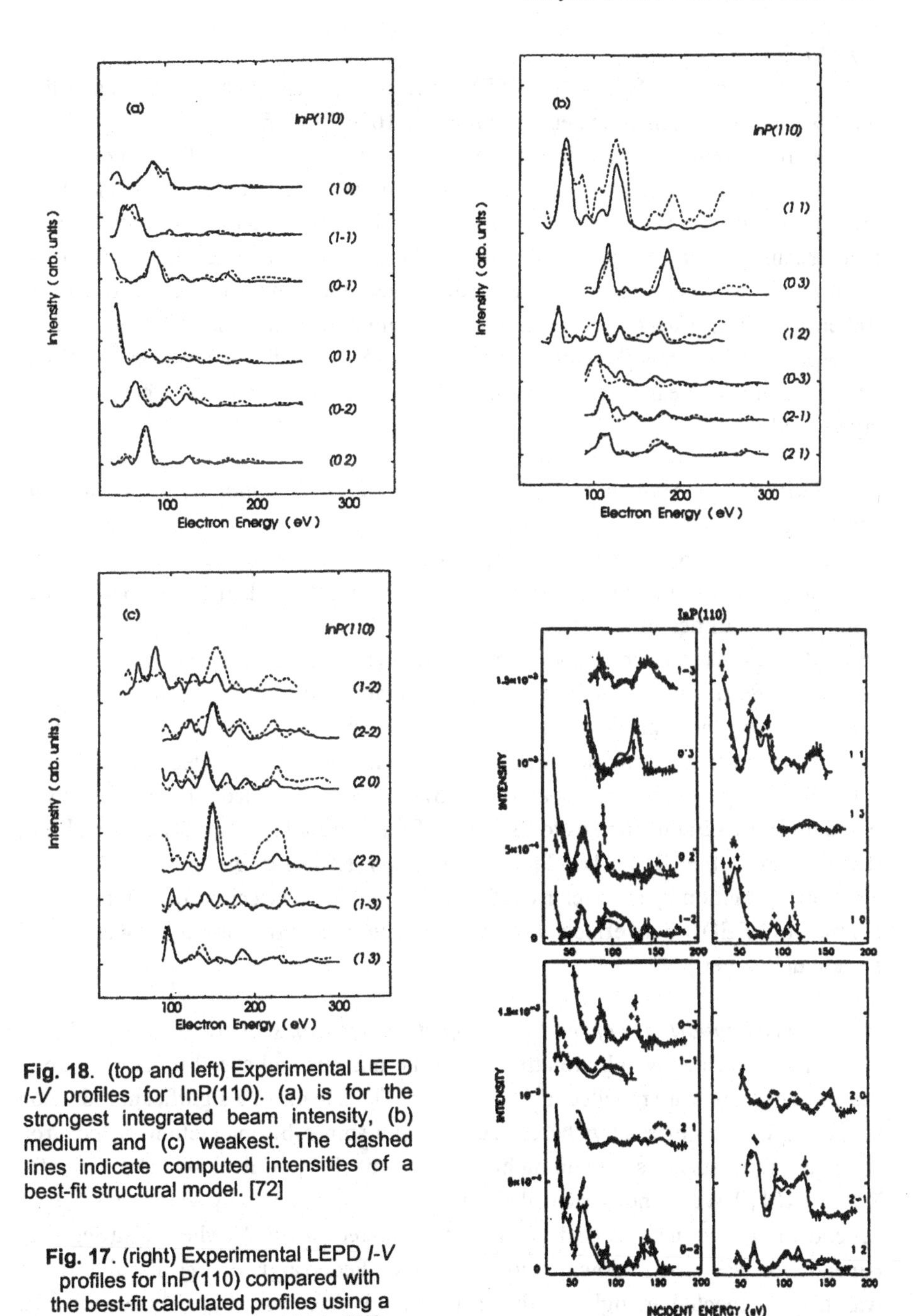


**Fig. 18.** (top and left) Experimental LEED *I-V* profiles for InP(110). (a) is for the strongest integrated beam intensity, (b) medium and (c) weakest. The dashed lines indicate computed intensities of a best-fit structural model. [72]

**Fig. 17.** (right) Experimental LEPD *I-V* profiles for InP(110) compared with the best-fit calculated profiles using a global normalization constant. [52]

*(ii) LEPD is more surface sensitive*

The inelastic mean free path (IMPF) for electrons is as much as 40% higher than the IMFP for positrons in the energy range 30-100eV [54, 58-60]. The longer IMFP for electrons stems from the fact that when one electron scatters from another, the final states of both particles are excluded from the filled states below the Fermi level (a significant fraction of the total available phase space). In contrast, since there is, in general, only one positron in the system at a time; the positron can scatter into a wider range of final states extending down to the bottom of its (otherwise empty) band. The IMPF sets an important length scale in determining the probe depth. Thus the shorter inelastic mean free path for positrons results in LEPD having a greater sensitivity than LEED to the structure of the outermost layers [58].

*(iii) Reduced importance of uncertainties in the correlation term in positron scattering:*

Uncertainties regarding the appropriate functional form of the electron-electron correlation term in the single particle Hamiltonian used in LEED calculations can lead to uncertainties in the structural parameters deduced from these calculations. In LEPD calculations, the importance of the uncertainties in the positron–electron correlation term is greatly reduced by the fact that the low energy positron is largely excluded from the high electron density region surrounding the nucleus due to its coulomb repulsion from the nucleus [73]. It has been pointed out [59] that calculations reported by Jona et al. [60] provided clear evidence for the insensitivity of calculated LEPD intensities to the form of the correlation term. They found [60] that changing the Hedin-Lundqvist correlation term for the positron-ion core electron interaction by as much as 400% had a negligible effect on $I_{th}(V)$ in marked contrast to LEED where effects of electron-ion core electron correlation were found to be much larger [74].

*(iv) Reduced importance of relativistic effects in positron scattering:*

Unlike electrons, which are attracted to the nucleus and speed up as they enter the core, positrons are repelled and slow down. This leads to a significant reduction in the importance of relativistic effects (e.g. the spin-orbit interaction) in positron scattering from surfaces containing high Z elements [62]. Calculations by Feder for W(001) [61] have demonstrated that LEPD intensities are only very weakly spin dependent at low energies ( <100 eV ) as compared to LEED where a strong spin dependence was found down to low energies. Feder argued [62] that this result was to be expected in light of the fact that the effect of spin dependence was proportional to the average of the term $V_{s0}$ ~(1/r)(dV(r)/dr) weighted by the modulus of the scattering particle's wave function. Positrons are repelled from the core causing the modulus of their wave function to be small in those regions were the term $V_{s0}$ is significant. As a result, the spin orbit interaction for positrons is small. The opposite is the case for electrons.

*(v) The muffin-tin approximation works better for positrons:*

The dominance of the Hartree and the centrifugal terms in the scattering potential for positrons have been suggested as reasons why the muffin-tin model appears to work better for positrons than for electrons when considering scattering from covalently bonded semiconductors [55]. The lack of spin-exchange repulsion of the positron by the nonspherically distributed and spin-singlet correlated valence electrons reduces the influence of the interstitial valence electrons on positron scattering as compared to electron scattering.

### *3.4 Advantages of LEPD over LEED for surface structure determination*

As discussed above, calculated LEPD intensities are less sensitive to so called non-structural parameters such as the functional form of the correlation term used in the scattering potential. This reduced model dependence leads to an enhanced confidence in the LEPD determination of structural parameters particularly in those cases, e.g. compound semiconductors, where LEED has difficulty.

The work of the Brandeis group on compound semiconductors provides a good example of the advantages of LEPD over LEED in structural determinations. They have shown that the use of LEPD resulted in significantly better agreement between experimental and theoretical IV curves than could be obtained using LEED for those compound semiconductor surfaces investigated by both techniques to date, namely the cleavage faces of CdSe [50,51,54], the (110) faces of GaAs [52,55], InP [52] and CdTe [53].

The structural determination of the InP (110) surface provides a particularly good example of the power of LEPD [52]. The commonly used structural feature used to characterize relaxations of zinc-blende semiconductors is the top anionic dimer pair rotation angle $\omega_1$(see Fig. 15), which for InP(110) is found to be 24.3 ± 1.5 degrees using LEPD [52] (see Figs. 16 and 17) in good agreement with an independent measure using photo-electron diffraction in which it was found that of $\omega_1$ = 23 ± 3 degrees.[75]. In contrast, the agreement between experimental and theoretical *I-V* profiles using LEED was significantly worse and produced a value of $\omega_1$ = 31 ±3. [76] (see Figs. 16 and 18). A quantitative measure of the improvement resulting from the use of LEPD can be obtained from an evaluation of the x-ray R-factor, $R_{x,}$ a functional which provides a measure of goodness-of-fit defined such that a smaller value of $R_x$ corresponds to better agreement between theoretical and experimental *I-V* profiles [77]. The value $R_X$ = 0.0.5 obtained for the LEPD InP(110) data set was more than 3 times better than the best value (of $R_X$ = 0.16 ) obtained to date by LEED [52] (see Fig. 16).

### *3.5 LEPD theory*

The first LEPD calculations were performed for a W(001) surface by Feder [62], (who was unaware of the as yet unpublished experimental work by the Brandeis group for Cu). Feder adapted spin dependent, relativistic, muffin-tin LEED calculations by reversing the sign of the electrostatic Hartree potential and eliminating the exchange term. Feder's calculations demonstrated that LEPD intensities were only very weakly spin dependent; he also noted that the computation time for the LEPD calculations was ~1/2 that of the LEED calculations due to faster convergence.

After the first LEPD *I-V* measurements became available in April of 1980 [58], Jona et al. [60] provided the first comparison of calculations with measured LEPD intensities for the Cu(111) surface (see Fig. 13). Their calculations produced quite good agreement with the measurements of Rosenberg et al. [58] demonstrating the effectiveness of the use of modified LEED muffin-tin based multiple scattering codes in calculating LEPD intensities. Jona et al. found that, unlike the case in LEED, the calculated LEPD intensities were highly insensitive to changes in the correlation term in the scattering potential. As noted above, this insensitivity gives LEPD advantages in cases where uncertainties regarding the form of the correlation term give rise to uncertainties in determining the surface structure using LEPD. Subsequent theoretical work using similar codes was carried out for Cu(100) [65].

Duke and collaborators have carried out an extensive series of LEPD theoretical work [50-57] in close collaboration with the Brandeis experimental group. Early work [59] considered different treatments of the inner potential and of the imaginary part of the inner potential. It was found that best agreement with data obtained from Cu could be obtained with an attenuation length that varied as $E^{1/2}$ for LEPD but was constant in the case of LEED indicating a shorter inelastic mean free path for positrons at low energies. Good agreement was found using $\alpha = 2/3$ for electrons in the local density approximation to the exchange-correlation potential and $\alpha = 0$ (zero exchange-correlation) in the case of positrons. The calculations of Duke and collaborators were extended to include the first calculations of LEPD intensities for compound semiconductors including CdSe [50,51,54], InP [52], GaAs [52,57], and CdTe [53]. They found, in general, that

LEED codes could be used, *mutatis mutandis*, to calculate LEPD intensities that were in excellent agreement with experiment. Further they found that, these codes could be used to produced agreement between calculation and experiment which was often significantly better for LEPD than LEED. In a number of cases (e.g. GaAs [52,55], CdSe [50,51,54] and CdTe [53]) the LEPD analysis resulted in surface structure determinations that were close to those found previously by LEED. In these cases the improved agreement between experiment and theory led increased confidence in the LEED results for which agreement had been relatively poor. In other cases (e.g. CdTe [53]) the LEPD results produced surface structures tha differed from those obtained by LEED. Since agreement between calculated and experimental *I-V* curves was significantly better in the case of LEPD, this strongly suggests that the surface structures derived from LEPD are more accurate than those previously found from the LEED (see Figs.16-18).

Recently, a new theoretical approach to calculating diffracted intensities has been developed by Joly which holds out the promise of even more accurate and detailed surface structure determinations using LEPD. This approach (termed the finite difference method (FDM), as applied to LEPD or LEED) is based on methods developed by the Finnish Group for calculating positron wave functions in the bulk [79]. In the FDM, the single particle Schrödinger equation is solved numer-ically using the relaxation method on a mesh [80,81]. Unlike the formalisms typ-ically used in LEED, the FDM does not require that the crystal potential be modeled as a collection of spherically symmetric scattering centers (ie a muffin-tin poten-tial). This makes the FDM better suited to semiconductor systems in which consid-erable charge density is located in the covalent bonds connecting the atoms. The FDM has already been used by Joly to achieve improved agreement between calculated and measured *I-V* profiles in InP(110) and it is likely to find wide applicability in the future.

### *3.6 Future LEPD research*

The advantages of LEPD have already been demonstrated for a number of compound semiconductors. There are reasons to believe that LEPD should be advantageous for most other ordered multi-component surfaces as well. This follows from the relative insensitivity of positron scattering to Z (for a neutral atom) which results in larger interference effects between positrons scattered from different elements leading to a greater sensitivity of LEPD diffracted intensities to small deviations in the surface structure [50,51].

Other areas in which LEPD could make significant contributions are in the determination of the surface structures of high Z metals and the structures associated with hydrogen adsorbed on semiconductors and metals. While LEED theory works well for most elemental metals (leaving relatively little room for improvement using LEPD), uncertainties regarding the proper treatment of electron scattering from the high electron density cores of high Z metals limits the accuracy of LEED in determining surface structure [78]. These uncertainties are largely eliminated in LEPD due to core repulsion implying that LEPD may have significant advantages for the study of high-Z metal surfaces. LEED has had great difficulty in determining surface structures associated with adsorbed hydrogen due to the fact that electron scattering from hydrogen is weak compared to that from most substrates. For example, in LEED studies of the Si(100) (1x1) H surface, [82] good agreement was obtained between theoretical and experimental *I-V* profiles even though scattering from the H atoms is completely ignored. The differences in scattering cross-section between H and other atoms are significantly less pronounced in the case of positrons. Preliminary calculations indicated that unlike LEED, LEPD intensities will be strongly influenced by adsorbed H, making it well suited in answering important questions regarding the position of H on surfaces. [71]

Finally, Tong [57] has made the intriguing suggestion that positrons may have significant advantages over electrons in so-called direct methods such as holographic wave-front reconstruction. The reasons given include the fact that the because of the net repulsive interaction the positron's scattering from an atom has a relatively simple angular dependence. This is in contrast to electron scattering where the attractive potential results in sharp resonance lobes in the differential scattering cross section (involving jumps of $\sim\pi$ in the scattering phase) which greatly complicate holographic analysis. Tong has also proposed that because the relatively weak scattering of positrons by atoms, the structure of large biological molecules can be studied by coherent beam positron diffraction.

## 4 Positron Annihilation Induced Auger Electron Spectroscopy (PAES)

### *4.1 Introduction*

Positron annihilation induced Auger Electron Spectroscopy (PAES) makes use of a fundamentally new method for the excitation of Auger electron emission in which a beam of low energy positrons is used to create the necessary core hole excitations by matter-antimatter annihilation [83]. PAES has been shown to have significant advantages over conventional Auger methods [84,85] including: 1. increased surface selectivity which permits PAES to be used to characterize the elemental content of the topmost atomic layer, 2. the ability to obtain Auger spectra using energy doses to the surface that are five orders of magnitude lower than in electron induced Auger spectroscopy (EAES) and 3. the ability to eliminate the very large secondary electron background which makes it impossible to unambiguously determine Auger line shapes using conventional Auger techniques.

The PAES mechanism was first demonstrated in 1987 [83]. Subsequently a number of studies at the University of Texas at Arlington (UTA) and Brookhaven National Lab (BNL) demonstrated that PAES could be used to eliminate the collisionally induced secondary electron background [86], and that the PAES signal originates from the annihilation of positrons in the surface state with the consequence that PAES probes the top-most atomic layer [87,88]. Early theoretical work was carried out by Jensen and collaborators [87,89-92], and subsequently extended by Fazleev, Fry and co-workers [93-98]. PAES has been applied by the UTA group in the study of the growth of ultra-thin metal and semiconductor films on metal and semiconductor substrates [99-105] and in studies of gas adsorption [101,105]. Recently, a group at the Electrotechnical Laboratory (ETL) in Tsukuba has utilized a pulsed positron beam to perform time of flight PAES spectroscopy [106-111]. Because of their high flux and the high efficiency of the t-o-f technique, the ETL group were able to observe the relatively weak PAES peaks associated with C and O and were able to demonstrate that PAES is highly sensitive to small quantities of impurities adsorbed on $MoS_2$. Currently, PAES experiments are actively being pur-sued at UTA in the USA and at the ETL in Japan, Tsukuba, Japan and are in the planning stages at laboratories in the UK and Germany. This section provides an update of previous reviews [84-5] and highlights some more recent PAES research.

### *4.1.1 The PAES mechanism*

The Auger process is a non-radiative transition in which an atom with an inner shell hole relaxes by filling this hole with a less tightly bound electron while simultaneously emitting another electron (the Auger electron) which carries off the excess energy. The energy of the Auger electron is given to a first approximation by the equation, $E_{XYZ} = E_X - E_Y - E_Z^*$ where $E_X$ and $E_Y$ are the binding energies of the electron removed to form the original inner shell hole and the electron that fills the hole respectively, and $E_Z^*$ is the binding energy of the outgoing electron where the binding energy has been calculated for an atom with a pre-existing hole in the Y level. Because the energy levels of different elements are in general unique, the elemental identity of an atom may be deduced from the energies of the Auger electrons emitted as a result of core hole excitations. This fact, along with the short escape depth of low energy electrons, has been exploited in the widely used surface analysis tool, Auger Electron Spectroscopy (AES) [112,113].

The PAES mechanism can be outlined as follows:

1. A positron implanted at low energy diffuses to and gets trapped at the surface.
2. A few percent of the trapped positrons annihilate with core electrons leaving the atom in an excited state.
3. The atom relaxes via emission of an Auger electron. The PAES mechanism is contrasted with that of electron induced Auger Spectroscopy (EAES) in Fig. 19.

### *4.1.2 PAES experimental apparatus*

PAES experiments require a positron beam to make low energy positrons incident onto a surface, an electron energy analyzer for measuring the spectra of annihilation induced Auger spectra and a surface preparation and analysis chamber which can be maintained under ultra-high vacuum (UHV) conditions. To date most PAES measurements have been made using a magnetically guided positron beam system equipped with a trochoidal energy analyzer [114,115]. Such systems are highly efficient, collecting Auger electrons from a solid angle of ~2 Radians, but in general have relatively poor energy resolution. Figure 20 provides a schematic diagram of the UTA High Resolution PAES apparatus [116] which makes use of an electrostatic positron beam and a cylindrical mirror analyzer to achieve a resolution of E/E < 2% at the cost of collecting only over ~10% of 2 Radians. Recently, the ETL group has developed a highly efficient PAES system that uses a LINAC based pulsed high intensity positron beam coupled to a time of flight (T-O-F) energy analyzer

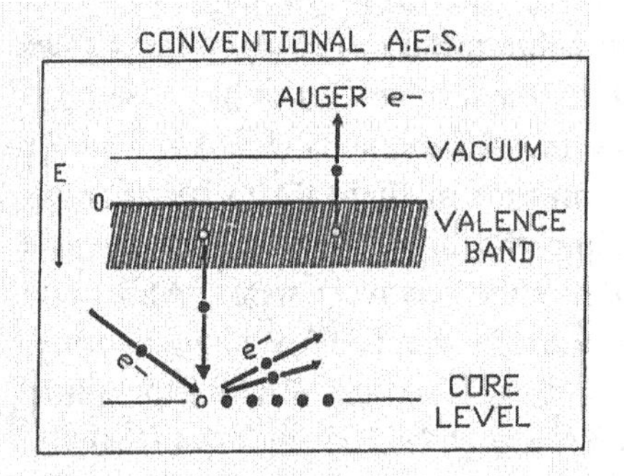


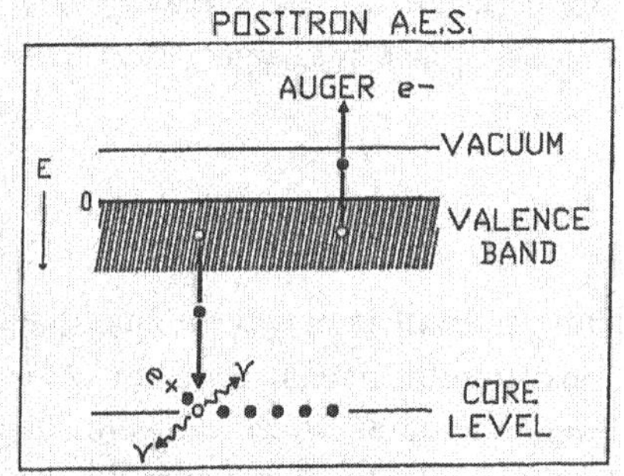


**Fig. 19.**Comparison of the core-hole creation mechanism in Conventional AES (EAES) and PAES. In EAES core electrons are removed by collisions with an energetic electron. The incident electron beam energy must exceed the binding energy of the core electron. In PAES the core hole is created via a matter-antimatter annihilation process. The incident positron beam energy can be made arbitrarily low. Auger electron emission follows the creation of the core hole in both cases [86].

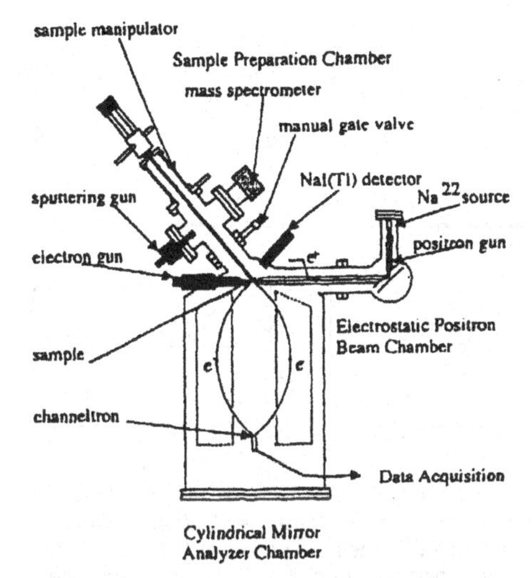


**Fig. 20**. Schematic Overview of the UTA high-resolution PAES system. An electrostatic beam is used to transport positrons from a $^{22}$Na source to the sample surface. A large cylindrical mirror analyzer (CMA) is used to select the energy of the annihilation induced Auger electrons that reach the channeltron and are detected [from Ref. 116].

[117]. The T-O-F analyzer allows for parallel collection of Auger electron over entire energy range at one time leading to a factor of 10 to 100 increase in the data collection rate as compared to a sequentially scanned analyzer.

### *4.1.3 Secondary Electron Background Elimination*

The secondary electron background in the energy range of the Auger electron can be eliminated by using an incident beam energy that is less than the Auger electron energy. This follows from the fact that secondary electrons cannot be created through collisional processes with energies in excess of an energy $E_k$ given by:

$$E_{sec} \leq E_k = E_p - \phi^- + \phi^+ \tag{4}$$

where $E_{sec}$ is the kinetic energy with which the secondary electrons leave the surface of the sample, $E_p$ is the kinetic energy of the primary beam at the sample surface, and $\phi^-$ and $\phi^+$ are the positron and electron work functions, respectively [86].

The enhanced surface selectivity of PAES stems from the fact that positrons implanted into a metal or semiconductor at low energies have a high probability of diffusing to the surface and becoming trapped in an "image-correlation-well" before they annihilate [118]. The positrons in this well are localized at the surface and annihilate almost exclusively with atoms at the surface. Consequently, almost all of the Auger electrons originate from the top-most atomic layer [87]. This is in contrast to conventional Auger techniques in which the Auger electrons originate from an excitation volume that extends hundreds of atomic layers below the surface, limiting the surface selectivity to the 4-20Å escape depth of the Auger electron [112,113]. Figure 21 provides a comparison of PAES and EAES spectra which demonstrates the ability of PAES to eliminate the large secondary electron background associated with electron and photon excited Auger

Electron emission [86]. The EAES signal (shown in Figs 21A and 21C) can be seen siting on top of a large secondary electron background. This background is almost entirely absent in the Positron annihilation Induced Auger spectra shown in Figs. 21B and 21D. Note the rapid increase in the number of detected electrons below the energy $E_k \sim 25$eV due to the onset of collisionally excited secondary electrons.

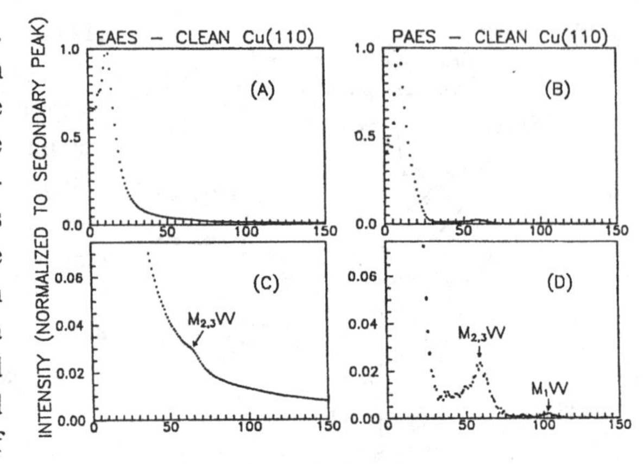


**Fig. 21.** Comparison of (A) electron induced vs (B) positron annihilation-induced Auger spectra from Cu(110). The same data, magnified 14.7 times are shown in (C) and (D) for electron- and positron-induced AES respectively. The incident beam energies were 3 keV and 25 eV for the electron and positron beam respectively. Both spectra were obtained using the same sample and analyzer and have been normalized to their respective secondary electron peaks. [from ref.86.]

The ability of PAES to eliminate the secondary electron background is particularly useful in determining the line shape of the low energy Auger lines (those most surface sensitive). In addition, the large improvement in signal to background permits PAES measurements to be made with energy doses ~5 orders of magnitude lower than doses required for to EAES [86]. The large reductions in charge and especially energy dose may make PAES useful for determining the elemental content of insulators and fragile adsorbate systems damaged by conventional AES methods.

*4.1.4 Surface sensitivity*

The enhanced surface selectivity of PAES was first demonstrated by a series of measurements performed on single crystal Cu with varying coverage of S [87]. As can be seen in Fig. 22. a half monolayer of S caused a four fold reduction in the PAES intensity, a much larger reduction than was observed using EAES. The reduction in the PAES signal is consistent with the fact that the S overlayer pushes the positron wave function away from the surface, thereby reducing its overlap with the top layer of Cu atoms as can be seen in the calculated positron wave functions shown in Fig. 23 [87]. Detailed calculations of the probability of core-hole formation due to positrons trapped at clean and S covered Cu surfaces predict attenuation in the Cu PAES signal in agreement with the observations.

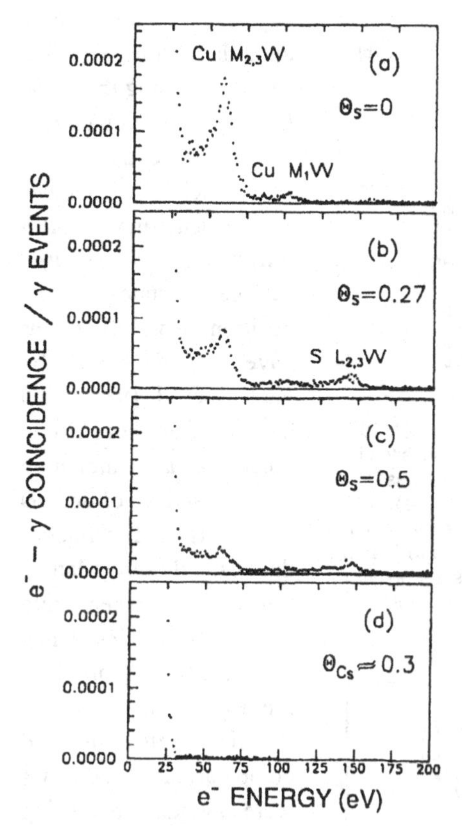


**Fig. 22.** Positron-annihilation-induced Auger spectra of (a) clean Cu, (b) Cu with 0.27ML S, and (c) 0.5ML of S. Panel (d) shows the PAES spectrum with ~0.3 ML of Cs on Cu. Note that the Cs causes the PAES signal to vanish by lowering the electron work function to the point where most positrons leave the surface as positronium before they can annihilate with a core electron and excite a PAES transition. The sample was maintained at a voltage of ~5 V with respect to ground. The positrons were incident with a kinetic energy of ~25eV [from reference 87].

### *4.2 Theoretical calculations*

Computer codes have been developed to perform detailed calculations of PAES intensities [89,97,98]. In these calculations, the image-potential-induced positron surface state wave function is found by numerical solution of a single particle Schrödinger equation. The potential is calculated using the corrugated mirror model [119,120]. The charge density is determined from a superposition of atomic charge densities. The surface dipole term is taken into account by the use of a short range ramp potential [89] or by modifying the atomic charge density in such a way as to give the correct potential step at the surface after superposition [121]. After calculation of the positron surface state wave function, the annihilation rates λn,1 for different electronic levels were calculated using the Independent Particle Model (IPM). The core annihilation rate is taken to be:

$$\lambda_{n,l} = \pi r_0^{\,2} c \int d\underline{r}\, |\psi_+(\underline{r})|^2 \left| \sum \psi^i_{n,l}(\underline{r}) |^2 \right| \tag{5}$$

where $r_0$ is the classical electron radius, c is the velocity of light, $\Psi_+$ is the positron wavefunction and $\Psi^i_{n,l}$ denotes core electron wave functions represented by free atom wavefunctions [89] where *n* and *l* are the principal and orbital quantum numbers, respectively. The PAES intensities are taken to be proportional to the annihilation probabilities $p_{n,l}$ obtained by dividing

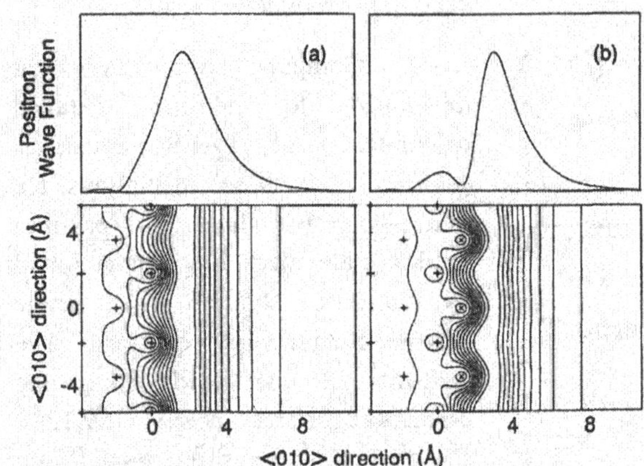


**Fig. 23.** Positron wave functions at (a) a clean Cu(100) surface and (b)Cu(100) with 0.5 ML of S at fourfold sites in a c(2x2) structure shown as contour plots (lower panels) and cross sections through a fourfold site (upper panels). The crossed dots denote positions of Cu atoms in the plane of the figure. The dots marked with a x in (b) correspond to S atoms. Note how the presence of the S atoms pushes the positron wave function away from the Cu atoms [from reference 87].

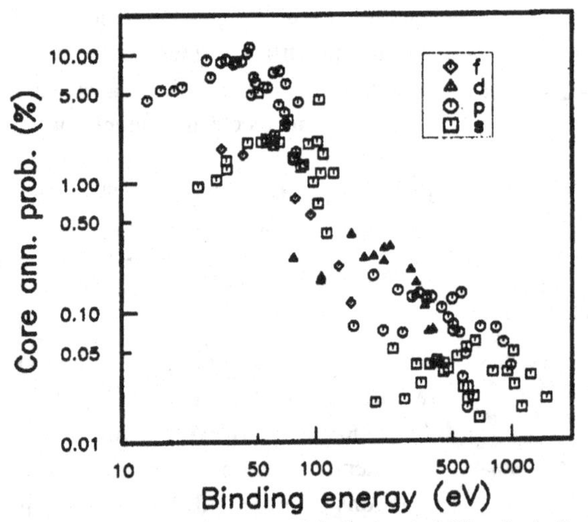


**Fig. 24.** Calculated core-annihilation probabilities for bulk positron states as functions of the binding energy of the core levels of 30 different elements. The probabilities represent the combined annihilation probabilities of all electrons in a particular level. Different symbols are used for s,p,d and f levels as indicated [from reference 89].

$\lambda_{n,l}$ by the total annihilation rate, which is calculated using the Local Density Approximation.

Recently, new, first-principles methods for calculating PAES intensities, positron surface states and positron work functions have been developed by Fazleev et al. [98] which have wide applicability in other areas including positron studies of defects in metals and semiconductors. Discrete-lattice effects are taken into account and inconsistency associated with previous methods in which different potentials and computational schemes were avoided by using the same scheme in calculating the positron surface state binding energy and positron work function. Calculations, using the new methods demonstrated for the first time that the positron surface state was the ground state in all of the alkalis. Li was found to have a stable surface state localized strongly at the surface with an energy ~

0.5 eV below the bulk positron band. The surface states on the other alkalis were found to penetrate deeper into the bulk and to have energies much closer to the bulk ground state energies. These results indicate that experiments which require the positron to be trapped in a surface state (e.g. PAES) can be performed on the free electron like alkali metals.

### *4.3 PAES intensities*

PAES spectra have been obtained from a number of metals including Cu [82, 87,88, 90-93, 104, 122], Al [124], Ni [82], Fe [104,122], Pd [91,122], Au [92, 100, 101, 102, 122], and Ag [99,125]; from semiconductors Si [97,100-103,105,126,127] and Ge [103,128]; C on Ge [128], C (graphite) [106,117], Ga (GaAs) [129] and As (GaAs) [129], and from sputtered insulators [130].

Experimental estimates based on these measurements indicate that the fraction of surface state positrons that annihilate with the outer core electrons (those with binding energies between ~10 and ~100 eV) are typically on the order

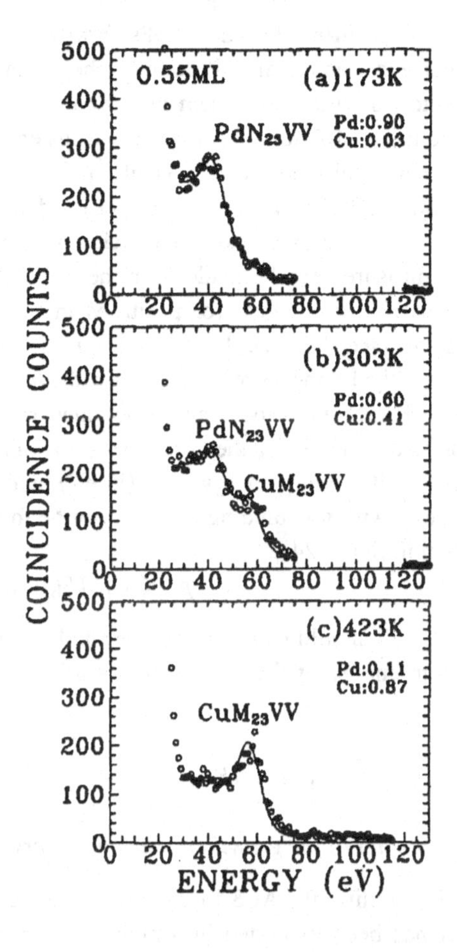


**Fig. 25.** PAES spectra for ~0.5 ML of Pd deposited on Cu(100). Panel (a) shows the "as-deposited" spectrum at 173 K, (b) showsthe PAES spectrum after the sample was heated to 303 K, (c) shows the PAES spectrum after the sample was heated to 423 K. The Pd $N_{23}VV$ (at ~40eV) and the Cu $M_{23}VV$ (at ~60eV) peaks are identified. The relative intensities of the Pd and Cu peaks obtained from fits to the data (solid lines), are indicated in the right hand side of each spectrum [from reference 87].

of $10^{-2}$ to $10^{-1}$ (see table 1, reference 84). Taking into account the fraction of incident positrons lost due to Ps formation (typically ~ 1/2), and the fraction of Auger electrons that fail to exit the surface due to inelastic processes or to trajectories which take them into the bulk, the efficiency of the PAES process is found to be $\sim 10^{-3}$ to $\sim 10^{-4}$ Auger electrons emitted from the surface, (integrated over the Auger peak), per incident positron. This is comparable or higher than the Auger excitation efficiency of electrons, per incident particle [113].

Estimates of expected PAES intensities for surfaces which have not yet been measured can be made from the calculations of Jensen and Weiss [89] of the core annihilation rates (for positrons in the bulk ground state) for 30 different elements (see Fig. 24). In all cases, the core hole annihilation cross sections for the less tightly bound core levels is on the order of several percent indicating that usable PAES intensities can be obtained from most elements. An approximate expression (eq. 6) for the core electron annihilation probability, $p$, in percent, as a function of binding energy, $E_b$, (in eV), and the number of core electrons, $N(E_b)$, per atom with this binding energy, can be found from a fit to the calculated values shown in figure 24:

$$p = 600[N(E_b)]\,(E_b)^{-1.6}\,. \qquad (6)$$

It should be noted that the annihilation probability decreases rapidly as the binding energy of the core increases as a consequence of positron repulsion from the core.

### 4.4 *Applications of PAES*

#### 4.4.1 *Studies of the growth of ultrathin layers on metals and semiconductors*

The ability of PAES to determine the elemental content of the topmost atomic layer has been exploited in a number of investigations aimed at the study of the growth and stability of ultrathin metal layers on metal substrates including: Pd on Cu [97], Au on Cu [91], Rh on Ag [99], Au on Si [100] and Cs on Cu [90]. These systems are important because of their unique magnetic and catalytic surface properties. A few examples are discussed in more detail below:

Pd Films on Cu(100)

The first confirmation of top-layer selectivity of PAES in metal-on-metal systems was obtained by Koymen et al., [90] who used PAES, EAES and LEED to study the composition and structure of vapor-deposited Pd films on Cu(100). Their results indicate that the PAES intensity from Pd saturates while that from Cu attenuates to near zero by one monolayer of Pd deposition at 173K. While the Pd/Cu system has been studied extensively using other techniques, PAES measurements provided a direct indication of the intermixing of Cu and Pd in the top most layer as a function of temperature. Combined PAES and EAES measurements where able to show that a Pd film deposited at 173K remains on top at 173K, forms a mixed Cu/Pd top layer when the same surface is warmed to 303K and that a Cu overlayer forms and the Pd goes one layer below the top layer at 423K.

PAES spectra at three different temperatures for ~0.69 monolayers Pd on Cu(100) at 173K are shown in Fig. 25. PAES spectrum for the as deposited surface at 173K (Fig. 25a) shows predominantly the Pd $N_{23}N_{45}N_{45}$ (40eV) Auger peak. Fig. 25b shows the PAES spectrum after the sample was heated to 303K without further Pd deposition. Here the appearance of the Cu $M_{23}M_{45}M_{45}$ (60eV) Auger peak is clearly observed as well as the decrease in the Pd Auger peak. As shown in Fig. 25c upon heating the sample to 423K the Pd Auger peak almost disappears and the Cu Auger peak gets larger, showing that Pd is moving below the top surface layer. The corresponding EAES spectra show only small variation with temperature [90]. A two parameter linear least square fit was used to extract the fractions of Pd and Cu contributing to the PAES spectra [90].

<u>Au on Si</u>

Experiments carried out by the UTA group have demonstrated the utility of PAES in probing the interface formed during metalization of semiconductors [100-102]. The properties of such interfaces have become increasingly important as devices are made smaller. When Si surfaces with thicker ( 2-6ML) layers of Au are warmed from 173K to 460K the Au PAES signal increases by ~ 50% with no indication of a PAES signal from Si. In contrast the EAES signal from Au decreases and the EAES signal from Si increases. These results have been interpreted [102] as indicating the formation of a silicide capped by a Au overlayer.

*4.4.2 Adsorption studies*

<u>PAES Study of the adsorption of Hydrogen and Oxygen on Si(100)</u>

Kim et al.[105] demonstrated that the exposure of a Si surface to Hydrogen or Oxygen gas results in large decreases in the PAES signal from the Si substrate (see Fig 26). They used this effect to measure the sticking coefficient by measuring the

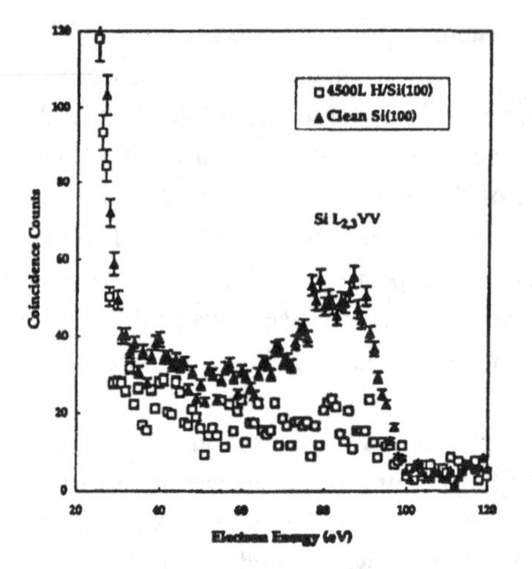


Fig. 26. PAES spectra obtained from an annealed Si(100) surface before and after exposure to 4500 Langmuirs (L) of hydrogen. Note the reduction of the Si $L_{2,3}VV$ Auger peak after exposure. Kim et al. posit that this reduction is due to adsorbed hydrogen pushing the positron wave function away from the Si substrate reducing the annihilation with the Si ion cores [from reference 105].

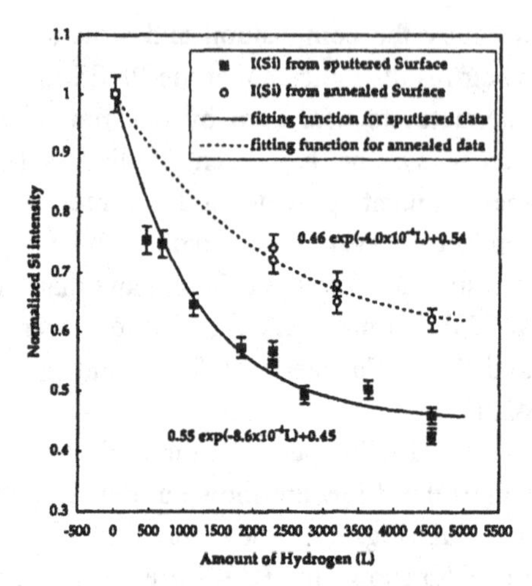


Fig. 27 Si $L_{2,3}VV$ PAES intensities measured as a function of the amount of Hydrogen exposure at 300 K of a sputtered and annealed Si(100) surface (open circles) and a sputtered Si(100) surface (filled squares). The solid and dotted lines were obtained from least-square fits to a model function from which the sticking coefficients of $S_0 = 9.8 \pm 2.2 \times 10^{-5}$ and $S_0 = 4.4 \pm 1.8 \times 10^{-5}$ were obtained for the sputtered and the sputtered-annealed surfaces respectively [from reference 105].

decrease in the PAES intensity for the Si substrate as a function of increasing gas exposure (Fig. 27). Theoretical calculations indicate that the decrease in the Si PAES intensity is due to the fact that the adsorbed gas on the surface pushes the positron wave function away from the Si substrate reducing the positron overlap (and consequently the probability of annihilation) with the Si core electrons. These results demonstrate that unlike most surface science techniques, PAES is sensitive to the presence of hydrogen on the surface and is therefore a valuable new tool in hydrogen adsorption and related measurements.

### *4.5 PAES sensitivity to impurities*

The presence of steps, defects or impurities on the surface can strongly affect the form of the wave function of the positron in the surface trap. In some cases, this may result in an enhanced sensitivity to surface impurities if the positron

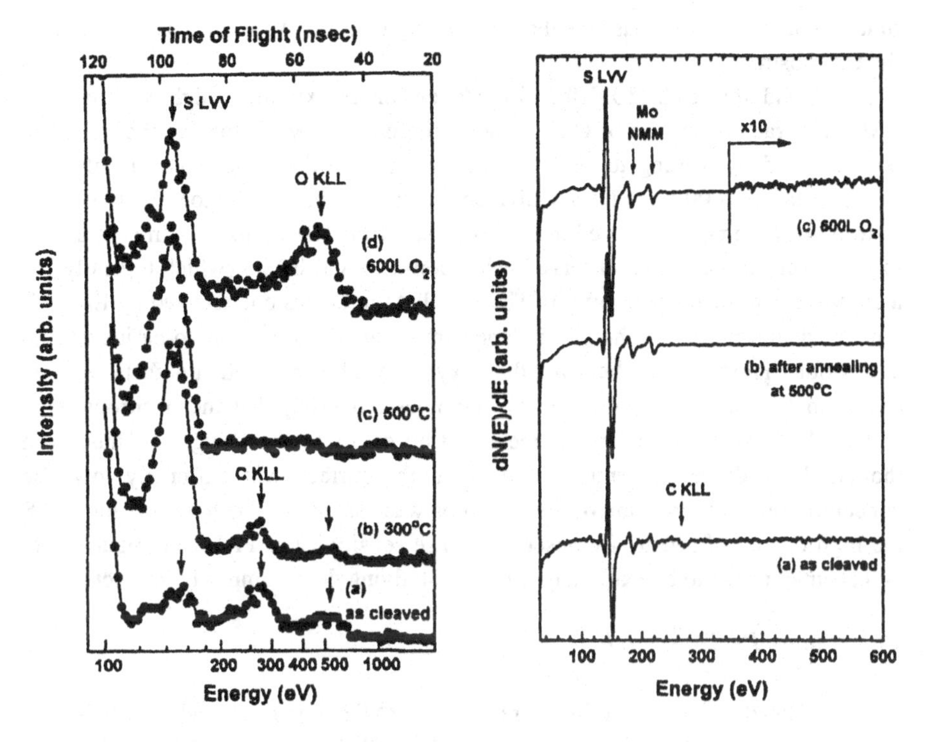


**Fig. 28**. Comparison of PAES, (left), an EAES, (right), spectra obtained from a $MoS_2$ (0001) surface. Starting from the bottom of each panel, the spectra correspond to the $MoS_2$ surface, as cleaved, after annealing, and after exposure to 600 L Oxygen. Note that a strong O KLL peak may be observed in the PAES spectrum taken after Oxygen exposure while none is visible in the EAES spectrum (even after 10X magnification) providing evidence that PAES has a significantly enhanced sensitivity to O impurity atoms adsorbed on the surface [from reference 109].

density is concentrated at the position of the impurity atoms rather than being evenly distributed over the surface.

Lee et al. found [92] strong evidence for such an enhanced sensitivity in measurements of the PAES intensity of Au as a function of the amount of Au vapor deposited on a Cu(100) surface. They found that the PAES signal from Au rose very rapidly for low coverages, reaching ~50% of its maximum value at ~0.1 ML and then rose more slowly, reaching ~100% at 1 ML Au. They were able to model this behavior by assuming that the positrons, instead of being delocalized on the surface were instead trapped on top of Au islands formed during deposition. They posited that differences in the positron surface affinities resulted in a potential well over the Au island. As a consequence of this well, the ground state wave function of the positron has an enhanced density over the Au islands as compared to the Cu substrate.

Ohdaira et al. [109] found evidence for an extremely high sensitivity of PAES for trace quantities of Oxygen adsorbed on a cleaved $MoS_2$ surface that had been cleaned by heating to 500° C. They reported that after exposure of a the $MoS_2$ surface to 600 L ( 1 L is equivalent to an exposure to $10^{-6}$ torr for 1 second), a strong oxygen peak appeared in the PAES spectrum (see Fig. 28). In contrast, an oxygen peak could not be observed using conventional EAES (which typically has a sensitivity in the percent range). The fact that the O was not detected by EAES is not surprising given that the basal plane surface of $MoS_2$ is know to be inert to $O_2$ adsorption. However it is believed that Oxygen can be adsorbed at defects and step edges on this surface. Ohdaira et al. hypothesized [109] that the positrons were localized at some of the same defects or step edges as the Oxygen. Thus, even though the average concentration of O on the surface was extremely low, the concentration at the location of the positrons was assumed to be high, giving PAES the high degree of sensitivity observed. The sensitivity of PAES to impurities at defects may prove to be extremely useful in studying defect-impurity interactions.

### *4.6 Future PAES research*

The development of high resolution PAES apparatus and especially the development of high intensity positron beams with fluxes ($\geq 10^8$ $e^+$/s ) which are two to three orders of magnitude greater than those previously used in PAES has made it possible to greatly extend the range of future PAES investigations. Two interesting areas for the future application of PAES are discussed below:

<u>Spin Polarized Positron Annihilation Induced Auger Electron Spectroscopy:</u>
The use of beams of polarized positrons will make possible a new spectroscopy for the study of the magnetic properties of surfaces: Polarized Positron annihilation induced Auger Spectroscopy (P-PAES) [84]. The ability of positrons to create polarized core holes stems from the fact that the annihilation rate for spin singlet collisions is many times faster than for spin triplet collisions at the velocities involved in PAES. As a consequence the positron will annihilate ~558 times faster with an electron with anti-parallel spin than an electron with parallel spin. It is therefore possible to use a polarized beam of positrons to create polarized core holes with a net polarization approximately equal in amount and opposite in direction of the incident positron beam.

Selection rules based on the conservation of total spin in Auger transitions will permit the study of the polarization of valence electrons by measuring the intensities of Auger transitions in which the core hole is filed by a valence electron. In the first step of the P-PAES mechanism a polarized positron produces a polarized core hole by annihilating an electron of anti-parallel spin. This process is followed by the filling of the core hole by a valence electron. Neglecting double spin flip processes, only electrons of like spins will fall back in the hole. In the absence of correlation effects the spectra would represent the convolution with one spin sub-band with the whole conduction band. The spin dependence of correlation effects will be studied and can be expected to provide additional information about magnetic surface systems and the Auger process.

PAES Microprobe: In order to use the PAES signal in conjunction with a positron microprobe it is necessary to overcome the problem presented by the fact that positrons at the relatively high beam energies >1000 keV required to form a micro-beam can generate high energy secondary electrons and Auger excitations in the bulk. Thus the main advantages of PAES, surface selectivity and low background, disappear. Two methods of avoiding these problems have been suggested. In one scheme due to Y. Kong [131], a large field gradient established using a superconducting magnet would be used to both focus the incident positron beam to an intense spot and to provide a magnified image of the outgoing Auger electrons. In another scheme [84], the high-energy secondary electrons and Auger electron resulting from the use of a highly focused, energetic positron beam would be eliminated using timing. The annihilation induced signal would be separated out from the prompt collisionally induced backgrounds by opening the gate for electron detection after a time comparable to the positron lifetime, (~400 hundred pico-seconds) completely eliminating the prompt backgrounds while reducing the PAES signal by only a factor of ~1/e . If these or other methods succeed then it will be possible to use PAES to obtain a map of the elemental composition of the surface with an in-plane resolution comparable to scanning EAES but with a depth resolution almost an order of magnitude higher!

## 5 Positron-induced secondary electron emission

### *5.1 Introduction*

Electrons leaving a surface under primary electron bombardment can be classified into two groups; (i) primary electrons which have either elastically or inelastically backscattered out of the surface, and (ii) true secondary electrons which were originally bound in the material, but have gained energy from the electron energy loss of the primary beam or from other bulk electrons and have managed to leave the surface before inelastic processes put them at energies below those needed to escape. These secondary electrons give rise to a large background under the signal spectra in re-emitted electron spectroscopies such as AES (which makes it difficult, for example, to extract the Auger lineshape from the experimental data). Electron-induced secondary emission has been studied extensively [*e.g.*, 132,133]; however, it is impossible to identify unambiguously true secondaries and redistributed primaries due to the indistinguishability of the electrons - a problem which is exacerbated at low incident energies. If a primary positron beam is used instead of an electron beam, then the primary and secondary particles may be experimentally distinguished and, because positrons have the same mass and magnitude of charge as an electron, their scattering properties are similar, and it is therefore possible to study the secondary electron distribution and the redistributed primary spectra independently.

Positron-induced secondary electron emission has been used to tag or time slow positrons, as described in Chapter 2. It is also essential to the detection of positrons by electron muliplier devices, also discussed in Chapter 2. In this chapter we shall focus on the spectroscopy of secondary electrons ejected from solid surfaces by positron bombardment.

## *5.2 Experimental studies of secondary electron emission*

The first observation of positron-induced secondary electron emission was made by Cherry in 1958 [134]. He compared the yields of secondary electrons from an activated Mg-Ag alloy bombarded by positrons and electrons. Over twenty years passed until Weiss and Canter [135] used a slow positron beam to measure the yield, energy spectra and angular distributions of secondary electrons ejected by positrons in the $10^2$ eV range. They found broad similarities between the secondary electron characteristics for incident positrons and electrons (Figure 29).

Mayer and Weiss [136] measured the total secondary electron yields $Y$ from Ni, Si and MgO bombarded with electrons and positrons at incident energies $E$ up to 500eV. They observed similar $Y(E)$ curves for the three targets, but the ratios $Y_{e-}/Y_{e+}$ were 0.79, 1.26 and 0.6 for Ni, Si and MgO, respectively. Mayer *et al.* [137] measured secondary electron energy spectra for electron- and positron-bombarded MgO and Ni crystals, and first demonstrated the similarity between secondary electron and re-emitted epithermal positron spectra.

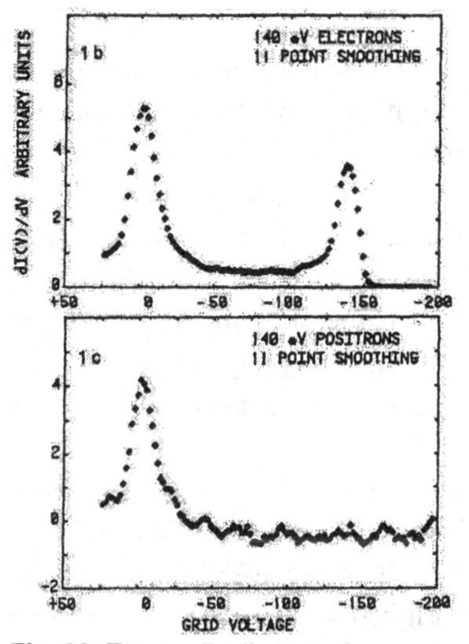


**Fig. 29** Energy distribution of secondary electrons produced by 140eV electrons (top) and positrons (bottom) [135]

Knights and Coleman compared the epithermal positron reemission spectra from an Ag(100) surface (which has a positive positron work function) with secondary electron spectra for incident beam energies of 0.1, 0.5, 1 and 2 keV [138]. The epithermal positron and secondary electron spectra are essentially identical at an incident energy of 100 eV; it is reasonable to assume that both epithermal positrons and secondary electrons experience energy loss over an equivalent distance. However, as the incident positron energy increases the epithermal positron spectra become narrower than the electron spectra (Figure 30) as the re-emerging positrons travel further and experience more inelastic collisions prior to reaching the surface - whereas the secondary electrons which leave the surface always originate from the same range of escape depths. This is a complementary example of the difference between true secondary particles and 'redistributed primary' particles.

Recent work on positron-induced secondary electron emission has focused on the electron energy spectra. Experiments of this type have been performed using the recently-completed high resolution PAES system at UT Arlington [139] and at the University of East Anglia [140].

An example of an electron-induced energy spectrum (redistributed primaries plus secondaries), together with positron-induced secondary electron and redistributed primary spectra, are shown in Figure 31 for

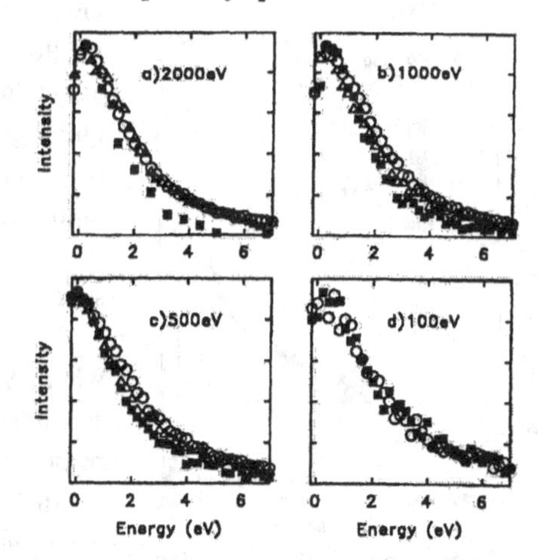


**Fig. 30** Energy spectra of secondary electrons from Ag(100) under positron (Δ) and electron (O) bombardment. Epithermal positron re-emission spectra are shown for comparison (■). [138]

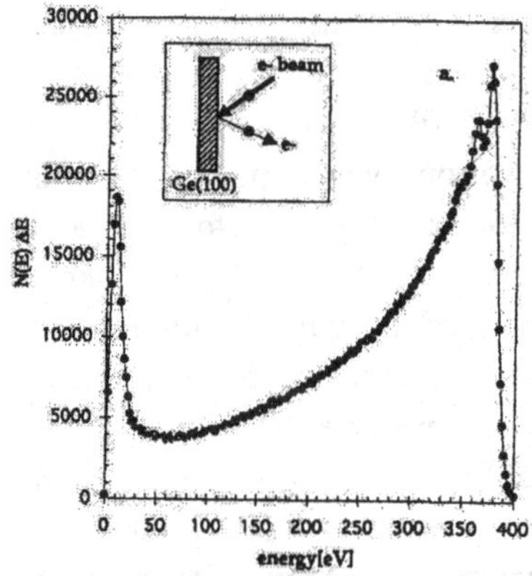


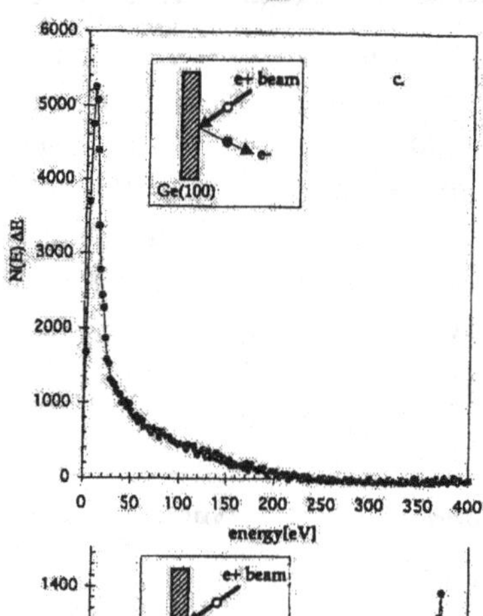


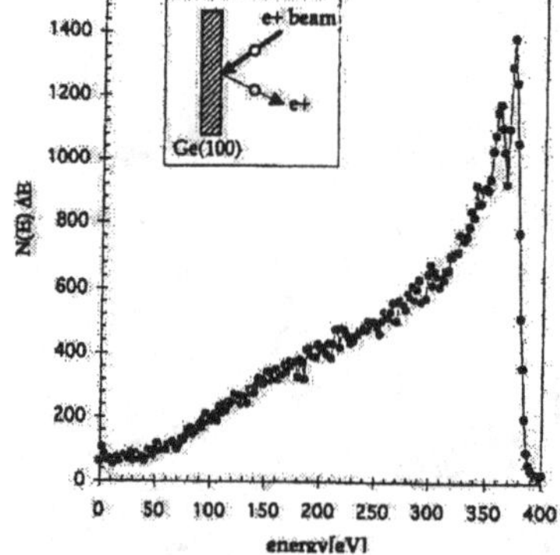


**Fig. 31** Energy spectra of $e^-/e^+$ emitted from Ge(100) bombarded with 375eV $e^-/e^+$. Top: $e^-$ in-$e^-$ out: middle: $e^+$ in-$e^-$ out: bottom: $e^+$ in-$e^+$ out [139]

Ge(100). The familiar electron-induced spectrum shows three different regions; region I is comprised of primary electrons which are elastically reflected at the incident electron beam energy, here 375 eV. Electrons which have lost characteristic amounts of energy to the crystal are referred to as redistributed primaries and are found in region II. Finally, the electrons in the low energy peak of region III are typically characterized as "true secondaries".

Regions II and III overlap and thus indistinguishability of the electrons makes it impossible to tell if the electron exiting the surface was originally from the incident beam or from the material being bombarded. The measured spectrum of secondary electrons ejected by 375eV incident positrons, with detector background subtracted, has a cutoff at the incident beam energy. The spectrum of redistributed primary positrons from the Ge(100) sample due to 375 eV incident positrons shows only two regions: region I is again the elastically-reflected peak at 375eV, and region II contains the redistributed primary positrons which have lost characteristic amounts of energy to the sample.

The electron beam-induced spectrum $B_e(E)$ can be expressed as

$$B_e(E) = S_e(E) + R_e(E) \qquad (7)$$

where $S_e(E)$ and $R_e(E)$ are the distributions of secondary and redistributed primary electrons, respectively. Analogously, $Sp(E)$ and $Rp(E)$ are the distributions of positron-induced secondary electrons and redistributed primary positrons.

Given the similarity in the way positrons and electrons interact with solids, it is reasonable to suppose that a weighted sum

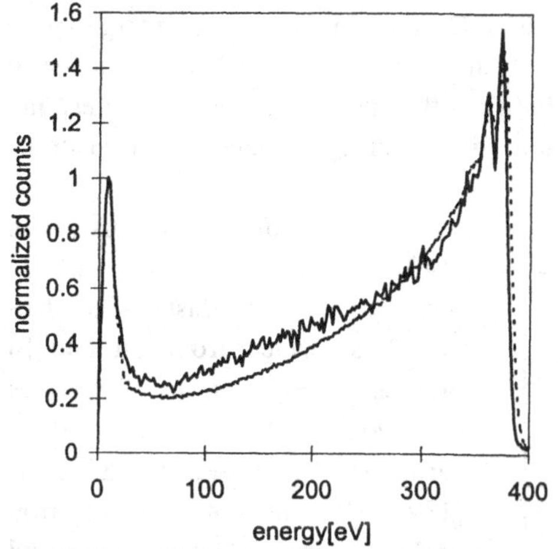


**Fig. 32** $B_{e\text{-model}}(E)$ fitted to the total re-emitted electron energy spectrum (see text) [150].

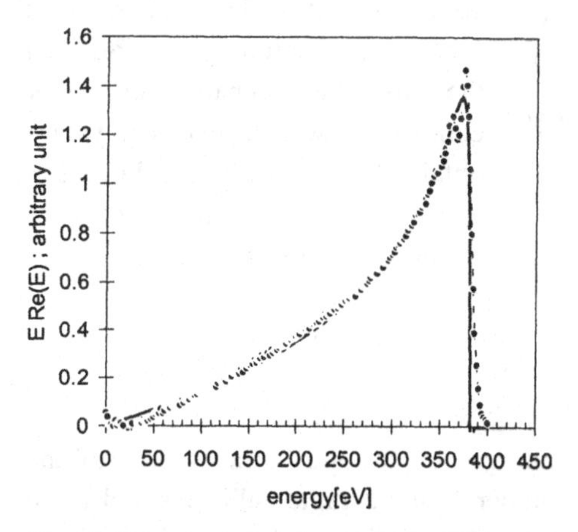


**Fig. 33** Ramaker's function (solid line) fitted to the experimentally-determined redistributed primary electron spectrum (circles) [150].

of the distributions $R_p(E)$ and $S_p(E)$ will yield a function which approximates $B_e(E)$, i.e.,

$$B_{e\text{-model}}(E) = \alpha_1 S_p(E) + \alpha_2 R_p(E) \quad (8)$$

$B_{e\text{-model}}(E)$ is compared with $B_e(E)$ in Figure 32 [150].

To separate the redistributed primary electrons ($R_e(E)$) from the electron induced electron energy spectra ($B_e(E)$), one can use the positron-induced secondary electrons ($S_p(E)$) to approximate the electron-induced secondary electron spectra ($S_e(E)$). An estimate of $R_e(E)$ may thus be obtained by subtracting an appropriately scaled $S_p(E)$ from $B_e(E)$. The result of this procedure, illustrated in Figure 33, shows good agreement with the theoretical model of Ramaker [150,151].

The energy spectrum of positrons re-emitted from Ge(100) bombarded with 375 eV positrons shows three regions – an elastic peak, a redistributed primary distribution, and a low-energy region containing near-thermalized, or epithermal, positrons. Epithermal positron emission from solid surfaces can have significant effects on the interpretation of experimental data involving the recording of an annihilation line shape parameter as a function of implantation energy such as those collected in defect profiling measurements [152]. It is likely that epithermally re-emitted electrons are present in the electron-induced spectrum at low energies, but it is impossible to separate them from the secondary electron contribution.

The energy spectra of electron- and positron-induced redistributed primaries (backscattered particles) (Figs. 31 and 33) are similar at high energies; the energy loss peaks (due to the excitation of the surface plasmons) are the same in both spectra. The principal difference between the spectra is at low energies, and this is probably indicative of differences in the scattering mechanisms of positrons and electrons.

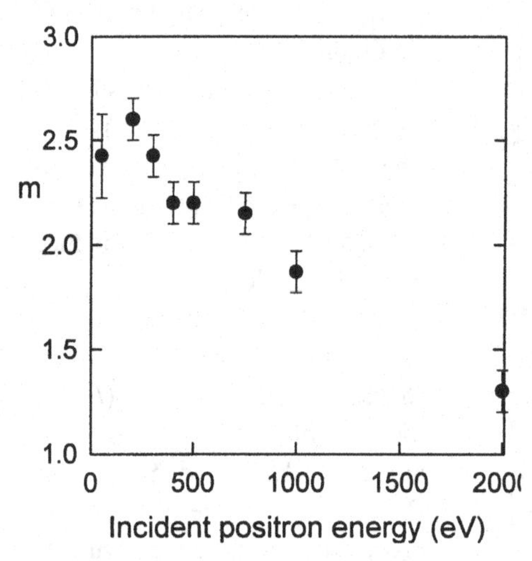


**Fig. 34** Sickafus index *m* for secondary electrons ejected from Cu by positrons (see text) [140].

Overton and Coleman [140] measured the energy distributions of fast secondary electrons ejected from Cu and Si bombarded by positrons of energies 0.3, 1 and 2 keV, and fit them to the form $AE^{-m}$ proposed by Sickafus [132]. The elimination of electron backscattering, discussed above, and the reduction of cascade effects common in electron-stimulated secondary electron spectra allowed the direct determination of *m* for comparison with theory. The values obtained lay in the range 2.0 – 2.6, and are all outside that found for incident electrons (0.5-1.5) (Fig. 34).

## 6 Positron backscattering

When a beam of energetic positrons impinges on a solid surface some of the incident particles will undergo a small number of large-angle collisions and leave the target with a fraction – possibly a high fraction – of their original energy. These are termed *backscattered*, or alternatively *redistributed primary*, positrons. Knowledge of 1) the angular and 2) the energy distributions of backscattered positrons, and/or 3) the total backscattering probability (expressed as the coefficient $\eta_+$), is essential if a full understanding and a correct interpretation is to be made of positron studies of surface and near-surface phenomena. Measurement of these three quantities also provides a stringent test of theoretical calculations and simulations of positron – and also electron – collisions processes in solids, thereby underpinning and strengthening the interpretation not only of positron data but also that from the large array of spectroscopies in which electrons are used to probe

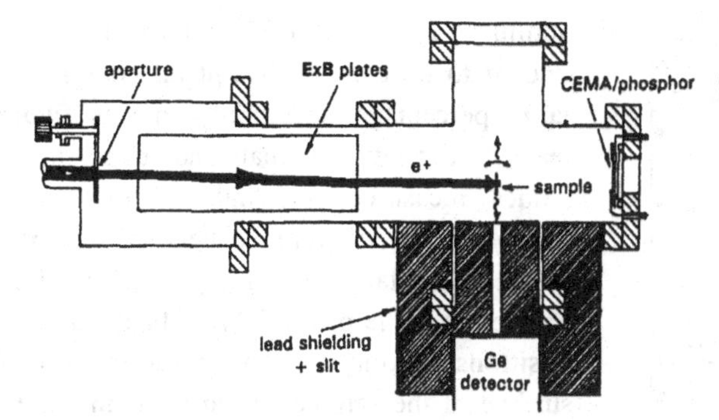


**Fig. 35** Schematic diagram of apparatus for measurement of total positron backscattering coefficients. Deflection by the ***ExB*** plates prevent positrons from returning to sample: annihilation photons from the sample target pass through a 1cm-wide slit in the lead shielding surrounding the Ge detector [158].

solids. Knowledge of positron backscattering probabilities is also of practical use if positron annihilation is used to probe the surface and subsurface regions of solids; those positrons which backscatter from a sample can be annihilated at solid surfaces – for example, chamber walls – in sight of the gamma signal detector, thereby contaminating the experimental data with unwanted contributions. Backscattering is also used to maximise positron emission into a vacuum chamber or gas-filled container by backing a primary radioactive positron source, or lining a container, with high-$Z$ (and thus generally high $\eta_+$) material.

The first positron backscattering measurements were performed for the entire beta spectrum of energies from a radioactive source (e.g., 0 to 540 keV for $^{22}$Na) [153,154]; one attempt was made to select energy windows from the beta spectrum using a magnetic spectrometer [155]. The first report of backscattering probabilities for truly monoenergetic positrons was for sub-3keV positrons interacting with thin Al foils [156], which formed part of a study of the penetration of positrons through thin metallic foils.

Accurate measurements of total backscattering coefficients $\eta_-$ for electrons have required detection of the scattered electrons, for example by a spherical particle detector surrounding the target. In contrast, $\eta_+$ for positrons may be made relatively straightforwardly by exploiting one of the properties not shared with electrons – annihilation. By measuring the annihilation gamma count rate from the sample target one is essentially measuring $(1-\eta_+)$ directly, as long as the total count rate is known for the incident beam. This method was employed in 1988 by Baker and Coleman [157], who measured total backscattering coefficients for monoenergetic positrons incident on Al, Cu, Ag and W targets at energies between 0.5 and 30keV (Figure 35). They ensured that backscattered positrons were not returned to the sample by passing the incident beam through ***ExB*** deflector plates, and that only annihilation gamma rays from the sample were detected by using a lead viewing slit. They measured the incident beam intensity by allowing the positrons to be

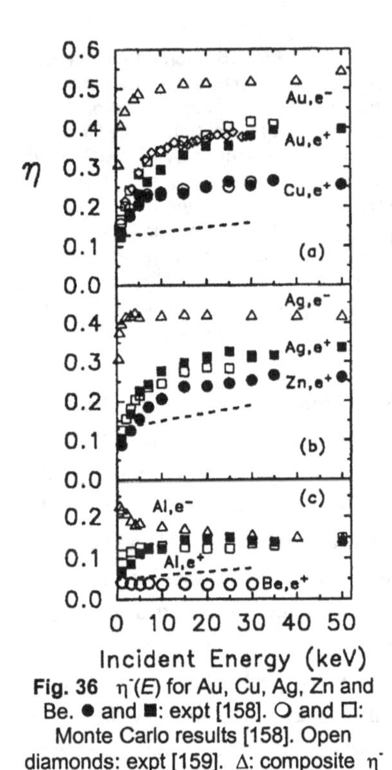


**Fig. 36** $\eta^{-}(E)$ for Au, Cu, Ag, Zn and Be. ● and ■: expt [158]. O and □: Monte Carlo results [158]. Open diamonds: expt [159]. Δ: composite $\eta^{-}(E)$ for electrons.

annihilated in a Be sample, for which $\eta_+$ was known to be close to 3% at all energies (a large percentage uncertainty in this figure leads to a relatively small uncertainty in $\eta_+$ values measured for higher-$Z$ materials). The pure samples were cleaned and were not single crystals (avoiding channelling problems). The probability of backscattered positrons picking up an electron at the exit surface of the sample and annihilating as p-Ps in sight of the gamma detector was found to be negligible.

The measurements of Baker and Coleman were extended by Coleman et al. [158] who measured $\eta_+$ for 21 different elemental solids for incident energies up to 50keV and at different incident angles. At about the same time similar measurements were made by Mäkinen et al. [159], and the two sets of results showed reasonable agreement with each other and with Monte Carlo simulations. An example of the results is shown in Figure 36. At high incident energies the stopping cross section for positrons of energy $E$ in a material of atomic number $Z$ can be described approximately by the Bethe-Bloch formula (thus $\propto Z/E$) and the transport cross section by the modified Rutherford cross section ($\propto Z^2/E^2$). Substituting these cross sections into analytical transport models such as that of Vicanek and Urbassek [160] leads to the conclusion that $\eta_+$ varies approximately as $Z^{1/2}$, which is borne out by the results exemplified by those in figure 36. At lower energies, however, more sophisticated modelling of positron transport is required.

The ratio of $\eta_-$ to $\eta_+$ is shown in Figure 37 as a function of incident particle energy. The ratio tends to a constant value of ~1.3 at higher energies; the increase at low energies is partly due to the differences in scattering cross sections, but is probably due also in part to the contribution to $\eta_-$, of energetic secondary electrons being indistinguishable and thus inseparable from true backscattered primary electrons (again, a problem not experienced in positron measurements).

Measurement of $\eta_+$ at low incident energies is made difficult by the effect of

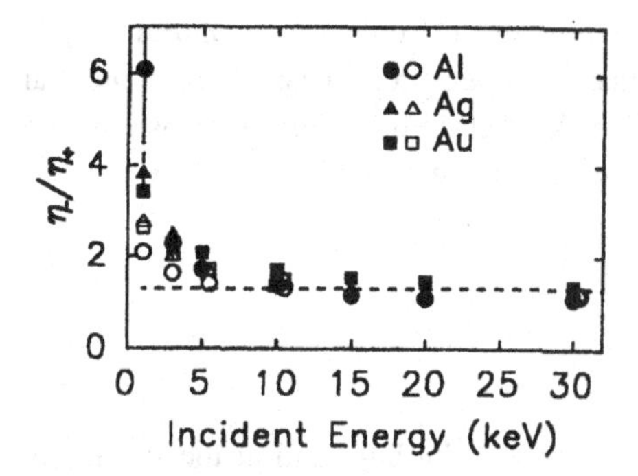


**Fig. 37** $\eta^-/\eta^+$ vs $E$ for Al (●), Ag (△) and Au (■). Open symbols are Monte Carlo results [158].

surface contamination, the emission of epithermal positrons, and the increased probability of positronium formation. However, these problems can all be overcome, and Knights and Coleman [161] observed structure in the dependence of $\eta_+$ on atomic number $Z$ for $E$ = 1keV, which they attributed to the variation of the effective electron density seen by the incident positron.

While magnetic-transport beams are ideal for measurements of $\eta_+$, information on the energy and angular distributions of backscattered positrons requires electrostatic beam systems, such as that used by Weiss and co-workers to study positron and electron backscattering below 1keV as reported in the secondary electron section of this chapter [162]. For $E$ = 35keV, however, Schultz and co-workers performed a series of measurements of doubly-differential positron backscattering yields using a moveable silicon barrier detector [163-165]. Examples of their results for angular and energy distributions are shown in Figures 38 and 39,

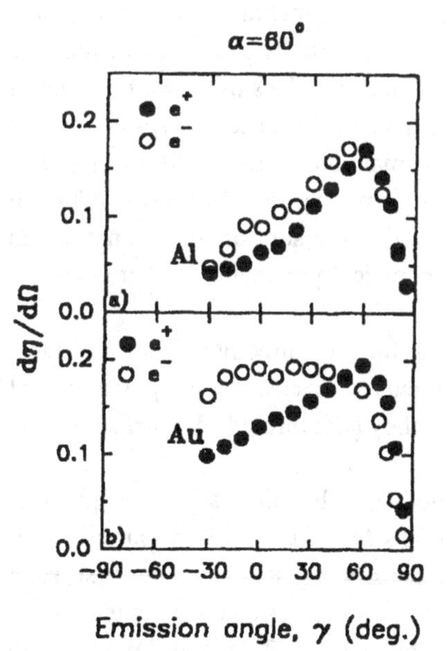


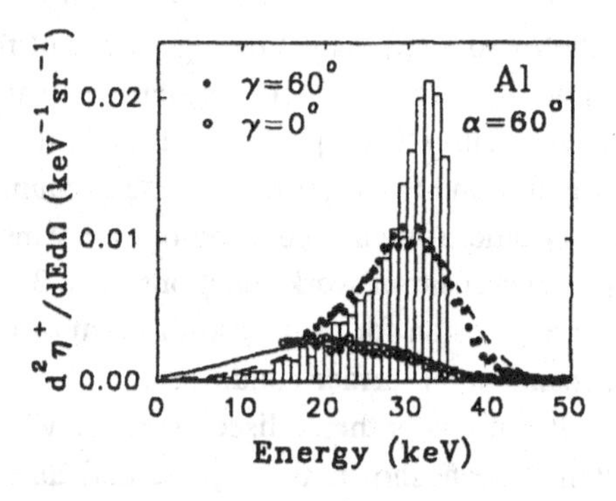


**Fig. 38** (left) $d\eta/d\Omega$ for $e^+$ and $e^-$ on Al and Au as a function of emission angle (incident angle 60°) [154].

**Fig. 39** (above) $d^2\eta^+/dEd\Omega$ for Al at 0 and 60° (incident angle 60°). Smooth curves: convoluted Monte Carlo results. Histogram: Monte Carlo results for 60° before convolution [154].

respectively. The results, after deconvolution of instrumental resolution, again agreed well with Monte Carlo simulations. The measured quasi-specular angular distributions are in contrast to the much broader distributions for electrons, and are attributed to positrons undergoing a large number of weak elastic collisions.

## 7 Positronium at surfaces

### *7.1 Direct positronium formation*

Positronium (Ps) is formed by slow positrons in the bulk and at the surface of insulators, but – because of screening effects – can only be formed at the surfaces of metals and semiconductors, where the electron density is sufficiently reduced to allow single positron-electron pairing [166-169]. [At this point $r_s = (3/4\pi n)^{1/3}/a_0 \approx 6$: $n$ = mean background conduction electron density, $a_0$ = Bohr radius]. It is this fact which has makes Ps formation a potentially useful probe of solid surfaces. It has also allowed many important experiments on fundamental aspects of Ps, but we shall not dwell here on basic Ps physics (see Chapter 3), focussing instead on those aspects of Ps formation and decay which can play a part in surface science [170]

Ps is formed directly at surfaces with a formation potential

$$\varepsilon_{Ps} = \phi_+ + \phi_- - 6.8 \text{ eV} \tag{9}$$

where $\phi_+$ and $\phi_-$ are the positron and electron work functions, respectively, and 6.8eV (=0.5 $R_\infty$) is the binding energy of free ground-state Ps ($\varepsilon_{Ps}$ is often called the Ps work function [171]). One can see that for typical $\phi_-$ values of ~ 5eV that even if $\phi_+$ is positive (i.e., positrons are not re-emitted from the surface) $\varepsilon_{Ps}$ can still be negative and Ps is ejected into the vacuum. The maximum energy of the emitted Ps is $-\varepsilon_{Ps}$ and, if measured, provides a means of determining experimentally values of positive positron work functions [172-3]. Note that the sum of work functions in the expression for $\varepsilon_{Ps}$ is also the sum of bulk chemical potentials, so that $\varepsilon_{Ps}$ is not dependent on surface conditions.

It is not only thermalised positrons which can form Ps directly at a solid surface; it has been shown that epithermal and low-energy backscattered positrons (of energies ~ $10^0$-$10^2$ eV) can pick up electrons as they pass through the surface region and leave as energetic Ps [174].

Assuming that direct Ps formation is a sudden, rather than adiabatic, process, then measurement of the velocity spectrum of Ps formed directly at surfaces can give information on the surface electronic density of states. The first such measurement was performed by Mills et al. in 1983 [175] and was extended by Howell et al. [176] and theoretically treated by Ishii [177] and by Walker and

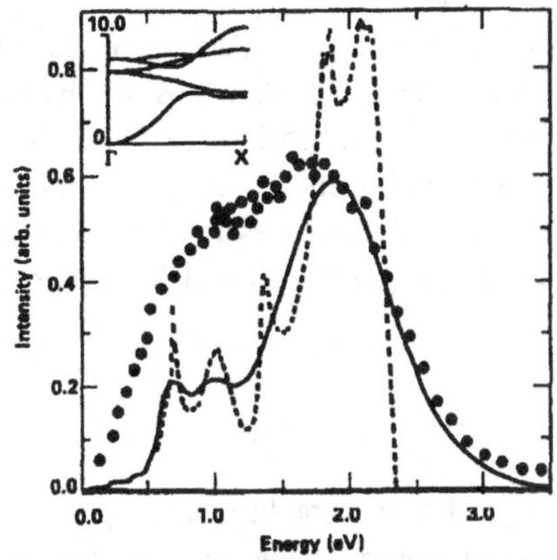


**Fig. 40** Ps energy distribution from Ni(100). Broken line: theory (band structure, inset). Solid line: theory smoothed with Gaussian resolution function [176].

Nieminen [178]. Experimental results to date imply that Ps formation leaves a single hole in the conduction band of the metal, but there are significant discrepancies between theory and experiment at low Ps energies (Figure 40).

Direct Ps formation and emission from Al(100) was studied by Chen et al. [179] and Howell et al. [180] using an intense positron beam coupled with a two-dimensional angular correlation (2D-ACAR) apparatus, which measures the momentum distribution of the annihilating pairs. In ref. [179] Ps momentum spectra were measured for three crystal orientations of Al, and compared with theoretical results for the nearly-free electron model; the results demonstrated sensitivity to electronic structure (Figure 41).

*7.2 Thermal desorption of positronium from surfaces*

One of the major three channels open to the thermally diffusing positron as it encounters the exit surface of a metal or semiconductor is to fall into the surface potential well. There has long been two pictures of this positronic surface state: that of a positron strongly bound by the image-correlation potential, and that of a physisorbed Ps atom [181]. (Both models give a similar surface state lifetime of ~500ps, the spin-averaged Ps lifetime in vacuo.) An attempt to investigate the

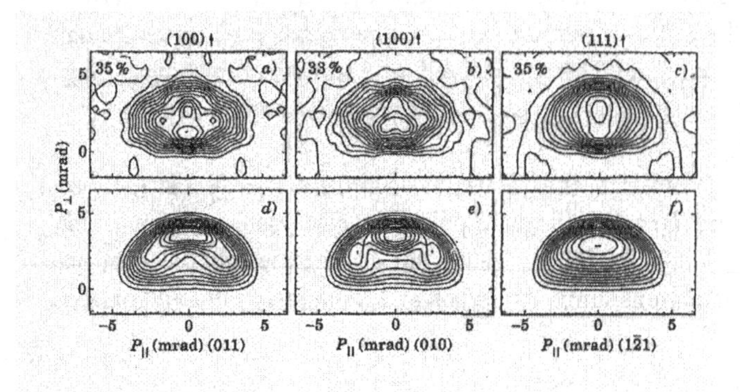


**Fig. 41** a-c: 2D-ACAR results for momentum spectra of Ps emitted from the (100), (110) and (111) faces of Al.
d-e: results from nearly-free electron model [179].

relative veracity of these two models was made by Lynn et al. [182], who by using 2D-ACAR found a symmetrical momentum distribution for annihilating pairs in the Al(100) surface – a result not consistent with either Ps or bare positron-based surface state picture. At elevated temperatures (~$10^2$ °C) the positrons are desorbed into the vacuum as thermal-energy Ps, this channel being energetically favourable [183-9]. This desorption process has provided thermal-energy Ps for fundamental studies, and has also been shown to exhibit sensitivity to surface oxidation (see 10.3.6 below) [190].

### *7.3 Ps studies of non-metallic surfaces and thin films*

7.3.1 Graphite: Direct Ps formation at the surface of graphite is not possible because of the high momentum of electrons parallel to the surface. However, in 1988 Sferlazzo et al. [191] reported on experimental evidence, based on 2DACAR momentum measurements and the temperature dependence of Ps formation, that Ps can be formed if a phonon is absorbed or emitted, thereby taking up the initially high parallel momentum (see Figure 42). Rice-Evans and co-workers [192,193] have conducted a series of studies on the formation of Ps at graphite surfaces covered with physisorbed gases at submonolayer levels (Figure 43); they interpret their results in terms of the excess electron momentum being taken up by molecular recoil – maximum Ps formation being observed at semilayer coverage. They propose that this technique is a sensitive tool for studying film fluidity.

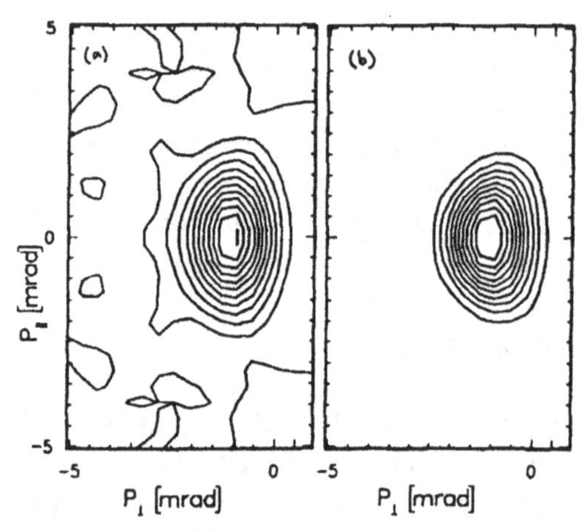


**Fig. 42** 2D-ACAR projections of momenta of Ps emitted from graphite. (a) experimental results, (b) theory based on phonon-assisted emission [191].

7.3.2 Alkali halides: Tuomissari et al. [194] measured the energy distributions of Ps from LiF, NaF, KCl, KBr and KI using a pulsed intense positron beam. Their results were consistent with a model in which hot Ps diffuses to the surface, experiencing inelastic interactions, until it reaches and penetrates the exit surface.

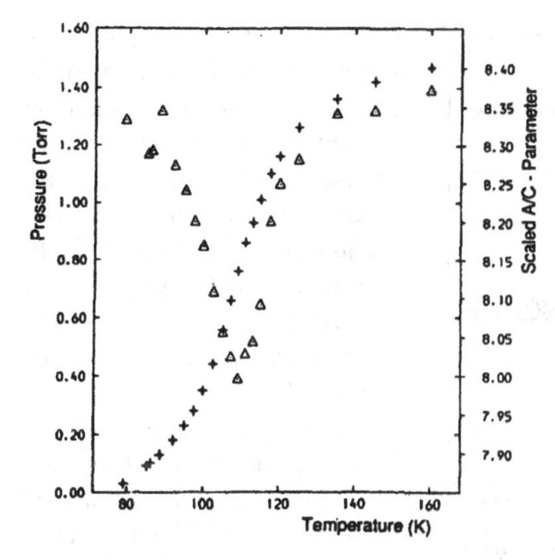


Fig. 43 Ps formation at a graphite surface (Δ) ( low A/C parameter ≡ high Ps) and $CH_4$ pressure (+) vs. temperature (and hence monolayer development)[179]

LiF showed two peaks in the energy spectrum of emitted Ps atoms, the second being attributed to fast Ps formation at the surface.

7.3.3 $SiO_2$: The temperature dependence of Ps emission from the surface of crystalline $SiO_2$ (quartz) suggests that there is, in addition to the diffusion of hot Ps from beneath the surface, the desorption of Ps physisorbed on the surface [195]. This result was recently confirmed by time-of-flight measurements at a pulsed, intense positron beam by Nagashima et al. [196].

7.3.4 Polymers: In a series of experimental studies Gidley and co-workers have used Ps formation and decay as a probe of the physical state of thin polymer films, using a positron beam time-tagged by detecting secondary electrons ejected from the film essentially at the moment of implantation [197]. In a study of the glass transition temperature $T_g$ for thin polystyrene films, DeMaggio et al. [198] found that the reduction in $T_g$ with decreasing film thickness is accompanied by a decrease in void volume expansion; their results pointed to the existence of a near-surface region with reduced $T_g$. (Fig. 44.)

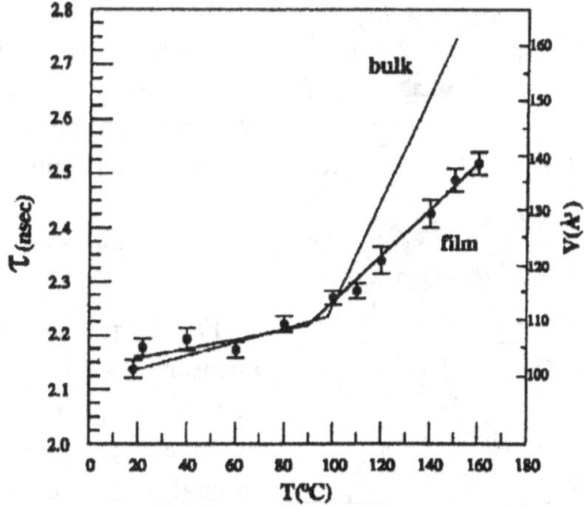


Fig. 44 Ps lifetime (left axis) and spherical hole volume (right axis) for a 115Å polystyrene (PS) film vs temperature. Dashed line – fit to bulk PS data [198].

7.3.5 Ice: Eldrup et al. [199] studied the formation and

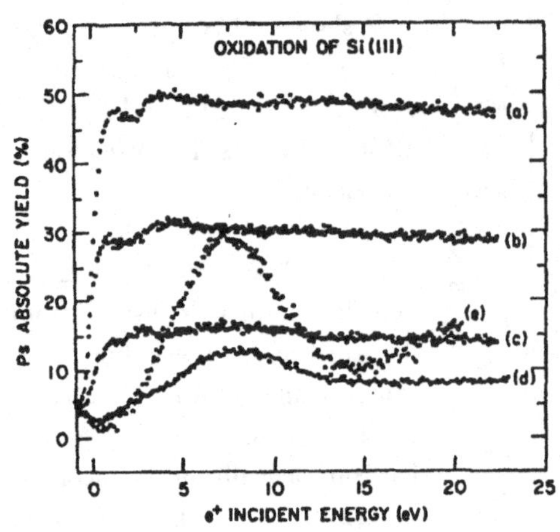


**Fig. 45** Ps fraction vs energy of positrons incident on (a) clean Si(111) and with oxide layers of thicknesses (b) 2.5Å, (c) 5.4Å, (d) 14Å and (e) 3500Å [200].

diffusion of Ps in amorphous ice, and its subsequent emission from the surface, as a function of incident positron energy $E$. They observed structure in the dependence of the ortho-Ps fraction on $E$, consistent with Ps formation in the Ore gap ($5<E<10$ eV). Similar structure was seen in Ps fraction measurements for oxide films (see below).

7.3.6 Oxide films: Oxidation of solid surfaces has been studied by measuring Ps fractions as a function of incident positron energy for a range of oxide thicknesses [200]. The results were sensitive to the different stages of oxidation (Figure 45).

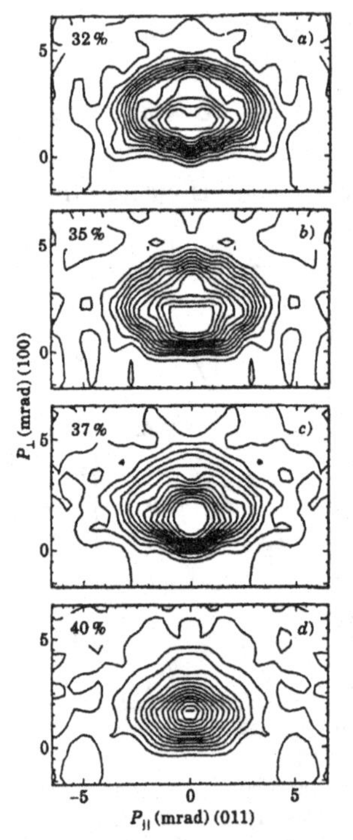


**Fig. 46** 2D-ACAR projections of momentum spectra of Ps emitted from (a) clean Al(100) and after oxygen contamination for (b) 3h, (c) 6h, (d) 9h [201].

2D-ACAR measurements also show the sensitivity of the Ps momentum distribution to oxide contamination [201]; the anisotropy observed in the projected distribution for a clean Al(100) surface, associated with surface electronic structure, progressively disappears as the oxide grows (Figure 46).

## *7.4 Ps beam studies*

Fast Ps production has been observed by passing positron beams through thin C films [202], neutralizing positron beams incident metallic surfaces at glancing angles [203], and passing positrons through a gas cell [204]. The last method has been used extensively to produce near-monoenergetic Ps beams for Ps-atom collision cross sections

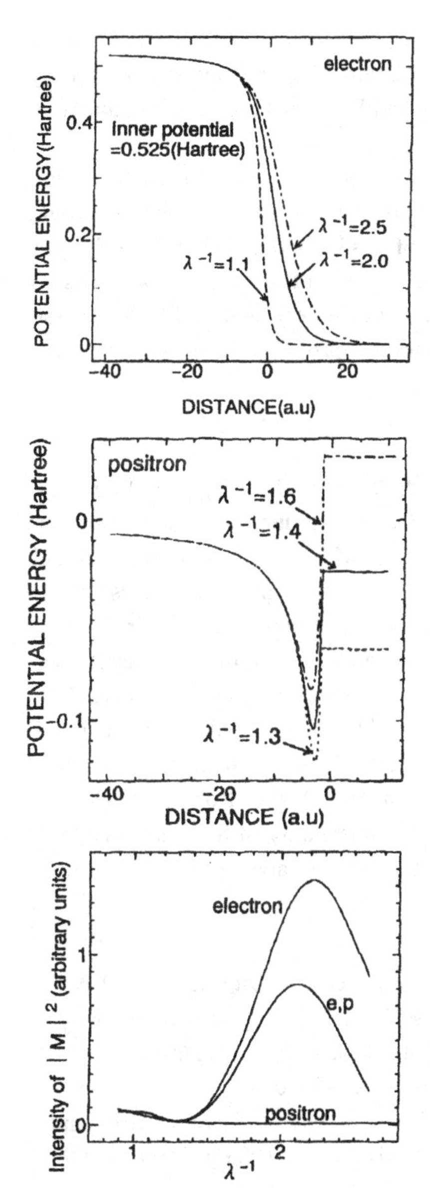


**Fig. 47** (top) Electronic surface barrier potential for different shape parameters $\lambda$. (middle) positronic surface barrier potential. (bottom) Ps formation intensity vs $\lambda^{-1}$ (incident positron energy 40.8 eV: 1 Hartree = 27.2eV). [209]

measurements, and for more information on these the reader is referred to Chapter 3. It has also been used to produce a Ps beam for Ps diffraction studies [205]. The use of Ps for diffraction measurements is proposed primarily for the reason that, being neutral, it interacts mainly with the surface atoms. While $H_2$ and He beams also have this advantage, they are necessarily of very low energy and so cannot probe small-scale surface structure.

## 7.5 Other Ps surface spectroscopies

Ishii [206-209] has performed extensive theoretical work showing how Ps formation can be used to provide information on solid surfaces in novel ways. Briefly, they are:

(a) *Inverse Ps Formation Spectroscopy* [206]. In this technique a beam of Ps atoms impinges on a surface; the electron is given up to an unfilled state and the positron takes away information on the state. Ishii states that this method will be more sensitive to the topmost surface layer than the two existing spectroscopies of unoccupied states, inverse photoelectron spectroscopy and two-photon photoemission. Another advantage should be the strength of the signal.

(b) *Adsorbate studies using Ps formation Spectroscopy* [207]. Ishii has shown that the dependence on incident positron energy of Ps formation probability at adsorbate-covered surface is sesnstive to the atomic positions of the adatoms. In

particular, he shows that the method is sensitive even to hydrogen adatoms, in contrast to the standard angle-resolved photoemission. This method is extended by Ishii to randomly-positioned adsorbate atoms when Ps formation is accompanied by desorption of the ionised adatom [208].
(c) *Surface barrier potential studies* [209]. The dependence of the Ps formation probability on incident positron energy is sensitive to the shape of the electronic surface barrier (see Figure 47). Measurements of this kind, coupled with LEED (for the surface atomic configuration), angle-resolved ultraviolet photoemission spectroscopy ARUPS (for information on electronic states), and LEPD (using the LEED and ARUPS data) for the positronic potential, provides the only method for measuring the electronic surface barrier potential.

**Acknowledgements**
It is a pleasure to acknowledge the significant help of many people in the preparation of this manuscript. In particular, thanks are due to K. Canter and D. Gidley for their many suggestions, for providing us with reprints and preprints and for reviewing and commenting on sections of the manuscript. Thanks also to R. Suzuki and T. Ohdaira for providing us with reprints and preprints of their work on PAES and pointing us to other relevant literature. Useful discussion are acknowledged with K.G. Lynn, R. Howell, A. Koymen, N. Fazleev, J. Fry, R.N. West, R. Nieminen and P. Simpson. A.W would like to thank J.-H. Kim, S Starnes, W.-C. Chen, N. Jiang, S. Xie, and R. Nayak, for help in preparing this manuscript. AW also acknowledges the support of the Robert A. Welch Foundation, NSF contract #DMR9812628 and the hospitality of S. Hulbert and the National Synchrotron Light Source, Brookhaven National Laboratory.

**References**

[1] A.P. Mills, Jr., 1983, *Positron Solid State Physics*, eds. A. Dupasquier and W. Brandt (Amsterdam:North Holland), p. 455. Also 1993, *Positron Spectroscopy of Solids,* eds. A. Dupasquier and A.P. Mills, Jr. (Amsterdam: IOS), p. 209.
[2] P.J. Schultz and K.G. Lynn, 1988, *Rev. Mod. Phys.* **60**, 701.
[3] R.M. Nieminen, 1983, *Positron Scattering in Gases,* eds. J.W. Humberston and M.R.C. McDowell (New York: Plenum), p. 139. Also as [1], 1983 p. 359 and 1993 p. 443.
[4] W. Cherry, 1958, *Ph.D. dissertation* (Princeton University).
[5] J.M.J. Madey, 1969, *Phys. Rev. Lett.* **22**, 784.
[6] D.E. Groce, D.G. Costello, J.Wm. McGowan and D.F. Herring, 1968, *Bull Am Phys Soc* 13, 1397.

[7] K.F. Canter, A.P. Mills, Jr. and S. Berko, 1974 *Phys. Rev. Lett.* **33**, 7.
[8] K.F. Canter, A.P. Mills, Jr. and S. Berko, 1974 *Phys. Rev. Lett.* **34**, 177.
[9] A.P. Mills, Jr., P.M. Platzman and B.L. Brown, 1978, *Phys. Rev. lett.* **41**, 1076.
[10] C.A. Murray, A.P. Mills, Jr. and J.E. Rowe, 1980, *Surf. Sci.* **100**, 647.
[11] A.P. Mills, Jr. and C.A. Murray, 1980, *Bull.Am.Phys.Soc.* **25**, 392.
[12] C.H. Hodges and M.J. Stott, 1973, *Phys. Rev. B* **7**, 73.
[13] R.M. Nieminen and C.H. Hodges, 1978, *Phys. Rev. B* **18**, 2568.
[14] A.P. Mills, Jr., 1979, *Solid State Commun.* **34**, 623.
[15] I.J. Rosenberg, A.H. Weiss and K.F. Canter, 1980, *Phys. Rev. Lett.* **44**, 1139.
[16] A.H. Weiss and K.F. Canter, 1982, *Positron Annihilation*, eds P.G. Coleman, S.C. Sharma and L.M. Diana (Amsterdam:North Holland) p 162.
[17] R. Mayer, C.S. Zhang, K.G. Lynn, W.E. Frieze, F. Jona and P.M. Marcus, 1987, *Phys. Rev. B* **35**, 3102.
[18] A.P. Mills and W.S. Crane, 1985, *Phys. Rev. B* **31**, 3988.
[19] M.H. Weber, S. Tang, S. Berko, B.L. Brown, K.F. canter, K.G. Lynn, A.P. Mills, L.O. Roellig and A.J. Viescas, 1988, *Phys. Rev. Lett.* **61**, 2542.
[20] A. Weiss, R. Mayer, M. Jibaly, C. lei, D. Mehl and K.G. Lynn, 1988, *Phys. Rev. Lett.* **61**, 2245.
[21] A.R. Koymen, K.H. Lee, D. Mehl, Alex Weiss and K.O. Jensen, 1992, *Phys. Rev. Lett.* **68** 2378.
[22] Cherry A. Murray, Allen P. Millls, Jr. and J. W. Rowe 1980, *Surf. Science* **100** 647
[23] D.A. Fischer, K.G. Lynn and D.W. Gidley, 1986, Phys. Rev. B**33**, 4479.
[24] A.P. Mills, Jr., 1983, in *Positron Solid-State Physics,* ed. A. Dupasquier and W. Brandt (Amsterdam:North-Hoplland) p. 506.
[25] P.J. Schultz and K.G. Lynn, 1988, Rev. Mod. Phys. **60**, 701.
[26] Alex Weiss, E. Jung, J.H. Kim, A. Nangia, R. Venkataraman, S. Starnes and G. Brauer, 1997, *Appl. Surf. Sci.* **116**, 311.
[27] N.G. Fazleev, J.L. Fry and A.H. Weiss, 1998, Phys. Rev. B **57**, 12506 (and references therein).
[28] M. Jibaly, A. Weiss, A.R. Koymen, D. Mehl, L. Stiborek and C. Lei, 1991, Phys. Rev. B **44**, 12166 (and references therein).
[29] G. Fletcher, J.L. Fry and P.C. Pattnaik. 1983, Phys. Rev. B **27**, 3987 (and references therein).
[30] R. Nieminen and C. Hodges, 1976, Solid State Commun. **18,** 1115.
[31] A. Nangia, J.H. Kim, A.H. Weiss and G. Brauer, 1997, Mat.Sci.Forum **255-7**, 711.
[32] J.H. Kim, A. Nangia, E. Jung and A.H. Weiss, 1997, Mat. Sci. Foruim **255-7**, 638.

[33] N.G. Fazleev, J.L. Fry and A.H. Weiss, Mat. Sci. Foruim **255-7**, 145.
[34] B.D. Wissman, D.W. Gidley and W.E. Frieze, 1992, Phys. Rev. B **46**. 16058.
[35] J.G. Ociepa, P.J. Schultz, K. Griffiths and P.R. Norton, 1990, Surf. Sci. **225**, 281.
[36] G.R. Brandes and A.P. Mills, Jr., 1998, Phys. Rev. B **58, 4952.**
[37] M.R. Poulsen, M. Charlton and G. Laricchia, 1993, J.Phys.:CM **5**, 5209.
[38] A.P. Knights and P.G. Coleman, 1996, Surf.Sci. **367**, 238.
[39] M. Farjam and H.B. Shore, 1987, Phys. Rev. B **36**, 5089.
[40] E.M. Gullikson, A.P. Mills, Jr. and J.M. Phillips, 1998, Surf.Sci.**195**, L150.
[41] J. Kuriplach, M. Sob, G. Brauer, W. Anwand, E-M Nicht, P.G. Coleman and N. Wagner, 1999, Phys. Rev. B **59**, 1948.
[42] David W. Gidley, 1989, Phys. Rev. Lett. **62**, 811.
[43] W.E. Frieze, D.W. Gidley and B.D. Wissman, 1990, Solid State Commun. **74**, 1079.
[44] D.W. Gidley and W.E. Frieze, 1988, Phys. Rev. Lett. **60**, 1193.
[45] B.D. Wissman, D.W. Gidley and W.E. Frieze, 1992, Phys. Rev. B **46**, 16058.
[46] J.G. Ociepa, P.J. Schultz, K. Griffiths and P.R. Norton, 1990, Surf. Sci. **225**, 281.
[47] G.W. Anderson, K.O. Jensen, T.D. Pope, K. Griffiths, P.R. Norton and P.J. Schultz, 1991, Phys. Rev. B **46**, 12880.
[48] G.W. Anderson, K.O. Jensen, T.D. Pope, K. Griffiths, P.R. Norton and P.J. Schultz, 1993, Phys. Rev. B **48**, 15283.
[49] D.A. Fischer, K.G. Lynn and W.E. Frieze, 1983, Phys. Rev. Lett. **50**, 1149.
[50] T. N. Horsky, G. R. Brandes, K. F. Canter, C. B. Duke, S. F. Horng, A. Kahn, D. L. Lessor, A. P. Mills, Jr., A. Paton, K. Stevens, and K. Stiles, 1989, Phys. Rev. Lett. **62**, 1876.
[51] T.N. Horsky, G.R. Brandes, K.F. Canter, C.B. Duke, A.Paton, D.L. Lessor, A. Kahn, S.F. Horng, K. Stevens, K. Stiles, and A.P. Mills, 1992, Phys. Rev. B **46**, 7011.
[52] X.M. Chen, K.F. Canter, C.B. Duke, A. Paton, D.L. Lessor, and W.K., 1993, Phys. Rev. B, **48**, 2400.
[53] C. B. Duke, A. Paton, and A. Lazarides, D, Vasumathi and K.F. Canter, 1997, Phys. Rev. B **55** 7181.
[54] C.B. Duke, D.E. Lessor, T.N. Horsky, G. Brandes, K.F. Canter, P.H. Lippel, A.P. Mills,Jr., A. Paton and Y.R. Wang, 1989, J. Vac. Sci. Technol, **A7** 2030.
[55] D. L. Lessor, C. B. Duke, X. M. Chen, G.R. Brandeis, K.F. Canter, and W. K. Ford, J. Vac. Sci. Technol. 1992, **A10** 2585.
[56] C. B. Duke in "Surface Science : The First Thirty Years" Charles B. Duke (Editor) (January 1994), North-Holland; Amsterdam pp. 24-33.

[57] S. Y. Tong in "Applications and advances of positron beam spectroscopy," UCRL-JC-130205: preprint - http://www.llnl.gov/tid/lof/documents/pdf/233541.pdf
[58] I. J. Rosenberg, A. H. Weiss, and K. F. Canter, 1980, Phys. Rev. Lett. **44**, 1139.
[59] A.H. Weiss, I.J. Rosenberg, K.F. Canter, C.B. Duke, A. Paton, 1983, Phys. Rev. B **27**, 867.
[60] F. Jona, D. W. Jepsen, P. M. Marcus, I. J. Rosenberg, Alex H. Weiss, and K. F. Canter, 1980, *Sol. St. Comm.* **36**, 957.
[61] Allen P. Mills, Jr. and P.M. Platzman, 1980, Sol. St. Commun., **35**, 321.
[62] R. Feder, 1980, Sol. St. Comm. **34**, 541.
[63] W.E. Frieze, D.W. Gidley, and K.G. Lynn, 1985, Phys. Rev. B **31** 5628.
[64] D.R. Cook, T.N. Horsky, and P.G. Coleman, 1984, Appl. Phys. A**34**, 237.
[65] A.P. Mills, Jr. and W.S. Crane, 1985, Phys. Rev. B **31**, 3988.
[66] R. Mayer, C.S. Zhang, K.G. Lynn, J. Throwe, P.M. Marcus, D.W. Gidley, F. Jona, Phys. Rev. B **36** 5659.
[67] R. Mayer, C.-S. Zhang, K.G. Lynn, W.E. Frieze, F. Jona, P.M. Marcus, 1987, Phys. Rev. B **35** 3102.
[68] See for instance, J.B. Pendry, 1974, Low Energy Electron Diffraction (Academic Press, London).
[69] L.D. Hullett, Jr. , B.L. Brown, A.L. Denison, R. Gonzalez, R.H. Howell, Y.C. Jean, P.L. Jones, W.J. Kossler, K.G. Lynn, A.P. Mills,S. Okada, Jr., D.M Schrader, L.C. Smedskjaer, R.N. West, A.H. Weiss and Y. Chen, 1995, Advanced Materials Research Vol. 3, Scitec Publications, Zug, Switzerland.
[70] C.B. Duke and D.L. Lessor, 1990, Surf. Sci. **225**, 8.
[71] K. Canter, private communication.
[72] W.K. Ford, T. Guo, K.-J. Wan, C.B. Duke, 1992, Phys. Rev. B, **45**, 11896.
[73] K. G. Lynn, J. R. MacDonald, R. A. Bole, L. C. Feldman, J. D. Gabbe, M. F. Robbins, E. Bonderup, and J. Golovchenco, 1977, *Phys. Rev. Lett.* **38**, 241.
[74] R. J. Meyer, C. B. Duke, A. Paton, A. Kahn, E. So, J. L. Yeh, and P. Mark, 1979, *Phys. Rev. B* **19**, 5194.
[75] S. Gota, R. Gunnella, Zi-Yu Wu, G. Jezequel, C.R. Natoli, D. Sebilleau, E. L. Bullock, F. Proix, C. Guillot, and A. Quemerais, 1993, Phys. Rev. Lett., **71**, 3387.
[76] W.K. Ford, T. Guo, K.J. Wan, and C.B. Duke, 1992, Phys. Rev. B **45**, 11896.
[77] E. Zanazzi and F. Jona, 1977, Surf. Sci. **62**, 61.
[78] M. N. Read and G. J. Russell, 1979, Surf. Sci. **88**, 95.
[79] R. M . Nieminen and M. J. Puska, 1983, Phys. Rev. Lett. **50**, 281.
[80] Y. Joly, 1992, Phys. Rev. Lett. **68**, 950.
[81] Y. Joly, 1994, Phys. Rev. Lett. **72**, 392, and 1996, Phys. Rev. **B53**, 13029.

[82] S.J. White, D.P. Woodruff, B.W. Holland, and R. S. Zimmer, 1978, Surf. Sci. **74**, 34.
[83] A. Weiss, R. Mayer, M. Jibaly, C. Lei, D. Mehl, and K. G. Lynn, 1988, Phys. Rev. Lett. **61**, 2245.
[84] A.H. Weiss, 1995, Positron Annihilation Induced Auger Spectroscopy, Chapt. 45 in *The Handbook of Surface Imaging and Visualization*, Arthur T. Hubbard, Ed., CRC Press, Boca Raton, Florida, pp. 617 - 633.
[85] A.H. Weiss, 1995, Positron-annihilation-induced Auger electron spectroscopy, in *Positron Spectroscopy of Solids*, Proceedings of the International School of Physics "Enrico Fermi," IOS Press, Amsterdam, pp. 259-284.
[86] Alex Weiss, David Mehl, Ali R. Koymen, K. H. Lee and Chun Lei, 1990, J. Vac. Sci. Technol A **8**, 2517.
[87] David Mehl, A. R. Koymen, Kjeld O. Jensen, Fred Gotwald, and Alex Weiss, 1990, Phys. Rev. B, **41**, 799.
[88] R. Mayer, A. Schwab, and Alex Weiss, 1990, Phys. Rev. B **42**, 1881.
[89] Kjeld Jensen and Alex Weiss, 1990, Phys. Rev. B., **41**, 3928.
[90] A.R. Koymen, K.H. Lee, D. Mehl, Alex Weiss, and K.O. Jensen, 1992, Phys. Rev. Lett. **68**, 2378.
[91] A.R. Koymen, K.H. Lee, G. Yang, K.O. Jensen and A.H. Weiss, 1993, Phys. Rev. B **48**, 2020.
[92] K.H. Lee, Gimo Yang, A.R. Koymen, K.O. Jensen, and A.H. Weiss, 1994, Phys. Rev. Lett., **72**, 1866.
[93] N.G. Fazleev, J.L. Fry, J.H. Kaiser, A.R. Koymen, K.H. Lee, T.D. Niedzwiecki, and A.H. Weiss, 1994, Phys. Rev. B, **49**, 10577.
[94] N.G. Fazleev, J.L. Fry, K. Kuttler, and A.H. Weiss, 1995, Appl. Surf. Sci. **85**, 22.
[95] N.G. Fazleev, J.L. Fry, K. Kuttler, A.R. Koymen, and A.H. Weiss, 1995, Appl. Surf. Sci. **85**, 26.
[96] N.G. Fazleev, J.L. Fry, K.H. Kuttler, A.R. Koymen, and A. H. Weiss, Phys. Rev. B:, **52**, 5351.
[97] K.H. Kuttler, J.L. Fry and A.H. Weiss, 1997, Appl. Surf. Sci. **116**, 304.
[98] N.G. Fazleev, J.L.Fry, and A.H. Weiss, 1998, Phys. Rev. B **57**, 12506.
[99] G. Yang, S. Yang, J.H. Kim, K.H. Lee, A.R. Koymen, G. A. Mulhollan, and A.H. Weiss, 1994, J. Vac. Sci. Technol. A**12**, 411.
[100] G. Yang, J.H. Kim, K.H. Lee, S. Yang, A.R. Koymen and A.H. Weiss, 1994, J. Materials Research. **9**, 2534.
[101] G. Yang, J.H. Kim, S. Yang and A.H. Weiss, 1995, Appl. Surf. Sci. **85**, 77.
[102] G. Yang, J. H. Kim, S. Yang, and A.H., 1995, Surf. Sci. **367**, 45.

[103] J.H. Kim, S. Wheeler, A. Nangia, E. Jung, and A.H. Weiss, 1997, Appl. Surf. Sci., **116** , 314.
[104] K.H. Lee, J.H. Kim, G. Yang, A.R. Koymen, K.O. Jensen and A.H. Weiss, 1998, submitted to J. Vac. Sci.Tech.
[105] J.H. Kim, G. Yang, and A.H. Weiss, 1998, Surf. Sci. **396**, 388.
[106] R. Suzuki, Y. Kobayashi, T. Mikado, H. Ohgaki, M. Chiwaki, and T. Yamazaki, 1994, Hyperfine Interactions **84** , 345.
[107] T. Ohdaira, R. Suzuki, T. Mikado, and T. Yamazaki, 1998, J. Electron Spectroscopy, **88-91**, 677.
[108] T. Yamazaki, R. Suzuki, T. Ohdaira, T. Mikado, Y. Kobayashi, 1997, Radiation Phys. Chem. **49**, 651.
[109] T. Ohdaira, R. Suzuki, T. Mikado, H. Ohgaki, M. Chiwaki, T. Yamazaki and M. Hasegawa, 1996, Appl. Surf. Sci. **100-101**, 73.
[110] T. Ohdaira, R. Suzuki, T. Mikado, H. Ohgaki, M. Chiwaki and T. Yamazaki, 1997, Appl. Surf. Sci. **116**, 177.
[111] T. Ohdaira, R. Suzuki, T. Mikado, and T. Yamazaki, 1997, Mat. Sci. Forum **255-257**, 769.
[112] Joshi, L. E. Davis, and P. W. Palmburg, 1975, in *Methods of Surface Analysis*, ed. A. W. Czanderna (Elsevier, Amsterdam).
[113] Practical Surface Analysis, Second Edition, Vol. 1, D. Briggs and M.P.Seah, 1990 (Wiley, Chichester, UK).
[114] Chun Lei, David Mehl, A.R. Koymen, Fred Gotwald and Alex Weiss, 1989, Rev. Sci. Instrum. **60**, 3656.
[115] R. Mayer, D. Becker, and A. Schwab, and Alex Weiss, 1990, Rev. Sci. Instrum., **61** , 42.
[116] S. Yang, H.Q. Zhou, E. Jung, A. H. Weiss and P.H. Citrin, 1997, Rev. Sci. Instrum. **68** 3893.
[117] R. Suzuki, T. Ohdaira, T. Mikado, H. Ohgaki, M. Chiwaki, T. Yamazaki, 1996, Appl. Surf. Sci. **100/101** 297.
[118] See reviews by A. P. Mills, 1983, in *Positron Solid state Physics, International School of Physics "Enrico Fermi," Course 83* ed. W. Brandt and A. Dupasquier (North-Holland, Amsterdam) and Peter J. Schultz and K. G. Lynn, 1988, Rev. Mod. Phys. **60**, 701, and references therein.
[119] R. M. Nieminen and M. J. Puska, 1983, Phys. Rev. Lett. **50**, 281.
[120] R. M. Nieminen and K. O Jensen, 1988, Phys. Rev. B **38**, 5764.
[121] M. Weinert and R.E. Watson, 1984, Phys. Rev. B **29**, 3001.
[122] K.H. Lee, Ali R. Koymen, David Mehl, K.O. Jensen, and A. Weiss, 1992, Surf. Sci. **264** , 127.

[123] A.H. Weiss, S. Yang, H.Q. Zhou, E. Jung, and S. Wheeler, 1995, J. Elect. Spec. and Rel. Phenom.**72**, 305.
[124] W-C. Chen, R. Nayak, A. H. Weiss, 1999, to be published.
[125] E. Jung, H.Q. Zhou, J.H. Kim, S. Starnes, R. Venkataraman and A. H. Weiss, 1997, Appl. Surf. Sci. **116**, 318.
[126] David Mehl, 1990, Ph.D. Dissertation, University of Texas at Arlington, unpublished.
[127] T. Ohdaira, R. Suzuki and T. Mikado, 1999, J. Surf. Anal. **8**, 158.
[128] E. Soininen, A. Schwab and K.G. Lynn, 1991, Phys. Rev. B**43**, 10051.
[129] J.H. Kim, A. Nangia, E. Jung and A.H. Weiss,1997, Mat. Sci. Forum **255-257**, 638.
[130] K.G. Lynn and Y. Kong, 1992, in *Positron at Metallic Surfaces*, A. Ishii, Editor, Solid State Phenomena **28-29**, pp. 275-292, Trans Tech, Aedermannsdorf, Switzerland.
[131] Yuan Kong and K.G. Lynn, 1993, Nucl. Instrum. Meth. B **79**, 341.
[132] E.N. Sickafus, 1977, *Phys. Rev. B* **16**, 1436.
[133] D.C. Peacock and J.P. Duraud, 1986, *Surf. Interface Analysis* **8**, 1.
[134] W. Cherry, 1958, *Ph.D. dissertation* (Princeton University).
[135] A.H. Weiss and K.F. Canter, 1982, *Positron Annihilation*, eds P.G. Coleman, S.C. Sharma and L.M. Diana (Amsterdam:North Holland) p 162.
[136] R. Mayer and A.H. Weiss, 1988, *Phys. Rev. B* **38**, 11927.
[137] R. Mayer, E. Gramsch and A.H. Weiss, 1989, *Phys. Rev. B* **40**, 11287.
[138] A.P. Knights and P.G. Coleman, 1995, *Appl. Surf. Sci.* **85**, 43.
[139] Alex Weiss, E. Jung, J.H. Kim, A. Nangia, R. Venkataraman, S. Starnes and G. Brauer, 1997, *Appl. Surf. Sci.* **116**, 311.
[140] N. Overton and P.G. Coleman, 1997, *Phys.Rev.Lett.* **79**, 305.
[150] E. Jung, R.Venkataraman, S. Starnes and A. H. Weiss, 1997, *Mat. Sci. Forum* **255**, 708.
[151] D.E. Ramaker, J.S. Murday and N.H. Turner, 1979, *J. Elec. Spect. and Rel. Phenomena* **17**, 45.
[152] H. Huomo, E. Soininen and A. Vehanen, 1989, *Appl. Phys. A* **49**, 647.
[153] I.K. MacKenzie, C.W. Schulte, T. Jackman and J.L. Campbell, 1973, Phys. Rev. A**7**, 135.
[154] V.A. Kuzminikh and S.A. Vorobiev, 1979, Nucl. Instrum. Methods **167**, 483.
[155] P.U. Arifov, A.R. Grupper and H. Alimkulov, 1982, Positron Annihilation, eds. P.G. Coleman, S.C. Sharma and L.M. Diana (Amsterdam: North Holland) p. 699.
[156] A.P. Mills, Jr. and R.J. Wilson, 1982, Phys. Rev. A**26**, 490.

[157] J.A. Baker and P.G. Coleman, 1988, J.Phys.C:Solid State Phys. **21**, L875.
[158] P.G. Coleman, L. Albrecht, K.O. Jensen and A.B. Walker, 1992, J.Phys.:Condens. Matter **4**, 10311.
[159] J. Mäkinen, S. Palko, J. Martikainen and P. Hautojärvi, 1992, J.Phys.:Condens. Matter **4**, L503.
[160] M. Vicanek and H.M. Urbassek, 1991, Phys. Rev. B **44**, 7234.
[161] A.P. Knights and P.G. Coleman, J.Phys.:Condens. Matter **7**, 3485.
[162] Alex Weiss, E. Jung, J.H. Kim, A. Nangia, R. Venkataraman, S. Starnes and G. Brauer, 1997, Appl. Surf. Sci. **116**, 311.
[163] G.R. Massoumi, N. Hozhabri, W.N. Lennard and P.J. Schultz, 1991, Phys. Rev. B **44**, 3486.
[164] G.R. Massoumi, N. Hozhabri, K.O. Jensen, W.N. Lennard, M.S. Lorenzo, P.J. Schultz and A.B. Walker, 1992, Phys. Rev. Lett. **68**, 3873.
[165] G.R. Massoumi, W.N. Lennard, P.J. Schultz, A.B. Walker and K.O. Jensen, 1993, Phys. Rev. B**47**, 11007.
[166] K.F. Canter, A.P. Mills, Jr. and S. Berko, 1974, Phys. Rev. Lett. **33**, 7.
[167] A.P. Mills, Jr., 1978, Phys. Rev. Lett. **41**, 1828.
[168] K.G. Lynn, 1979, J.Phys.C **12**, L435.
[169] R.M. Nieminen and J. Oliva, 1980, Phys. Rev. B **22**, 2226.
[170] A.P. Mills, Jr., 1993, Positron Spectroscopy of Solids, eds. A. Dupasquier and A.P. Mills, Jr. (Amsterdam:IOS Press) p. 237.
[171] I.J. Rosenberg, R.H. Howell and M.J. Fluss, 1987, Phys. Rev. B **35**, 2083.
[172] M.R. Poulsen, M. Charlton and G. Laricchia, 1993, J.Phys.:Condens. Matter **5**, 5209.
[173] R.H. Howell, I.J. Rosenberg, P. Meyer and M.J. Fluss, 1987, Phys.Rev.B **35**, 4555.
[174] R.H. Howell, I.J. Rosenberg and M.J. Fluss, 1986, Phys.Rev.B **34**, 3069.
[175] A.P. Mills, Jr., L. Pfeiffer and P.M. Platzman, 1983, Phys. Rev. Lett. **51**, 1085.
[176] R.H. Howell, I.J. Rosenberg, M.J. Fluss, R.E. Goldberg and R.B. Laughlin, 1987, Phys.Rev.B **35**, 5303.
[177] A. Ishii, 1984, Surface Sci. **147**, 295.
[178] A.B. Walker and R.M. Nieminen, 1986, J.Phys. F **16**, L295.
[179] D.M. Chen, S. Berko, K.F. Canter, K.G. Lynn, A.P. Mills, Jr., L.O. Roeelig and P. Sferlazzo, 1987, Phys. Rev. Lett. **58**, 921.
[180] R.H. Howell, P. Meyer, I.J. Rosenberg and M.J. Fluss, 1985, Phys. Rev. Lett. **54**, 1698.
[181] P.M. Platzman and N. Tzoar, 1986, Phys.Rev. B**33**, 5900.

[182] K.G. Lynn, A.P. Mills, Jr., R.N. West, S.Berko, K.F. Canter and L.O. Roellig, 1985, Phys.Rev.Lett. **54**, 1702.
[183] A.P. Mills, Jr. and L. Pfeiffer, 1979, Phys.Rev.Lett. **43**, 1961.
[184] K.G. Lynn, 1979, J.Phys.C **12**, L435.
[185] S. Chu, A.P. Mills, Jr. and C.A. Murray, 1981, Phys.Rev.B **23**, 2060.
[186] A.P. Mills and L. Pfeiffer, 1985, Phys. Rev. B **32**, 53.
[187] K.G. Lynn and D.O. Welch, 1980, Phys. Rev. B **22**, 99.
[188] I.J. Rosenberg, A.H. Weiss and K.F. Canter, 1980, J.Vac.Sci.Technol. **17**, 253.
[189] M.R. Poulsen, M. Charlton, J. Chevallier, B.I. Deutch, L.V. Jørgensen and G. Laricchia, 1991, J.Phys.:Condens. Matter **3**, 2849.
[190] P.J. Schultz, K.G. Lynn and B. Nielsen, 1985, Phys. Rev. B **32**, 4732.
[191] P. Sferlazzo, S.Berko, K.G. Lynn, A.P. Mills, Jr., L.O. Roellig, A.J. Viescas and R.N. West, 1988, Phys.Rev.Lett. **60**, 538.
[192] P. Rice-Evans and K.U. Rao, 1988, Phys. Rev. Lett. **61**, 581.
[193] P. Rice-Evans, M. Moussavi-Madani, K.U. Rao, D.T. Britton and B.P. Cowan, 1986, Phys.Rev.B **34**, 6117.
[194] M. Tuomisaari, R.H. Howell and T. McMullen, 1989, Phys. Rev. B **40**, 2060.
[195] P. Sferlazzo, S. Berko and K.F. Canter, 1985, Phys. Rev. B **32**, 6067.
[196] Y. Magashima, Y. Morinaka, T. Kurihara, Y. Nagai, T. Hyodo, T. Shidara and K. Nakahara, 1998, Phys. Rev. B **58**, 12676.
[197] L. Xie, G.B. deMaggio, W.E. Frieze, J. Devries, D.W. Gidley, H.A. Hristov and A.F. Yee, 1995, Phys. Rev. Lett. **74**, 4947.
[198] G.B. deMaggio, W.E. Frieze, D.W. Gidley, M. Zhu, H.A. Hristov and A.F. Yee, 1997, Phys. Rev. Lett. **78**, 1524.
[199] M. Eldrup, A. Vehanen, P.J. Scultz and K.G. Lynn, 1983, Phys. Rev. Lett **51**, 2007 and 1985, Phys. Rev. B **32**, 7048.
[200] Y-C. Chen, K.G. Lynn and B. Nielsen, 1988, Phys.Rev.B **37**, 3105.
[201] D.M. Chen, S. Berko, K.F. Canter, K.G. Lynn, A.P. Mills, Jr., L.O. Roellig, P. Sferlazzo, M. Weinert and R.N. West, 1987, Phys. Rev. Lett. **58**, 921.
[202] A.P. Mills, Jr. and W. S. Crane, 1985, Phys. Rev. A **31**, 593.
[203] D.W. Gidley, R. Mayer, W.E. Frieze and K.G. Lynn, 1987, Phys. Rev. Lett. **58**, 595.
[204] G. Laricchia, S.A. Davies, M. Charlton and T.C. Griffith, 1988, J.Phys. E **21**, 886.
[205] M.H. Weber, S. Tang, S. Berko, B.L. Brown, K.F. canter, K.G. Lynn, A.P. Mills, Jr., L.O. Roellig and A.J. Viescas, 1988, Phys. Rev. Lett. **61**, 2542.
[206] A. Ishii, 1990, Nucl. Instrum. Methods B **48**, 386.

[207] A. Ishii, 1993, Surface Sci. **287/8**, 811.
[208] A. Ishii, 1993, Surface Sci. **283**, 462.
[209] A. Ishii and T. Aisaka, 1995, Appl. Surf. Sci. **85**, 33.

CHAPTER 6

# DEPTH-PROFILING OF SUBSURFACE REGIONS, INTERFACES AND THIN FILMS

A. VAN VEEN, H. SCHUT and P.E. MIJNARENDS

*Interfaculty Reactor Institute, Delft University of Technology, Mekelweg 15, 2629 JB Delft, The Netherlands*

*E-mail: avveen@iri.tudelft.nl*

In positron beam analysis, mono-energetic positrons are implanted into a solid surface and the characteristics of the positron annihilation process are interpreted in terms of implantation depth, positron transport, and trapping at defects or at interface sites. In principle, all observation methods used for bulk studies with unmoderated positrons can be used also for beam studies, but the majority of them rely on Doppler broadening techniques. A few beam analysis systems have been designed to carry out depth sensitive positron lifetime analysis or analysis with 2D-ACAR. This chapter describes the observation techniques, including the recent refinement of the Doppler broadening method by employing two-detectors in coincidence techniques, and the use of a set of line-shape parameters (S,W) related to valence and core-electrons, respectively. Examples are given of analysis for point defects and defect clusters, precipitates, and layered structures with planar interfaces. Aspects related to the charged nature of the positron are described for electrically biased capacitor (e.g., MOS) systems.

## 1 Introduction

Positrons have a high affinity for trapping in open-volume defects. Consequently, positron annihilation and in particular the positron beam technique can provide information which is complementary to that provided by other techniques used for the analysis of composition and structure of materials. A wealth of techniques is available for these latter purposes. Among them are Rutherford Backscattering (RBS), Nuclear Reaction Analysis (NRA), and Secondary Ion Mass Spectrometry (SIMS), which are capable of depth profiling of impurity atoms (Feldman and Mayer 1986). These techniques, however, are not sensitive to the presence of open-volume defects. An indirect method for probing open volume is Small Angle Neutron or X-ray scattering (SANS/SAXS). Open volume is best probed by using probe particles that are trapped in the cavities. Suitable particles are gas atoms such as helium and hydrogen, or positrons. Other techniques using a probe atom are Mössbauer and PAC (Perturbed Angular Correlation), but these are better suited for probing very small cavities (1-5 vacancies). The use of positrons as a probe has the advantage that the material is not damaged by the irradiation with the probe particles. A recent review on defect profiling by positrons and other techniques has been given by Dupasquier and Ottaviani (1995).

In layered structures obtained by various deposition or implantation techniques positrons are used to detect open-volume defects or microcavities (Schut et al. 1991, Asoka-Kumar et al. 1993, van Veen et al. 1995). The defects can have various origins such as the low mobility of the deposited atoms (low-temperature deposition), imperfect regrowth during crystallization, stress relaxation in heterostructures, ion implantation damage etc. Microcavities can be observed easily in measurements of the Doppler broadening of the 511 keV annihilation line. With standard positron beams a layer of thickness ~ 1μm below the material surface can be probed in order to reveal information on the depth distribution of the defects. Positron beam analysis with millimeter lateral size beams has been developed since about 1982 (see the reviews by Schultz and Lynn (1988) and by Asoka-Kumar et al. (1994). In the near future it is expected that micrometer size beams will be available so that lateral resolution can be achieved with the technique (see Chapter 7).

In § 2 of the present chapter a short overview is given of the observables in positron annihilation. § 3 describes the Doppler broadening parameters $S$ and $W$ in some detail. It also discusses the recently re-discovered two-detector Doppler broadening technique. In § 4 a brief description is given of the two-dimensional angular correlation (2D-ACAR) technique used in combination with a slow-positron beam. Positron lifetime measurements with positron beams are treated in § 5, while the first part of the chapter is concluded with a discussion of the positronium fraction in § 6.

The second part of the chapter is devoted to the analysis of the behaviour of the implanted and thermalized positrons in various systems of interest. § 7 presents briefly the theory of positron diffusion and trapping. The diffusion equation thus obtained lies at the basis of several computer programs developed as an aid in this analysis. In § 8 the surface branching of positrons into emitted and surface trapped positrons and positronium is described. The §§ 9 through 11 are devoted to positron beam analysis of defects, precipitates, grain boundaries and planar interfaces in surface layers of solids. The effects of electric fields on positrons are the topic of § 12. Finally, in § 13 we present a summary of the chapter and an outlook on future developments.

## 2 Observables

The positron annihilation process possesses a number of signatures that contain information concerning the system under investigation. Various experimental methods have been developed during the past decades to measure these signatures. In this section we shall discuss how and which information can be extracted from the experimental data. It will be assumed that the positrons annihilate from the bulk and from (a number of) possible defects; for the history of the positrons up to the moment of their annihilation we refer to Chapter 4.

A thermalised positron and an electron annihilate under emission of mostly two $\gamma$ quanta. In the centre-of-mass system of the annihilating particles the $\gamma$ quanta are emitted in exactly opposite directions and the available energy of $2m_0c^2-E_B$ (where $E_B=E_B^- + E_B^+$ denotes the binding energy of the two particles in the solid) is divided equally over the two quanta. In the laboratory coordinate system the momentum $\boldsymbol{p}$ of the two particles (largely determined by the electron, the positron being thermal) manifests itself in two different ways. Firstly, the momentum component $p_{\parallel}$ along the direction of $\gamma$-ray emission results in an energy shift $\Delta E=\pm\frac{1}{2}\, p_{\parallel}c$ of the energy of the $\gamma$ quantum from its average value $m_0c^2-\frac{1}{2}E_B$. Since one half of the electrons moves towards the detector while the other half moves away from it, this shift results in a Doppler broadening by several keV's of the 511 keV line which can be measured with a solid-state detector (MacKenzie 1969). Since the distance between the detector and the sample is usually very small the Doppler broadening experiment integrates over virtually all values of the transverse component of the momentum.

This transverse component can be measured (in one or two dimensions) in an Angular Correlation of Annihilation Radiation (1D- or 2D-ACAR) experiment (Berko and Plaskett 1958, Berko et al. 1977). In the laboratory system of reference, the directions of $\gamma$-ray emission differ from $180^0$ by a small angle $\theta \sim p_{\perp}/m_0c$, typically of the order of 10 mrad and proportional to the component $p_{\perp}$ of the momentum perpendicular to the direction of $\gamma$-ray emission. In view of the smallness of $\theta$ the latter type of experiment requires large sample-detector separations in order to realise a good momentum resolution, and thus a strong radioisotope source or an intense positron beam is needed. In this section we shall concentrate on the Doppler broadening technique (however, see cf § 4 for applications of the angular correlation method to positron beam experiments).

A third signature of the annihilation process is the positron lifetime, which is the inverse of the rate at which positrons annihilate. The lifetime depends on the density of electrons at the position of the positron. It can be measured in a delayed coincidence experiment in which the time interval is measured between the emission of a positron by the positron source, c.q. the entrance of a positron into the sample, and its subsequent annihilation. Finally, positrons diffusing back to the surface or trapped at the surface of the sample may be re-emitted after having formed positronium (Ps), a hydrogen-like bound state of a positron and an electron. This so-called Ps fraction is usually not measured directly but derived in an indirect way from other measurements. In the following we shall examine the various signatures of the annihilation process more closely.

## 3 Doppler broadening

### 3.1 *S and W parameters*

The resolution of a solid-state detector used to measure the Doppler profile (typically ~1 keV) is not very narrow compared to the line broadening; hence, the observed resolution-broadened profiles are relatively smooth and featureless. Broadly speaking, two parts of the profile are of interest: the low-momentum central part of the peak and the high-momentum tails. Due to its positive charge, the positron is repelled by the atomic nuclei. It therefore samples mainly the interstitial region where it annihilates predominantly with the low-momentum valence electrons and thus contributes to the centre of the profile. The wings of the profile have a much lower intensity owing to the small overlap of the positron wavefunction with the core electron wavefunctions. A missing atom, i.e., a vacancy, represents an attractive potential well which may trap a positron. The absence of the atom with its core electrons manifests itself as a decrease of high-momentum content and thus a sharpening of the profile.

It is then profitable to characterise the shape of the profile quantitatively with the aid of two parameters: $S$ and $W$ (fig. 1). The parameter $S$ gives the area under the central part of the profile, divided by the total area under the profile; thus a high value of $S$ signals the presence of open-volume defects, whereas a 'defect-free' sample will show a low value of the $S$ parameter. The energy windows are set symmetrically around the centre of the peak in such a way as to maximise the sensitivity to vacancies. Many workers use an $S$ value of ~ 0.50. In order to make $S$ parameter measurements in different laboratories easier to compare, the $S$ values are usually quoted with respect to some standard material, often a highly perfect Si crystal. The orientation of this reference crystal and that of the crystal under investigation, if the latter also is a single crystal, should be specified as the $S$ parameter is an anisotropic quantity with the point

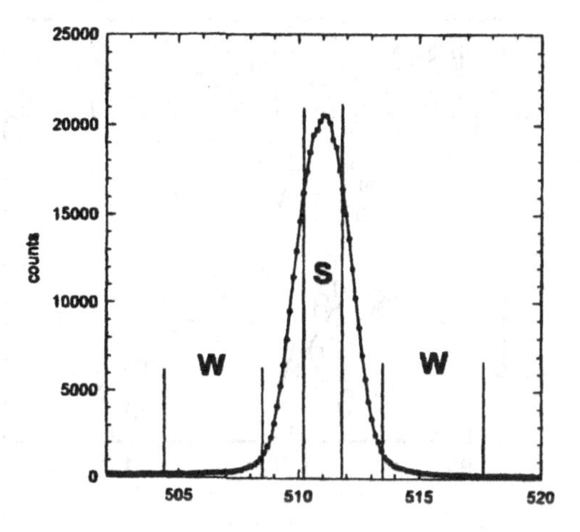


Fig. 1: A Doppler broadened 511 keV peak measured for perfect silicon. The intervals used for defining the S and W parameters are indicated.

symmetry of the crystal. The second parameter, *W*, gives the ratio between the area under the wings and the total area. The settings of the energy windows which define the *W* parameter differ more between different laboratories. Some workers choose their settings such that the value of *W* is approximately 0.25 of that of *S*, thereby obtaining a relatively high statistical precision, while others choose a lower fraction ($\leq 0.1$) to increase the sensitivity to defects. As *S* and *W* parameters are given by areas under parts of the Doppler broadening profile they can be measured in a short time with a good statistical precision without putting high demands on the positron beam intensity. This makes this type of measurement highly suitable for defect investigations in which some parameter such as the positron energy, sample temperature or previous heat treatment is varied.

### 3.2 *S - W mapping*

An important property of *S* and *W* is their linearity. By this we mean that if there are several possible modes of positron annihilation in a material with probabilities $f_i$ and corresponding *S* parameters $S_i$, then the measured *S* parameter will be found from

$$S=\sum_i f_i\, S_i / \sum_i f_i \tag{1}$$

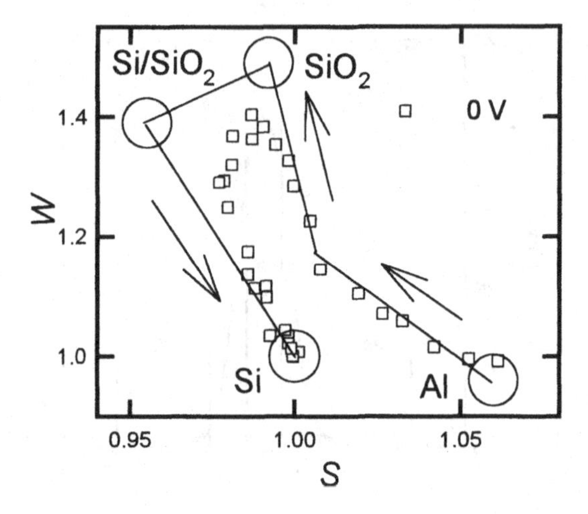


Fig. 2: S - W diagram for a MOS system. The arrow along the S - W trajectory shows the direction of increasing implantation energy (i.e., implantation depth) of the positrons. The large circles indicate the four possible annihilation sites present in the system. Note that relative S and W parameter values are plotted (relative with respect to bulk silicon).

with a corresponding expression for $W$. One can make use of this by plotting $S$ and $W$ of a profile in a graph of $S$ against $W$. Thus, each profile is represented by a point in the graph. If a number of profiles is measured under variation of some external parameter which affects positron trapping, special points may be distinguished in the $S$ - $W$ diagram which correspond to a situation in which all positrons annihilate via one and the same mode. This mode may be the annihilation of free positrons from the bulk of a defect-free sample, or the saturation annihilation from a particular type of defect. Intermediate situations, in which there is a mixture of annihilations from defects and from the bulk, will be represented by points on a straight line connecting these two special points. Figure 2 shows the $S$ - $W$ diagram for positrons implanted into a metal-oxide-silicon (MOS system). The running parameter is the positron energy which is directly related to the positron implantation depth. In this layered structure the positrons successively probe, with increasing implantation depth, the aluminium gate, the $SiO_2$ layer, the $SiO_2$-Si interface, and the silicon substrate. The trajectory in the $S$ - $W$ map reaches or approaches the characteristic $S$ - $W$ cluster points which are known for the four different annihilation sites. When the trajectory reaches the point (as in the case of the Si substrate) there is 100% annihilation at that point. In other cases the annihilation is divided over 2 or 3 annihilation sites. Note that the trajectory contains points with the same $S$ but different $W$. It is clear from this example that an $S$ - $W$ diagram allows one to distinguish annihilations from different types of defects that on the basis of measurements of the $S$ or $W$ parameter alone would go undetected.

### *3.3 Two-detector Doppler broadening measurements*

In 1977 Lynn et al. added a second Ge(Li) detector to the Doppler broadening setup in order to observe the second annihilation quantum in coincidence (Lynn et al. 1977). This resulted in an improvement by a factor ~100 of the peak to background ratio and a ~$\sqrt{2}$ better energy resolution. As a result it became possible to observe high- momentum annihilations with the core electrons without being hampered by the background. Recently this method, dormant for many years, has been revived and it

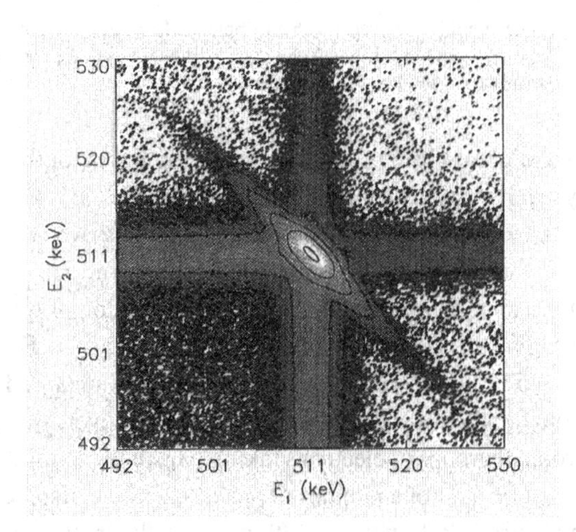


Fig. 3: Histogram of $E_1$-$E_2$ events in a two-detector coincidence measurement.

has been demonstrated that its elemental specificity can be employed in the identification of impurity atoms bound to vacancies (Szpala et al. 1996, Asoka-Kumar 1996).

The count rate in the two-detector experiment is a function of the energies $E_1$ and $E_2$ of the two detected photons. A histogram in the $E_1$ - $E_2$ plane shows a roughly elliptical profile (fig. 3) with its major axis parallel to the line $E_2 = 2 \times$ 511 keV- $E_B$ - $E_1$, stemming from positrons undergoing two-photon annihilation in the sample. In addition there are weak ridges along the lines $E_1$ = 511 keV and $E_2$ = 511 keV due to accidental coincidences between the two detectors. The low-energy parts are caused

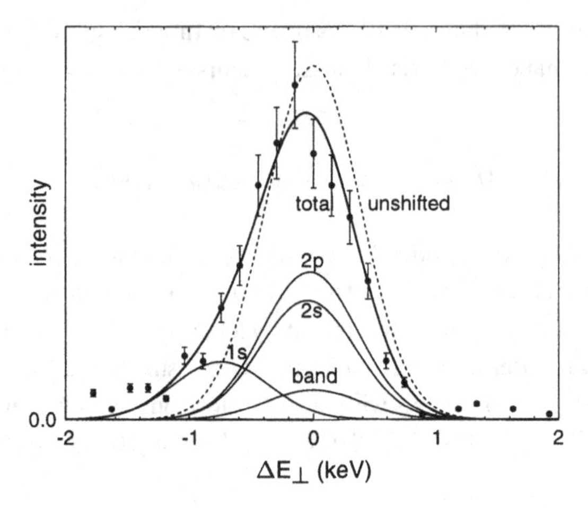


Fig. 4: Energy profile for gamma annihilation events with a Doppler shifted energy of 9 keV calculated and measured for aluminium. The contributions of different electron shells are indicated. The curve labelled 'unshifted' represents the result when the binding energy is ignored.

by incomplete charge collection or by Compton scattering in one of the detectors; the high-energy parts by cosmic rays or by pile-up events.

The crest of the three-dimensional profile yields the well-known Doppler broadening profile, but with a superior signal to background ratio ($10^5$:1) and a better resolution than in the case of a single detector. If the resolution of the two detectors were ideal, a cross-section through the profile along the line $E_2 = E_1$ would yield a $\delta$-function at $E_2 = E_1 = 511 - \frac{1}{2}E_B$ keV. In reality a peak is obtained with a width which is representative for the resolution of the setup. Moreover, this peak is made up of contributions from all groups of electrons taking part in the annihilation process. Thus, the peak consists of a sum of a number of peaks, each one displaced by half the binding energy of an electron shell, with a width given by the instrumental resolution and an intensity determined by the wavefunction overlap between the positron and the electrons in that shell. Figure 4 shows an example of this for aluminium. There is a relatively small contribution of the valence electrons centred at $\Delta E_\perp = 0$ (where $\Delta E_\perp$ is defined by $E_{1(2)}$=511 keV +(-)$\Delta E$ - $\Delta E_\perp$) since the binding energy of the valence electrons is negligibly small. The contributions of the 2s and 2p electrons are marginally shifted with respect to the centre of the peak of the conduction electrons, the binding energies of these two shells of electrons being 94 and 57 eV, respectively. Farther out in the wing there is a small contribution of the 1s electrons (at $\Delta E_\perp = \frac{1}{2}E_B$ = -0.744 keV). When all these contributions are added a curve results with a small but noticeable skewness which fits the experimental data markedly better than the unshifted curve (Mijnarends et al. 1998).

## 4 2D-ACAR

The observables discussed so far rely on a measurement of the longitudinal component $p_{\parallel}$ of the electron momentum. However, given a sufficiently intense positron beam it is also possible to observe the transverse component by measuring the angular correlation between the $\gamma$ rays. Position-sensitive detectors (Anger cameras) are placed on either side of the sample and register the positions at which the two annihilation quanta from an annihilating positron-electron pair hit the detectors in coincidence (West 1995). The energy resolution of the cameras is insufficient to resolve the longitudinal momentum component; thus, the system integrates over all values of this component. The low intensity of slow-positron beams generated by radio-isotope sources have made this type of experiment in combination with a beam very difficult until now, but now that intense positron beams are or soon will become available in more laboratories it is of interest to consider the information that can be derived from this experiment.

The pioneering 2D-ACAR work with beams was performed in the eighties in the laboratories of Lawrence Livermore and Brookhaven. The Livermore beam was made by moderation of high-energy positrons produced by pair production in a W target at the electron linac. The positrons were then moderated in a system of W vanes. The Brookhaven group used a slow positron source made by evaporating ~20 Ci of $^{64}$Cu obtained by irradiation in the High-Flux Beam Reactor onto a single-crystal W(110) substrate, where it was annealed to form an epitaxial Cu(111) layer. This thin Cu layer emitted and simultaneously moderated the positrons into a monoenergetic beam of slow positrons. Both beams could be accelerated to obtain the desired incident positron energy. These first 2D-ACAR experiments concentrated on the differences between the bulk and surface momentum distributions of single-crystal Al (BNL) (Lynn et al. 1985, Chen et al. 1987, 1989) and Cu (LLNL) (Howell et al. 1985b) samples and observed the momentum distribution of Ps emitted by the surface. More recent 2D-ACAR experiments have focussed on the contributions to the momentum density from various parts of a MOS system (see § 12).

## 5 Positron lifetime

As mentioned earlier, the positron lifetime is obtained by measuring the time interval between a suitably chosen start pulse which is as close as possible in time to

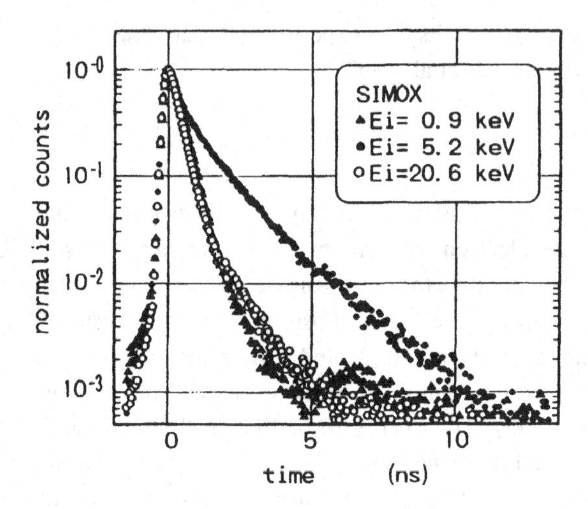


Fig. 5: Lifetime spectrum for an $SiO_2$ - Si system. From Suzuki (1994).

the entry of the positron into the sample. In conventional experiments $^{22}Na$ emits a 1.28 MeV $\gamma$ ray simultaneously with the positron, which is used to trigger the lifetime measurement. In a beam with one or more moderation stages, accelerating and retarding sections, and particle transport over an often considerable distance this is however not a feasible option. Several methods have been developed to overcome this problem. Early experiments used the emission of secondary electrons released from the target surface by the entrance of the positron and detected with the aid of a microchannel plate. They achieved a time resolution of ~600 ps FWHM (Lynn et al. 1984). The resolution is limited by positrons scattering off the sample surface and returning to it after some time and by the spread in response time of the detector. A different approach consists of 'bunching' the beam, i.e., collecting the positrons in the beam in groups ('bunches') equally separated in time, and subjecting these groups to fields designed to achieve a degree of time-focussing. Initial experiments along these lines produced resolutions of several ns FWHM (Mills et al. 1983, Howell et al. 1985a). Further development in Munich (Schödlbauer et al. 1988, Willutzki et al. 1995) and Japan (Suzuki et al. 1994) resulted in pulse widths of 150 ps with pulse repetition rates of $10^5$ Hz and a time resolution of ~ 200 ps which is comparable to that of good conventional lifetime circuitry. Also here backscattered positrons may interfere with the lifetime measurement, either by annihilations in the surrounding materials or by causing delayed start signals.

A lifetime spectrum obtained with a pulsed beam for an $SiO_2$ layer on silicon is shown in fig. 5. Usually the results of the measurements are reported as the averaged lifetime vs the positron implantation energy (Uhlmann et al. 1994). Resolving more than one lifetime component like in conventional lifetime measurements is possible. However, when the positrons are implanted at depths where substantial fractions can reach the surface the positron lifetimes are not well defined anymore. The time spectrum can no longer be described by a sum of exponential contributions. Methods to analyze the time spectrum in that case are discussed by Britton and Störmer (1995) and by Kögel (1996). In principle, the time dependent transport equation rather than the stationary equation (see § 7) has to be solved.

## 6 Positronium fraction

Most annihilations of positrons in an electron gas take place between particles of opposite spin (i.e., in a singlet state) under emission of two $\gamma$ quanta. A positron and an electron with parallel spin (triplet state) can only decay under emission of three $\gamma$ quanta. Taking into account the ratio between the cross- sections for $2\gamma$ and $3\gamma$ decay and the statistical weight of the triplet state, only a small fraction (1/372) of positrons decays under emission of three $\gamma$ quanta.

In situations where the electron density is sufficiently low, such as at surfaces, in voids or in the bulk of insulators and polymers, a positron and an electron may form a bound state called (quasi-)positronium (Ps) (the prefix *quasi* indicating that the Ps atom is affected by its environment) (Dupasquier 1983). In positron beam experiments the positrons have a certain probability, depending on their depth of implantation, of returning to the outer surface and forming positronium. For positrons arriving at the surface of the sample by diffusion there exist two other channels to escape from the bulk besides the transition to the bound surface state: *i*) re-emission as a positron; the available potential energy is converted into kinetic energy of the emitted positron, and *ii*) an electron is released from the valence band of the material and joins the positron to form Ps. In this process energy is used to release the electron, but a large part of the energy is provided by the binding energy (6.8 eV) of the electron in Ps. Depending on the spin orientation of the two particles, Ps decays under emission of two $\gamma$ quanta (opposite spins: para-Ps, lifetime 125 ps) or three $\gamma$ quanta (parallel spins: ortho-Ps). For Ps atoms in vacuum the ratio of $3\gamma$ to $2\gamma$ annihilation is 3:1; in matter this ratio is reduced by competing processes such as pick-off annihilation, in which the positron annihilates with an electron of opposite spin from one of the surrounding atoms. Hence, the lifetime of o-Ps can vary from 140 ns in vacuum to a few ns in matter.

It is therefore clear that the fraction of Ps formed constitutes valuable information. In 3$\gamma$ annihilation there is no unique correlation between the energies of the three emitted photons, and the energy spectrum of the emitted $\gamma$ radiation is therefore continuous with an end point at 511 keV. To quantify the differences between energy spectra with and without the formation of Ps one defines the ratio

$$R=(T-P)/P, \tag{2}$$

here $T$ denotes the total number of counts in the spectrum and $P$ the number of counts under the 511 keV peak. The quantity $R$ thus defined is independent of beam intensity and varies from a value $R_0$ if no Ps is formed to $R_1$ when all incident positrons form Ps. It can be shown that the positronium fraction $f_{Ps}$ is found from (Deutsch and Dulet, 1951)

$$f_{Ps}=[1+\frac{P_1}{P_0}\frac{R_1-R}{R-R_0}]^{-1} . \tag{3}$$

$f_{Ps}$ varies between 0 for no Ps formation and 1 for complete Ps formation. It will be clear from the above that $f_{Ps}$ refers only to the 3$\gamma$ decay mode; p-Ps decay and 2$\gamma$ pick-off annihilation of o-Ps are not included. Lynn and Welch (1980) have discussed some of the calibration problems of this measurement.

## 7 Diffusion and trapping

In experiments where the energy of the implanted positrons can be varied (beam experiments) it is important to know the depth dependent concentration profile of the thermalised positrons. From the concentration one can calculate the number of positrons per unit of time that *i*) is captured in defects and subsequently annihilates, *ii*) annihilates in the bulk, and *iii*) arrives at the surface. To this purpose the time averaged positron concentration is derived from the following balance equation:

$$D\frac{d^2c}{dz^2} - \frac{d}{dz}(v_d c) + I(z) - \sum_j k_j n_j c - \lambda_b c = 0 , \tag{4}$$

where

$c(z)$ = the depth dependent concentration,
$v_d(z)$ = $\mu\ E(z)$ = the drift velocity with $\mu$ the positron mobility and $E$ the electrical field strength in the material (in particular for activated semiconductor devices),
$I(z)$ = the positron implantation rate at depth $z$,
$n_j(z)$ = the density of defects of type $j$,
$k_j$ = specific capture rate for capture at defects of type $j$,
$\lambda_b = \tau_b^{-1}$ = annihilation rate in the bulk ($\tau_b$ is the bulk-lifetime),
$D$ = positron diffusivity.

The various terms in Eq. (4) describe the following processes: *diffusion, drift, implantation, capture* and *annihilation* of the positron. It should be noted that this balance equation is equally valid for other mobile particles implanted in materials. The equation may also be used for hydrogen and helium in metals in which case the electric field drift term and the annihilation term should be omitted.

When there is a uniform concentration of defects $n_t$ with specific trapping rate $\kappa_t$ and the surface is located at a sufficient distance from the depth of implantation $z_f$, the solution will read:

$$c(z) = c(z_f)\ \exp[-(z-z_f)/L_+] \tag{5}$$

where $L_+ = [\ D^+/(\kappa_t\ n_t\ +\lambda_b)]^{1/2}$ is the effective diffusion length, i.e., the averaged distance travelled by the positron before it is trapped or annihilated. From Eq. (5) follows the boundary condition at a large distance from the surface:

$$[\frac{dc}{dz}]_{z=z_f} = -c(z_f)/L_+ \tag{6}$$

For the outer surface a similar boundary condition can be defined, but now the absorption length $L_a$ instead of the effective diffusion length $L_+$ is involved:

$$[\frac{dc}{dz}]_{z=0} = -c(z=0)/L_a$$

The stationary diffusion equation can be solved numerically, for instance with the program VEPFIT (van Veen et al. 1990, Schut and van Veen 1995) or POSTRAP (Aers et al. 1995). These solutions constitute the basis for the analysis of positron annihilation measurements with variable energy positron beams. Layered structures or defect profiles can be modeled. The program calculates the depth distribution of trapped positrons and the fraction that is diffused to the surface for a range of implantation energies. Using assumed positron parameters $S$ or $W$, positron surface branching ratios (see § 8) and electric fields, predictions can be made of the $S(E)$, $W(E)$ and $f_{Ps}(E)$ curves. The curves can be made to match the measured data, thus yielding fitted values of $S$ and $W$, defect concentrations, interface positions, values of electric fields and surface properties.

## 8 Surface branching

In the previous sections it was mentioned that positrons may diffuse back to the surface and be emitted either as positronium or as positrons. A third possibility is the trapping of positrons in a surface state. Furthermore, in the first case fast and slow emitted positronium, positronium in an excited state or negatively charged Ps may be distinguished. The surface branching can be determined in a relatively simple way, when three branches are assumed: re-emitted positrons, surface trapped positrons and positronium. By varying the voltage of a grid in front of the sample from negative to positive the re-emitted positrons are affected. When the grid is negative, positrons are removed from the sample area while for a positive grid the positrons are returned to the surface and converted to either positronium or to positrons in a surface state. The measured positronium fractions $f_{Ps}^-(E)$ and $f_{Ps}^+(E)$ are related to the positronium fraction $f_{Ps}$ of all positrons (including those that are re-emitted) as follows

$$f_{Ps}^- = f_{Ps} \,/\, (1 - f_{e^+}) \qquad f_{Ps}^+ = f_{Ps} + \epsilon_r f_{e^+} \tag{8}$$

where $f_e^+(E)$ is the reemitted positron fraction and $\epsilon_r$ the fraction of re-emitted positrons that after returning to the surface end up as positronium. This positronium fraction can be derived from the gamma energy spectra taken for the negative and positive grid voltage, respectively, by substituting $\Delta R$ for $R$ in Eq. (2): $\Delta R = (\Delta T - \Delta P) / \Delta P$ with $\Delta T$ and $\Delta P$ given by $T^+ - T^-$ and $P^+ - P^-$, respectively. $T^+$ and $T^-$ are the total and peak contents for positive and negative grid voltages respectively. The fractions of Ps, reemitted positrons and surface trapped positrons are given by

$$f_{Ps} = f_{Ps}^{-}\,(\epsilon_r - f_{Ps}^{+})/(\epsilon_r - f_{Ps}^{-})\ ,$$
$$f_{e^+} = (f_{Ps}^{+} - f_{Ps}^{-})/(\epsilon_r - f_{Ps}^{-})\ , \qquad (9)$$
$$f_{ss} = f_{Ps}\,(1/f_{Ps}^{-}(E{=}0)-1)$$

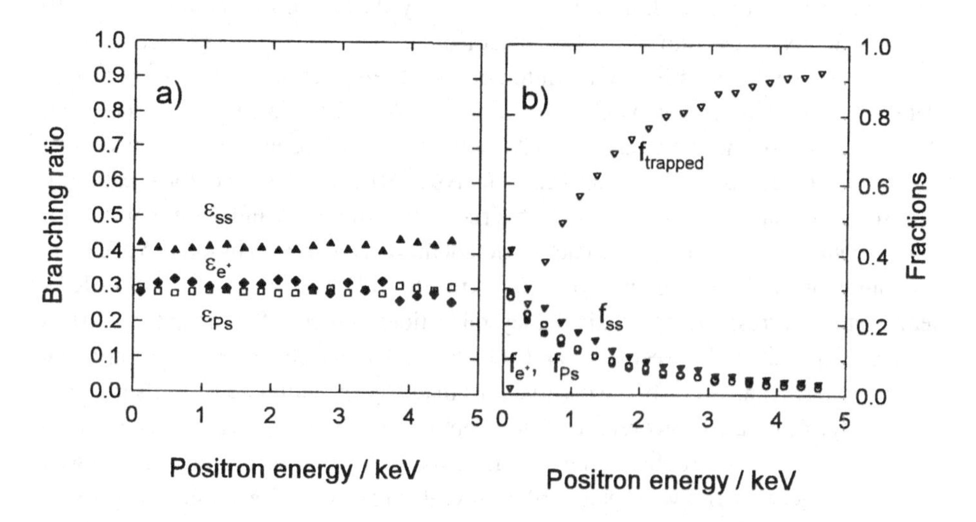


Fig. 6: a) Branching ratios for a β-SiC (100) layer on silicon; b) the various Ps and positron fractions obtained using Eq. (9). From Jørgensen et al. (1996).

The surface branching ratios for these three states are given by $\epsilon_{Ps} = f_{Ps}\ (E{=}0)$, $\epsilon_{e+}{=}f_{e+}(E{=}0)$, and $\epsilon_{ss}{=}f_{ss}\ (E{=}0)$, respectively. An example of surface branching ratios determined by this method for a SiC single crystalline surface by Jørgensen et al. (1996) is given in fig. 6. It is of interest to note that the experimentally derived branching ratios are independent of the positron implantation energy. These results support the validity of the approach followed. The surface branching is sensitive to coverage of the surface with gases. Howell et al. (1990) reported effects on the positronium fraction and reemitted positron fraction for oxygen coverage of Al and Ni single crystalline surfaces.

## 9 Point defects and clusters

- Ion implantation defects

Positron beam analysis has been proven to be a very sensitive technique to detect small amounts of defects created by ion implantation. Mäkinen et al. (1990) used the positron beam technique to detect vacancy clusters in silicon implanted with 12 MeV $Si^+$ ions. Uedono et al. (1991) studied vacancy type defects created by MeV donor and acceptor ions implanted into silicon. Nielsen et al. (1993) were able to detect defects in Si irradiated with MeV Si-ions at dose levels as low as $10^{10}$ ions $cm^{-2}$. These defect concentrations are below the appm level. Simpson et al. (1991) showed by the combined use of positron beam analysis, RBS (Rutherford Backscattering Spectroscopy) and infrared absorption that room temperature Si and helium implanted silicon contained divacancies. Some fraction of the defects created by 0.5 to 5 MeV ion implantation at doses varying from $10^{11}$ $cm^{-2}$ to $10^{16}$ $cm^{-2}$ annealed at 560 K, a temperature corresponding to divacancy migration. For the Si implanted crystals recovery took place between 870 and 970 K, probably due to recrystallization of amorphized silicon. Some of the observations made by the authors on the discrepancy between the calculated and observed defect depth profile in the case of helium could not be explained in that article. Later experiments on hydrogen and helium filling of defects in silicon by Hakvoort et al. (1995) revealed the reason for those observations. Hydrogen and helium when filling vacancy-type defects do not reduce the positron trapping but lead to an increase of annihilation with the Si matrix atoms. Therefore the highly gas decorated defects are difficult to distinguish from the bulk. Defects not filled with gas which are found in the zone between the surface and the gas implantation profile (which is usually a narrow distribution) will dominate the defect profile measured by positron beam analysis. Proof of the presence of the defects in the gas-rich zone is usually found by thermal annealing. At a certain temperature the gas will dissociate and be partly released. Then a strong positron signal indicating vacancy defects is observed. It is not always possible to derive defect densities from the measured positron sink strength because specific trapping rates are not known. However, for monovacancies and divacancies in a number of semiconductors detailed information is available. Hautojärvi (1995) has shown that the specific trapping rate for point defects (neutral mono-vacancies) in semiconductors increases in proportion to 1/T, and that negatively charged vacancies trap positrons 100 times faster than neutral vacancies (see also Nieminen, Chapter 4 ).

Fujinami (1996) reported on oxygen implantation induced defects in silicon. He found multioxygen-multivacany complexes which after low temperature annealing showed the normal positron annihilation behaviour known for vacancies and vacancy clusters. After annealing at 800 °C the excess vacancies dissociate from the defect complex and the annihilation behaviour shows strong positron interaction with the oxygen so that annihilation with high momentum oxygen electrons dominates. Therefore the value of the *S* parameter becomes much lower than the silicon *S* parameter. Similarly, in fused silica implanted with 1.5 MeV $Si^+$ ions defects were observed with a low *S* parameter (Knights et al. 1996). There might be a similarity in the annihilation behaviour: in both cases vacancies are present near oxygen atoms.

Hydrogen implantation effects are studied by many groups. Keinonen et al. (1988) have reported on the defects and their recovery for 35-100 keV $H^+$ implanted silicon. A detailed account of the hydrogen related defects in 15.5 keV hydrogen implanted silicon has been given by Brusa et al. (1994). The results reveal that hydrogen binds to vacancies and vacancy complexes, and that hydrogen in vacancies "passivates" the positron annihilation effects as discussed before.

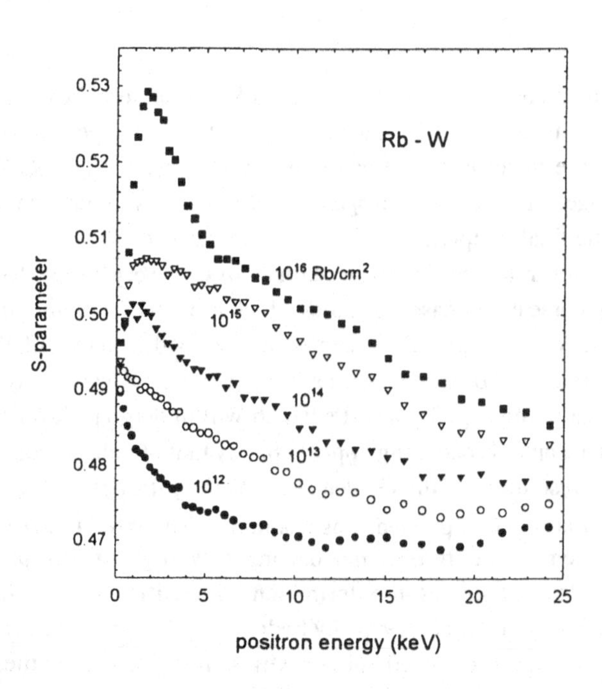


Fig. 7: S vs positron energy curves measured for a tungsten foil implanted with 110 keV $Rb^+$ at the indicated doses. From Schut et al. (1994).

Another interesting aspect of implantation depth profiling by positrons is that because of its high sensitivity defects are nearly always observed at depths much larger than the implantation depth of the projectiles (Smith 1990, Hakvoort et al. 1994). In many implantation experiments defect concentrations at the implantation depth are of the order of 1% so that saturation trapping occurs. Although at larger depths the concentrations decrease rapidly, positrons will still observe defects until defect levels drop 3 to 4 orders of magnitude. The deep defects might be due to channeling of projectile ions and to migration and clustering of point defects into dislocations.

Figure 7 shows the result of a defect tail observed for $Rb^+$ implantation of tungsten (Schut et al. 1994). Tungsten foils were irradiated with 110 keV $Rb^+$ ions with doses varying from $10^{12}$ to $10^{16}$ $cm^{-2}$. With increasing $Rb^+$ dose the $S$ parameter developed a maximum at 2 keV positron implantation energy, which corresponds to a mean implantation depth in tungsten of 6 nm. TRIM (Ziegler et al. 1985) calculations yield a range of 7 nm. Defects were observed to extend to depths larger than 250 nm (25 keV positron energy). After annealing to 1300 K bubbles developed, mostly at a low depth.

- *Microcavities*

Light-ion bombardment of silicon with keV ions at doses beyond $10^{16}$ $cm^{-2}$ and subsequent annealing up to a high temperature (1100 K) have been shown to generate microcavities in the material (Griffioen et al. 1987, Myers et al. 1993). The point defects cluster together with the hydrogen or helium atoms to form nanometer-sized bubbles. Near the final temperature of 1100 K the gas is dissolved and permeates to the surface where it is released. Positron beam analysis is very well suited to detect the cavities. The $S$ parameter increases to values which are 10-15 % higher than the value for defect-free silicon. These values can be understood in terms of Ps formation. Assuming that nearly all positrons form Ps in the cavity, 1/4 will decay as p-Ps with zero Doppler broadening and 3/4 as o-Ps which will undergo pick-off and therefore annihilate with a Doppler broadening approximately that of bulk silicon. Taking into account the finite resolution of the Ge detector, for the p-Ps peak $S/S_{silicon}$ = 1.59, the above values of $S$ are well explained. Positronium exists only in cavities larger than 0.8nm in size. When the cavity becomes decorated with gas a high pressure can be built up which also suppresses the formation of positronium. This has been demonstrated by Hakvoort et al. (1995) by hydrogen and helium ion irradiation and annealing of silicon which contained microcavities. In fig. 8 some of these results are shown. Cavities have been created by heliumimplantation and subsequent annealing to 800 °C. Refilling of the cavities leads to a decrease of the S-parameter. Repeated

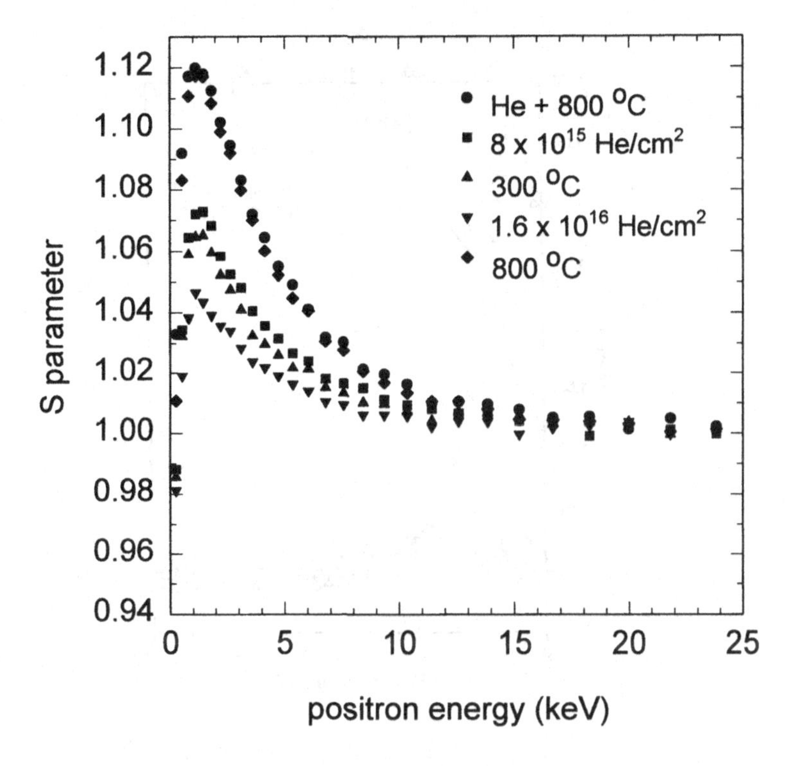


Fig. 8: Helium filling of microcavities in silicon. From Hakvoort et al. (1995).

annealing to 800 °C restores the old situation. The hydrogen results are generally in line with the hydrogen irradiation experiments on silicon (Brusa et al. 1994).

The formation of cavities is also observed by implantation with heavy ions. Simpson et al. (1992) found cavities in GaAs implanted with 220 keV $Si^+$ ions. The cavities were not visible after room temperature implantation but developed after annealing at 800°C. The threshold dose for void formation was $3\times10^{13}$ $cm^{-2}$.

Cavities at the $SiO_2$ metal interface have been observed by Simpson et al. (1994) in a system with $SiO_2$ passivated Al-Cu interconnect lines on silicon. The cavities were large enough to accommodate ortho-positronium for a long enough time to annihilate under $3\gamma$ decay .

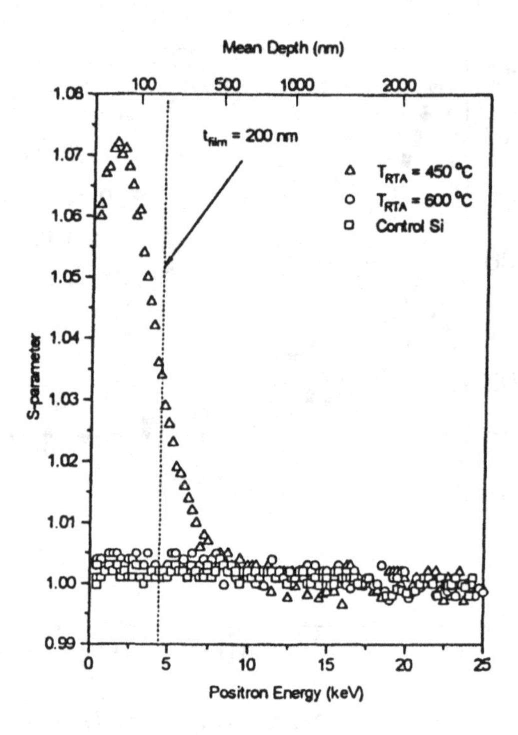


Fig. 9: Comparison of Doppler broadening measurements on LT-MBE films grown at 220 °C with different RTA temperatures and on a control sample of silicon. From Asoka-Kumar and Lynn (1995).

- *Defects in deposited layers*

During the deposition of thin layers by evaporation, sputtering or CVD processes, defects may be formed in the growing layer due to insufficient mobility of the deposited atoms and introduction of impurity atoms. Schut et al. (1991) have shown that amorphous layers of silicon deposited at room temperature may develop open volume defects during subsequent solid phase epitaxial (SPE) regrowth. Far fewer defects were present in MBE (Molecular Beam Epitaxy) layers grown at 1000 K. In § 12 it is shown that layers grown by evaporation at low temperature, e.g. aluminium contacts on MOS devices, show very large amounts of defects with high $S$ values. Soininen et al. (1992) measured for metal organic vapor-phase epitaxially (MOVPE) grown GaAs layers on a silicon substrate vacancy and vacancy cluster densities at the appm level. Divacancies were detected in low temperature (230-350 °C) grown GaAs (Umlor et al. 1993). Jørgensen et al. (1997a) were able to identify gallium vacancies in ECR-MBE deposited n-type GaN layers. TiN layers deposited by hollow-arc evaporation were examined by Uhlmann et al. (1994) with the positron beam lifetime technique. The defect positron lifetime of 225 ps could be assigned to the open volume at grain boundaries. Sputtering and ion beam assisted growth will cause far fewer open-volume defects because of the extra mobility caused by the energetic ions and atoms hitting the growing layer. The effects of ion bombardment of sputtered silicon layers (Greuter 1993, Greuter et al. 1993) and the degree of open volume in ion beam assisted carbon layers (Rossi et al. 1994a, 1994b, 1995) have been examined by positron beam analysis. Results obtained by Asoka-Kumar et al. (1993) for a silicon layer deposited at low temperature are shown in fig. 9. It is evident that for a silicon MBE layer of 200 nm deposited at 220 °C and subsequently thermally annealed at 450 °C (by Rapid Thermal Annealing, RTA) defects are found with $S = 1.07\ S_{silicon}$. The defects, presumably vacancy clusters in a concentration of $10^{18}\,cm^{-3}$, are not present in samples annealed at 600 °C or samples deposited at this temperature.

## 10 Precipitates

Precipitates consisting of particles of a different phase in alloys, nitrides, carbides and oxides have been the topic of many positron annihilation lifetime studies with unmoderated (fast) positrons (Dlubek et al. 1992). An interesting aspect is the trapping of positrons at the precipitates. At incoherent or semi-incoherent precipitates the defects at the interface will trap the positrons. Coherent precipitates will only trap positrons when the positron affinity ( $A_+ = \mu_- + \mu_+$, with $\mu_-$ and $\mu_+$ the electron and positron chemical potentials, respectively) is lower in the precipitate than in the metal or when the precipitates contain defects. Affinities for a number of transition metal carbides and nitrides have been calculated by Puska et al. (1994). Positron beam studies give access to precipitates which have been formed in thin layers only. The measurement of the effective diffusion length will yield the sink strength of the precipitates.

*Nitride precipitates in iron*

Nitriding is an important industrial process to provide surfaces of steel with a wear resistant layer. Nitriding occurs by dissolving nitrogen from gas mixtures at elevated temperatures. Nucleation centres for nitriding are dissolved, strongly nitride forming, atoms like aluminium or titanium. Once a precipitate has been formed the surrounding iron might form nitrides. Defects also play a role in the nitriding process. Figure 10 gives an overview of the (*S, W*) points observed for defects in FeAl2% which was deformed (creating defects, including vacancies), nitrided, and denitrided, respectively. Details are given by Schut et al. (1997). An interesting observation is that nitrogen decoration of open volume defects leads to a lower *S* because the volume is reduced, but also to a lower *W* because positrons annihilate at the core electrons of the light nitrogen atom which are less strongly bound than the core electrons in iron. When excess nitrogen is desorbed the change is in the opposite direction.

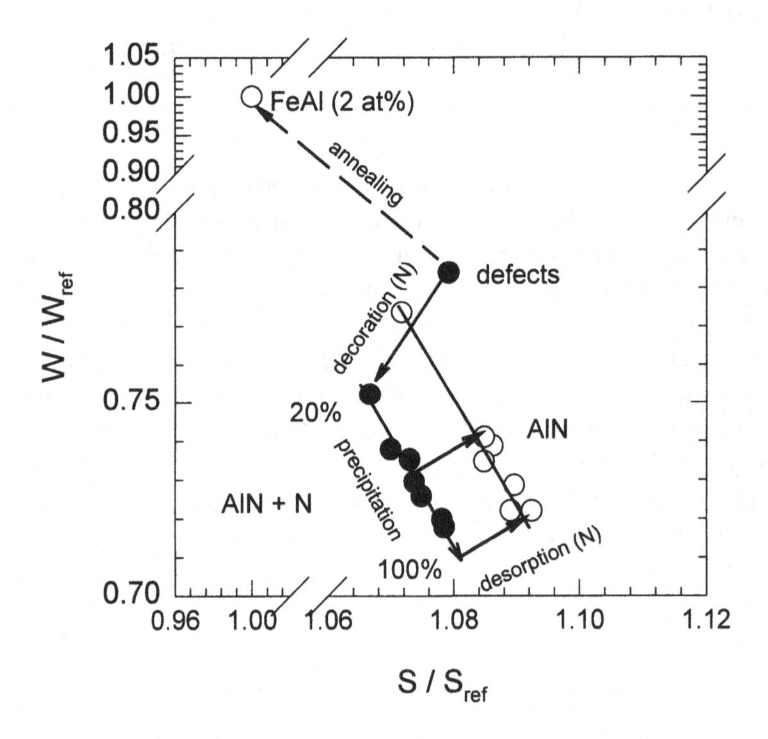


Fig. 10: S-W plot for nitride precipitates in iron. The effects of different treatments of the material are indicated. From Schut et al. (1997).

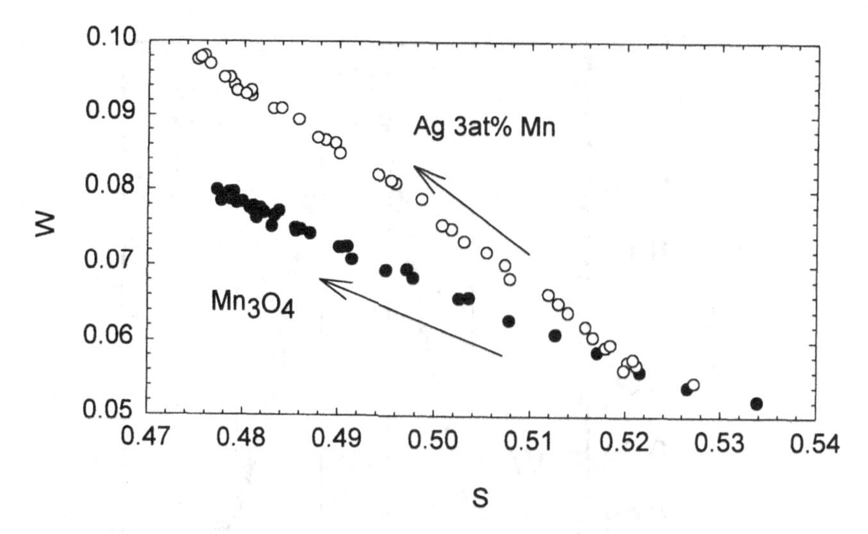


Fig. 11: S-W plots for silver with and without precipitates. The arrows indicate increasing positron implantation energy which varied from 0.05 to 30 keV. From van Veen et al. (1997).

*Oxide precipitates in silver*

Experiments are conducted on $Mn_3O_4$ precipitates in silver obtained by internal oxidation of manganese in a 3at% Mn solid solution in Ag. For a general overview of internal oxidation and metal-oxide interfaces see Ernst (1995). Figure 11 shows the $S$ - $W$ plots for Ag with $Mn_3O_4$ precipitates and for a Ag 3at% Mn solid solution annealed in vacuum. The trajectories with the energy (or positron implantation depth) as running parameter, start in an equal fashion in the surface layer but then divide into two straight branches which end at different cluster points: one for the alloy and one for the precipitates. It is of interest to note that the internal oxidation causes a relatively small change in $S$, but a considerable change in $W$. Apparently, the open volume available to the trapped positron is small, but the chemical environment of the positron has changed, presumably by the presence of the oxygen. The measured diffusion length was reduced from 109 to 30 nm by the internal oxidation. From this, one derives a trapping rate of $1.2\times10^{11}$ $s^{-1}$. Taking the precipitate density derived from HREM observations ($1.2\times10^{-7}$ at.fr.) and the diffusion limited specific trapping rate for precipitates of this size ($10^{18}$ $s^{-1}$), the predicted trapping rate is similar to the one found experimentally. Thus, the whole interface of the precipitate participates in the trapping process. However, it is also possible that positrons trapped at the interface reach the traps associated with the misfit ledges by interface diffusion. Results

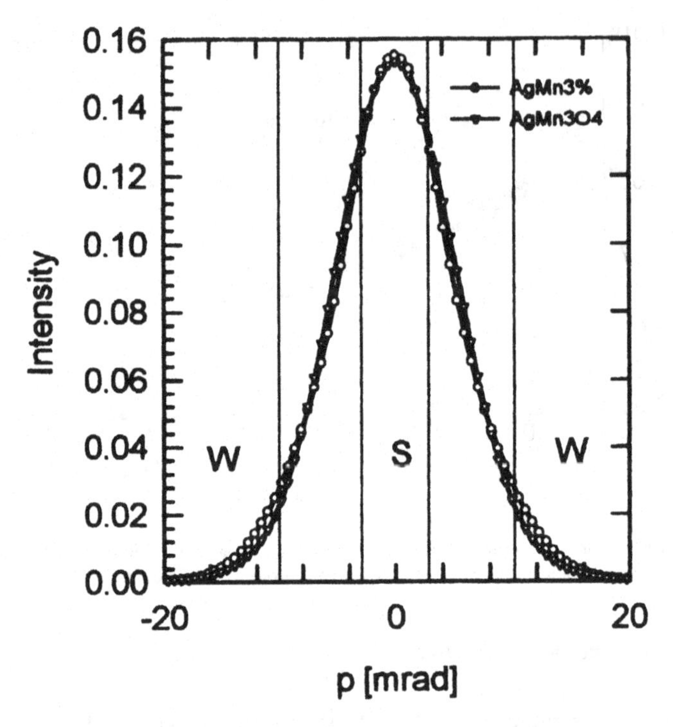


Fig 12. Momentum distributions for the AgMn alloy, vacuum annealed and internally oxidized (two-detector measurement). From van Veen et al. (1997).

of two-detector measurements are shown in fig. 12. The results for the sample with $Mn_3O_4$ precipitates show clearly why $S$ is compensated and shows little change, whereas $W$ changes considerably upon oxidation.

## 11 Planar interfaces

Positron beam analysis used with beam energies up to 30 keV can reach depths of about 2 μm in silicon and about 700 nm in many metals. This depth of analysis is sufficient for most of the thin films that are used in the semiconductor industry, while for metal coatings the thickness can be adapted to allow the coating film substrate interface to be monitored.

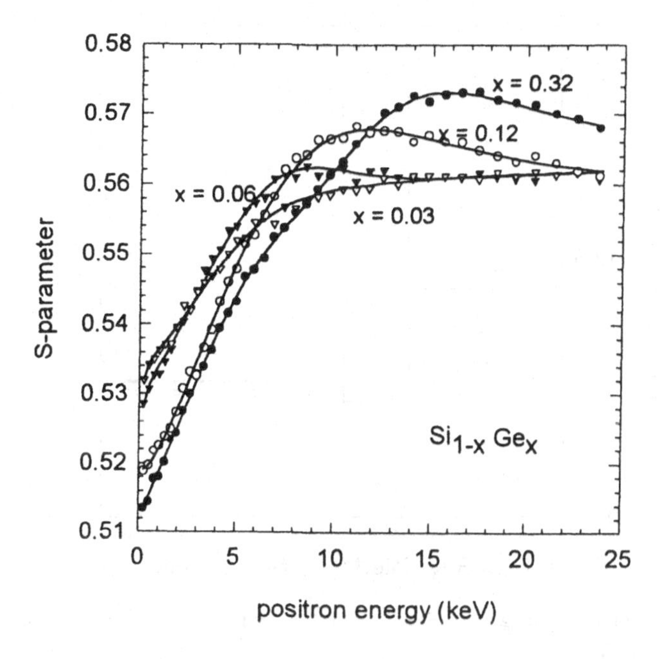


Fig. 13: S vs E curves for GeSi/Si heterostructures of different thickness.

In a thin film substrate system positrons can be used to study adhesion, i.e., the presence and development of defects at the surface, the effects of strain and strain relaxation in the thin film and the creation of defects in the substrate. In GeSi/Si heterostructures effects of strain and strain relaxation were found by van Veen et al. (1995) and by Baker and Coleman (1989). In particular, the relaxation of the system which caused misfit dislocations to be formed at the interface was accompanied by a considerable increase in the interface S value indicating cavity formation. Data obtained for this system are shown in fig.13. The results of the analysis, performed with the aid of VEPFIT, are shown in figs. 14a and b. It is observed that for $x > 0.12$ defects with high $S$ values, caused by strain relaxation, develop at the interface. In a metal ceramic system produced by MBE deposition of molybdenum on a MgO single crystal indications are found for the presence of cavities at the interface. Due to the misfit of both lattices a dislocation network will be formed to accommodate the misfit. The dislocations will not end only at the interface, leaving a core region with open volume, but also stand-off dislocations will be formed which give rise to an even larger core region. It is likely that for this system the positron affinity is lower than in the metal. However, the results do not provide evidence for this, probably because of the strong trapping at the interface.

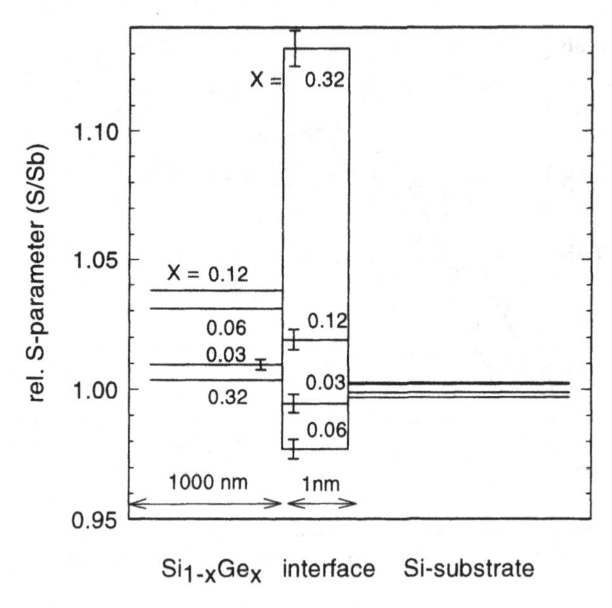


Fig. 14a: Results of model fitting with VEPFIT for the GeSi/Si layered structure of fig. 13. Relative S parameter vs depth.

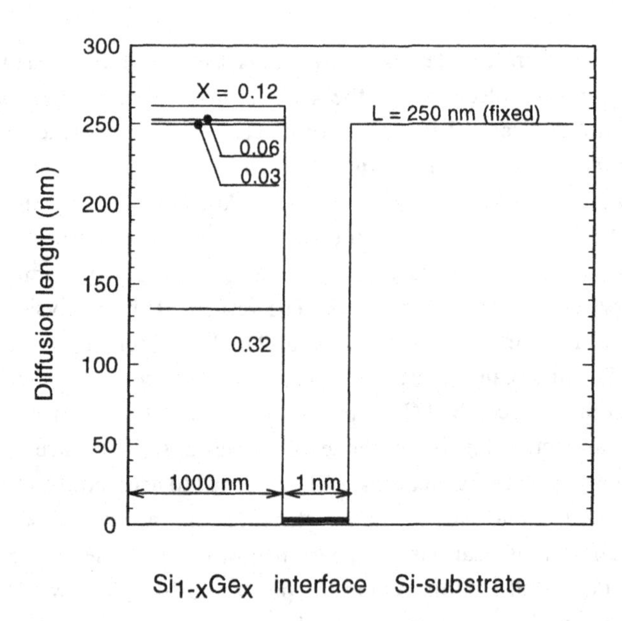


Fig. 14b: As fig. 14a, but showing the diffusion length vs depth.

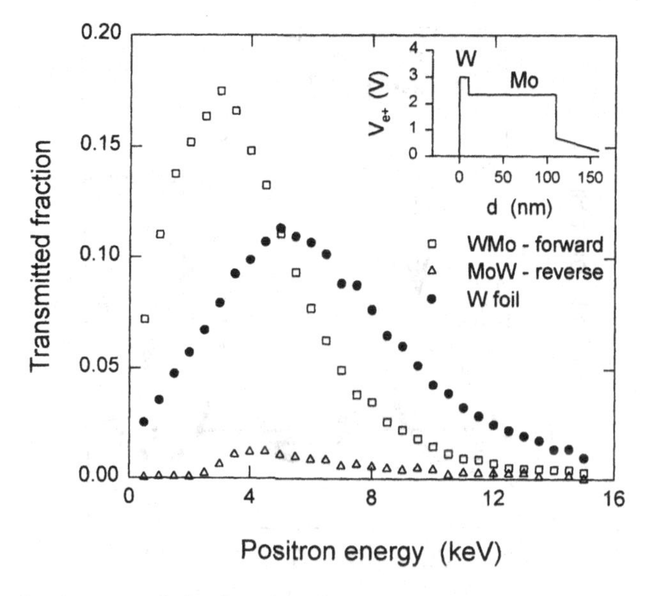


Fig. 15: Measured positron transmission through a foil consisting of W on Mo as a function of incident positron energy in the rectifying direction (positrons incident on the W side) and in the opposite direction. Also shown are transmission data for a 100 nm W foil. The insert illustrates the potentials involved as experienced by the positrons. From Jørgensen (1997b).

Recently, it was shown for metal-metal systems that when no trapping occurs the metals match perfectly (Jørgensen et al. 1997b). The positrons then experience only the energy difference between the two metals caused by the different positron affinities. The experiments were performed on a MoW thin film single crystalline assembly composed of 10 nm W on 100 nm Mo, both with (100) orientation. Figure 15 shows the positron transmission as a function of the energy in two directions. The results demonstrate that transmission in the direction from W to Mo is much higher than in the reverse direction. This so-called rectifying effect is ascribed to the internal potential barrier of about 0.6 eV that exists at the interface between Mo and W. For a similar sandwich of 10 nm Re evaporated on a W (110) foil the rectifying effect was not observed. The discovery of the rectifying effect is in fact based on pioneering work of Gidley and Frieze (1988) who performed systematic studies of the effect of thin metal overlayers (~1 nm) on the energy distribution of reemitted positrons.

Positron beam analysis has been used extensively to study $SiO_2$ layers on silicon; see the review by Asoka-Kumar et al. (1994). Recently, much attention is

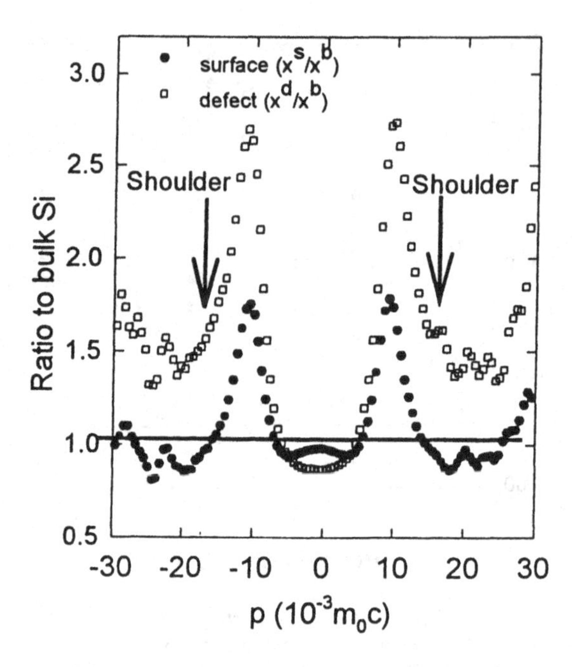


Fig. 16: Ratio curves of the defect and the surface contribution to bulk Si (from Kruseman et al. 1997).

devoted to defects at the $Si/SiO_2$ interface in MOS systems and in the buried oxide (BOX) layer of SIMOX material. SIMOX (separation by implanted oxygen) is produced by oxygen ion bombardment, typically 180 keV, followed by annealing at a very high temperature (1350 °C). Although cross-sectional TEM observations do not reveal any defects in the Si overlayer, positrons are able to detect small oxygen vacancy complexes which give rise to a low value of *S*. At the $Si/SiO_2$ interfaces defects are detected which have similar low *S* values. Kruseman et al. (1997) have studied the defects by performing two-detector measurements in SIMOX and in the MOS system. A typical result is shown in fig. 16. It appears that at a momentum of 15 mrad there is an extra contribution in the momentum distribution when the defects are present. This extra contribution causes a shoulder in the ratio curve obtained by dividing the observed Doppler broadening curve by a similar curve measured in a defect-free Si sample. The extra contribution is ascribed to overlap with oxygen electrons. Peng et al. (1996) employed a positron beam 2D-ACAR system to study the interface defects in the MOS-system. By using the variable positron energy to 'focus' on different depths they obtained four ACAR distributions which contained different percentages of the contributions from the surface, oxide, interface and bulk regions, respectively. This allowed them to separate these contributions, from which they infer positron trapping and annihilation at $P_b$ centres (dangling Si bonds) in the interface. Kauppinen et al. (1997) used the *S-W* method to investigate 38 $SiO_2$/Si interfaces formed in bulk or epitaxial Si covered by thin 10-50 Å oxide layers obtained by various processes. They found evidence that the positron sees the same interface state in all samples and that high concentrations of divacancies exist in Si domains at native interfaces.

## 12 Electric field effects

In semiconductor and insulating layers electric fields may be present due to surface or interface charges and charging of the bulk of the insulator (Coleman et al. 1990). In addition, electric fields may also be applied externally. It has experimentally been found that implanted and moderated positrons could be transported over much longer distances than diffusion would normally allow. Schultz and Lynn (1988) and Mäkinen et al. (1991) demonstrated that by applying electric fields in Si positrons could drift over distances larger than 1 μm. There are ongoing efforts to use electric fields to build efficient field assisted moderators (Beling et al. 1987, Brandes et al. 1997). Asoka-Kumar et al. (1994) and Uedono et al. (1988) have demonstrated that positrons implanted into an activated MOS system respond clearly to the applied bias

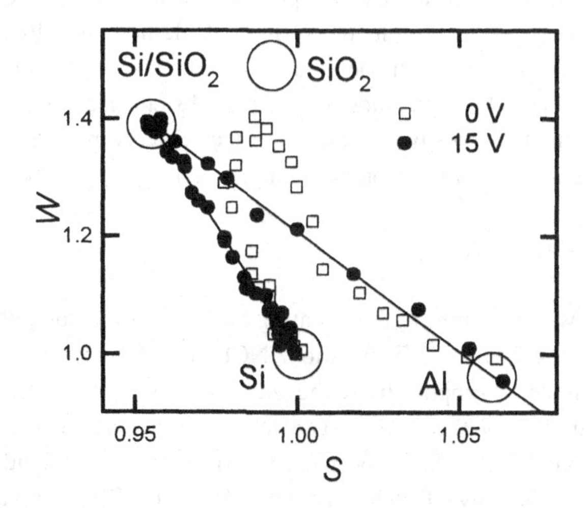


Fig. 17: S-W trajectories for the MOS-system of fig. 2, but with applied bias.

voltage. Clement et al. (1996,1997) explain the behaviour of positrons in an electrically activated MOS system by assuming that positrons in $SiO_2$ under the influence of fields of the order of MeV/cm are not trapped in the oxide but are transported over distances much larger than 100 nm. This constituted an experimental confirmation of the prediction by Simpson et al. (1987) that positrons in $SiO_2$ hardly cool down to zero energy and that non-thermalized positrons can easily be transported by applying an electric field. The oxide gate thickness in these experiments is 100 nm or smaller so that no trapping is observed in the gate oxide, but only at the interfaces of the oxide or in the metal gate or silicon substrate. With a positive gate voltage nearly all positrons implanted in the oxide could be collected at the interface to probe the defects present at the interface. For very thin oxides with a polysilicon gate it was shown that also positrons from the polysilicon gate could be collected at the interface (De Nijs and Clement 1998). In fig. 17 the effect of the fields is shown for both types of MOS-systems. This technique offers new opportunities for studying defects at the $Si/SiO_2$ interface of technologically important MOS systems with thin oxide layers. An example of the effect of post oxidation annealing on the interface defects at the $Si/SiO_2$ interface is given by Clement et al. (1995).

## 13 Summary

Techniques, including new developments, have been described which are used in positron beam analysis. Positron beam analysis has in the past relied mostly on the observation of the shape parameter of the Doppler broadened annihilation peak. It has now evolved into a technique employing nearly all methods developed in positron annihilation spectroscopy with unmoderated positrons. The control over the implantation depth and the possibility to guide the positrons by electric fields to interfaces in the material have made it possible to study interface defects in thin film structures. Examples have been given of positron beam applications for a number of systems.

## 14 Acknowledgements

This work was sponsored by the Stichting Nationale Supercomputer-faciliteiten (National Computing Facilities Foundation, NCF) for the use of supercomputer-facilities, with financial support from the Nederlandse Organisatie voor Wetenschappelijk Onderzoek (Netherlands Organization for Scientific Research, NWO). These investigations have been supported in part by the Netherlands Technology Foundation (STW). The Royal Dutch Academy of Sciences (KNAW) is acknowledged for providing travel grants and one of the authors (P.E.M.) wishes to thank NATO for a travel grant.

## 15 References

- Aers, G. C., Marshall, P. A., Leung, T. C. and Goldberg, R. D., 1995, *Appl. Surf. Sci.* **85**, 196-209.
- Asoka-Kumar, P., Alatalo, M., Ghosh, V. J., Kruseman, A. C., Nielsen, B. and Lynn, K. G., 1996, *Phys. Rev. Lett.* **77**, 2097-2100.
- Asoka-Kumar, P., Gossmann, H.-J., Unterwald, F. C., Feldman, L. C., Leung, T. C., Au, H. L., Talyanski, V., Nielsen, B. and Lynn, K. G., 1993, *Phys. Rev. B* **48**, 5345-5353.
- Asoka-Kumar, P., Lynn, K. G. and Welch, D. O., 1994, *J. Appl. Phys.* **76**, 4935-4982.
- Asoka-Kumar, P. and Lynn, K. G., 1995, *J. Physique IV, Coll. C1, Suppl. JP III*, **5**, 15-25.
- Baker, J. A. and Coleman, P. G., 1989, in: Positron Annihilation ICPA-8, edited by. L. Dorikens-vanPraet, M. Dorikens and D. Segers, (Singapore: World Scientific), p. 339-341.
- Beling, C. D., Simpson, R. I., Charlton, M., Jacobsen, F. M., Griffith, T. C., Moriarty, P. and Fung, S., 1987, *Appl. Phys. A* **42**, 111.
- Berko, S. and Plaskett, J. S., 1958, *Phys. Rev.* **112**, 1877-1887.
- Berko, S., Haghgooie, M. and Mader, J. J., 1977, *Phys. Lett.* **63A**, 335-338.
- Brandes, G. R., Canter, K., Krupyshev, A., Xie, R. and Mills, A. P. Jr.,1997, *Mat. Sci. Forum* **255-257**, 653-655.
- Britton, D. T. and Störmer, J., 1995, *Appl. Surf. Phys.* **85**, 1-7.
- Brusa, R. S., Duarte Naia, M., Zecca, A., Nobili, C., Ottaviani, G., Tonini, R. and Dupasquier, A., 1994, *Phys. Rev. B* **49**, 7271-7280.
- Chen, D. M., Berko, S., Canter, K. F., Lynn, K. G., Mills, Jr., A. P., Roellig, L. O., Sferlazzo, P., Weinert, M. and West, R. N., 1987, *Phys. Rev. Lett.* **58**, 921-924.
- Chen, D. M., Berko, S., Canter, K. F., Lynn, K. G., Mills, Jr., A. P., Roellig, L. O., Sferlazzo, P., Weinert, M. and West, R. N., 1989, *Phys. Rev. B* **39**, 3966-3989.
- Clement, M., de Nijs, J. M. M., van Veen, A., Schut, H. and Balk, P., 1995, *IEEE Trans. Nucl. Sci.* **NS-42**, 1717-1724.
- Clement, M., de Nijs, J. M. M., Balk, P., Schut, H. and van Veen, A., 1996, *J. Appl. Phys.* **79**, 9029-9036.
- Clement, M., de Nijs, J. M. M., Balk, P., Schut, H. and van Veen, A., 1997, *J. Appl. Phys.* **81**, 1943-1955.

- Coleman, P. G., Baker, J. A., and Chilton, N. B., 1990, *Vacuum* **41**, 1593-1594.
- De Nijs, J.M.M. and Clement, M., 1998, in: *Fundamental Aspects of Ultrathin Dielectrics on Si-based Devices*, edited by E. Garfunkel, E. Gusev and A. Vul (Dordrecht: Kluwer Academic Publishers) p. 25.
- Deutsch, M. and Dulet, E., 1951, *Phys. Rev.* **84**, 601.
- Dlubek, G., Krause, H., Krause, S., and Lademann, P., 1992, *Mat. Sci. Forum* **105-110**, 977-980.
- Dupasquier, A., 1983, in: *Positron Solid-State Physics,* edited by W. Brandt and A. Dupasquier (Amsterdam: North-Holland), p. 510-564.
- Dupasquier, A. and Ottaviani, G., 1995, in: *Positron Spectroscopy of Solids,* edited by A. Dupasquier and A. P. Mills, Jr (Amsterdam: IOS), p. 581-657.
- Ernst, F., 1995, Mat. Sci. and Engin. **R14**, 97-156 .
- Feldman, L. C. and Mayer, J. W., 1986, *Fundamentals of Surface and Thin Film Analysis,* (Amsterdam: Elsevier).
- Fujinami, F., 1996, *Phys. Rev. B* **53**, 13047-13050.
- Gidley, D. W., and Frieze, W., 1988, *Phys. Rev. Lett.* **60**, 1193-1196.
- Greuter, M. J. W., 1993, *Inert Gases in Metals and Semiconductors,* thesis, (Groningen: State University of Groningen).
- Greuter, M. J. W., Niesen, L., Hakvoort, R. A., de Roode, J., van Veen, A., Berntsen, A. J. M. and Sloof, W. G., 1993, *Hyp. Interact.* **79**, 669-674.
- Griffioen, G. C., Evans, J. H., de Jong, P. C., and van Veen, A., 1987, *Nucl. Instrum. and Meth. B* **27**, 417-420.
- Hakvoort, R. A., van Veen, A., Noordhuis, J. and de Hosson, J. Th. M., 1994, *Surface and Coating Technology* **66**, 393-397.
- Hakvoort, R. A., van Veen, A., Mijnarends, P. E. and Schut, H., 1995, *Appl. Surf. Sci.* **85**, 271-275.
- Hautojärvi, P., 1995, *J. Physique IV, Coll. C1, Suppl. JP III*, **5**, 3-14.
- Howell, R. H., Fluss, M. J., Rosenberg, I. J. and Meyer, P., 1985a, *Nucl. Instrum. and Meth. B* **10/11**, 373-377.
- Howell, R. H., Meyer, P., Rosenberg, I. J. and Fluss, M. J., 1985b, *Phys. Rev. Lett.* **54**, 1698-1701.
- Howell, R. H., Tuomisaari, M. and Jean, Y.C., 1990, *Phys. Rev. B.* **42**, 6921-6925.
- Jørgensen, L.V., van Veen, A., and Schut, H., 1996, *Nucl. Instrum. and Meth. in Phys. Res. B* **119**, 487-490.
- Jørgensen, L.V., Kruseman, A.C., Schut, H., van Veen, A., Fanciulli, M. and Moustakas, T.D.,1997a, *Mat. Res. Soc. Proc.* **449**, 853-856.
- Jørgensen, L.V., van Veen, A., Schut, H. and Chevallier, J., 1997b, *J. Appl. Phys.* **81**, 2725-2729.

- Kauppinen, H., Corbel, C., Liszkay, L., Laine, T., Oila, J., Saarinen, K., Hautojärvi, P., Barthe, M.-F. and Blondiaux, G., 1997, *J. Phys.: Condens. Matter* **9**, 10595-10601.
- Keinonen, J., Hautala, M., Rauhala, E., Karttunen, V., Kuronen, A., Räissänen, J., Lahtinen, J., Vehanen, A., Punkka, E. and Hautojärvi, P., 1988, *Phys. Rev. B* **37**, 8269-8277.
- Knights, A.P., Simpson, P.J., Allard, L.B., Brebner, J.L. and Albert, J.,1996, *J. Appl. Phys.* **79**, 9022-9028
- Kögel, G., 1996, *Appl. Phys. A* **63**, 227-235.
- Kruseman, A. C., Schut, H., Fujinami, M. and van Veen, A., 1997, *Mat. Sci. Forum* **255-257**, 793-795.
- Lynn, K. G., MacDonald, J. R., Boie, R. A., Feldman, L. C., Gabbe, J. D., Robbins, M. F., Bonderup, E. and Golovchenko, J., 1977, *Phys. Rev. Lett.* **38**, 241-244.
- Lynn, K. G. and Welch, D. O., 1980, *Phys. Rev. B* **22**, 99-110.
- Lynn, K. G., Frieze, W. E. and Schultz, P. J., 1984, *Phys. Rev. Lett.* **52**, 1137-1140.
- Lynn, K. G., Mills, Jr., A. P., West, R. N., Berko, S., Canter, K. F. and Roellig, L. O., 1985, *Phys. Rev. Lett.* **54**, 1702-1705.
- MacKenzie, I. K., 1969, *Phys. Lett.* **30 A**, 115-116.
- Mäkinen, J., Punkka, E., Vehanen, A., Hautojärvi, P., Keinonen, J., Hautala, M. and Rauhala, E., 1990, *J. Appl. Phys.* **67**, 990-995.
- Mäkinen, J., Corbel, C., Hautojärvi, P. and Mathiot, D., 1991, *Phys. Rev. B* **43**, 12114-12117.
- Myers, S. M, Follstaedt, D. M., Stein, H. J. and Wampler, W. R., 1993, *Phys. Rev. B* **47**, 13381-13394.
- Mijnarends, P. E., Kruseman, A. C., van Veen, A., Schut, H. and Bansil, A., 1998, *J. Phys.: Condens. Matter* **10**, 10383-10390.
- Mills, Jr., A. P., Pfeiffer, L. and Platzman, P. M., 1983, *Phys. Rev. Lett.* **51**, 1085-1088.
- Nielsen, B., Holland, O. W., Leung, T. C., and Lynn, K. G., 1993, *J. Appl. Phys.* **74**, 1636-1639.
- Peng, J. P., Lynn, K. G., Asoka-Kumar, P., Becker, D. P. and Harshman, D. R., 1996, *Phys. Rev. Lett.* **76**, 2157-2160.
- Puska, M. J., Šob, M., Brauer, G. and Korhonen, T., 1994, *Phys. Rev. B* **49**, 10947-10957.
- Rossi, F., André, B., van Veen, A., Mijnarends, P.E., Schut, H., Delplancke, M. P., - Gissler, W., Haupt, J., Lucazeau, G. and Abello, L., 1994a, *J. Appl. Phys.* **75**, 3121-3129.
- Rossi, F., André, B., van Veen, A., Mijnarends, P. E., Schut, H., Labohm, F., Dunlop, H., Delplancke, M. P. and Hubbard, K., 1994b, *J. Mater. Res.* **9**, 2240-2449.

- Rossi, F., André, B., van Veen, A., Mijnarends, P. E., Schut, H., Delplancke, M. P., Lucazeau, G. and Abello, L., 1995, *J. Physique IV, Coll. C1, Suppl. JP III*, **5**, 179-191.
- Schödlbauer, D., Sperr, P., Kögel, G. and Triftshäuser, W., 1988, *Nucl. Instrum. and Meth. B* **34**, 258-268.
- Schultz, P. and Lynn, K.G., 1988, *Rev. Mod. Phys.* **60**, 701-779.
- Schut, H., van Veen, A., van de Walle, G.F.A. and van Gorkum, A.A., 1991, *J. Appl. Phys.* **70**, 3003-3006.
- Schut, H., van Veen, A., Hakvoort, R. A., Greuter, M. J. W. and Niesen, L., 1994, in: *Slow Positron Beam Techniques for Solids and Surfaces*, edited by E. Ottewitte and A. H. Weiss, AIP Conf. Proc. **303**, 73-77.
- Schut, H. and van Veen, A., 1995, *Appl. Surf. Sci.* **85**, 225-228.
- Schut, H., van Veen, A., Jørgensen, L. V., van der Zwaag, S. and Geerlofs, N., 1997, *Mat. Sci. Forum* **255-257**, 427-429.
- Simpson, P.J., Vos, M., Mitchell, I.V., Wu, C. and Schultz, P.J., 1991, *Phys. Rev. B*, **44**, 12180-12188.
- Simpson, P. J., Schultz, P. J., Lee, S.-Tong, Chen, S. and Braunstein, G., 1992, *J. Appl. Phys.* **72**, 1799-1804.
- Simpson, P. J., Umlor, M. T. and Lynn, K. G., 1994, *Appl. Phys. Lett.* **65**, 52-54.
- Simpson, R.I., Beling, C.D., and Charlton, M., 1987, in: *The Physics and Technolgy of Amorphous* $SiO_2$, edited by R.A.B. Devine (New York: Plenum Press), p. 569.
- Smith, D. L., Rice-Evans, P. C., Britton, D.T., Evans, J.H. and Allen, A., 1990, *Phil. Mag. A* **61**, 839.
- Soininen, E, Mäkinen, J., Hautojärvi, P., Corbel, C., Freundlich, A. and Grenet, J.C., 1992, *Phys. Rev. B* **46**, 12394-12401.
- Suzuki, R., Mikado, T., Ohgaki, H., Chiwaki, M., Yamazaki, T. and Kobayashi, Y., 1994, in *Slow Positron Beam Techniques for Solids and Surfaces*, edited by E. Ottewitte and A. H. Weiss, AIP Conf. Proc. **303**, 526-534.
- Szpala, S., Asoka-Kumar, P., Nielsen, B., Peng, J. P., Hayakawa, S., Lynn, K. G. and Gossman, H.-J., 1996, *Phys. Rev. B* **54**, 4722-4731.
- Uedono, A., Tanigawa, S. and Ohji, Y., 1988, *Phys. Lett. A* **133** ,82-84.
- Uedono, A., Wei, L., Dosho, C., Kondo, H., Tanigawa, S. and Tamura, M., 1991, *Jap. J. Appl. Phys.* **30**, 1597-1603.
- Uhlmann, K., Härting, M. and Britton, D. T., 1994, *J. Phys.: Condens. Matter* **6**, 2943-2948.
- Umlor, M. T., Keeble, D. J., Cooke, P. W., Asoka-Kumar, P. and Lynn, K. G., 1993, *J. Electronic Mater.*, **22**, 1405-1408.
- Van Veen, A., Schut, H., de Vries, J., Hakvoort, R. A. and IJpma, M. R., 1990, AIP Conf. Proc. **218** (New York: AIP) 171-196.

- Van Veen, A., Hakvoort, R. A., Schut, H. and Mijnarends, P. E., 1995, *J. Physique IV, Coll. C1, Suppl. JP III*, **5**, 37-47.
- Van Veen, A., Kruseman, A. C., Schut, H., Mijnarends, P. E., Kooi, B. J. and de Hosson, J. Th. M., 1997, *Mat. Sci. Forum* **255-257**, 76-80.
- West, R. N., 1995, in: *Positron Spectroscopy of Solids*, edited by A. Dupasquier and A. P. Mills, Jr. (Amsterdam: IOS), p. 75-143.
- Willutzki, P., Störmer, J., Kögel, G., Sperr, P., Britton, D. T., Steindl, R. and Triftshäuser, W., 1995, *Mat. Sci. Forum* **175-178**, 237-240.
Ziegler, J.F., Biersack, J.P. and Litmark, U., 1985, *The Transport of Ions in Matter* (New York: Pergamon Press).

CHAPTER 7

# POSITRON MICROSCOPES AND MICROPROBES

P.G.COLEMAN
*Department of Physics*
*University of Bath*
*Claverton Down, Bath BA2 7AY, UK*
*E-mail: p.g.coleman@bath.ac.uk*

The considerable potential for novel surface and near-surface imaging by positron re-emission microscopy, and for extending the capability of traditional one-dimensional positron implantation studies by rastering a microbeam of controllable energy across a sample, is discussed. The advantages and limitations of the techniques are presented, as are the possibilities of future developments.

## 1 Positron microbeams

An ideal positron microbeam is parallel, of ~ μm diameter, and of useable intensity. These desirable properties are all encompassed by the *brightness B* of the beam of energy *E*, defined by Mills [1] as $S(\theta^2 d^2 E)^{-1}$, where *S* is the beam intensity (positrons $s^{-1}$), θ the angular divergence and *d* the beam diameter. To improve *B* - an important priority in positron optics - one has to overcome Liouville's theorem, which states that in the presence of conservative forces the positron beam occupies constant volume in phase space: *ie* $(\theta^2 d^2 E)^{1/2}$, and hence *B*, is conserved. This is done by repeated stages of focussing and re-moderation of the positron beam; the re-moderation process, in which incident positrons lose memory of their original paths and are re-emitted

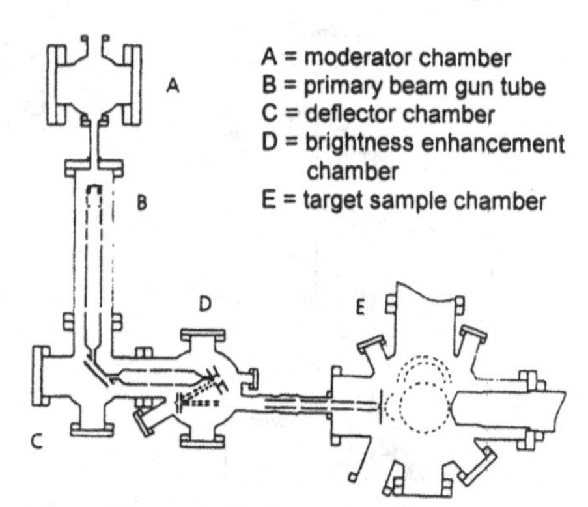


**Fig. 1.** First generation high-brightness beam at Brandeis University [2].

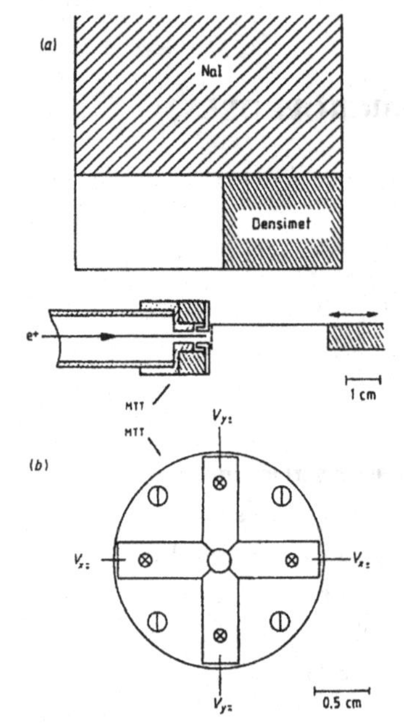


**Fig. 2** (a) Scanning positron microbeam test configuration. (b) segmented MTT lens, for rastering [4].

essentially normally to the remoderating surface, is non-conservative. Examples of brightness enhancement by repeated remoderation are given and discussed in chapter 2 and therefore will not be repeated here.

An early example of a working positron microbeam, capable of ~ 5μm resolution, was that developed at Brandeis University by Canter and co-workers [2] and shown in figure 1. The beam was used for a number of experimental studies including low-energy positron diffraction (see chapter 5). The possibilities for using such a microbeam to study (by positron annihilation) very small samples, or samples with small features of interest, was discussed by Brandes *et al* [3]. The rastering optics are shown in fig. 2.

The first microbeam at Brandeis used two-stage brightness enhancement of x500 and generated a beam of 12μm FWHM diameter and ~50 mr half-angle dispersion at 5keV [4].

In the microbeam being developed at the Interfaculty Reactor Institute at Delft, a brightness-enhanced positron beam is guided into the optics of a scanning electron microscope (figure 3) [5]. A similar approach has been taken by Greif *et al* at Bonn (figure 4) [6,7].

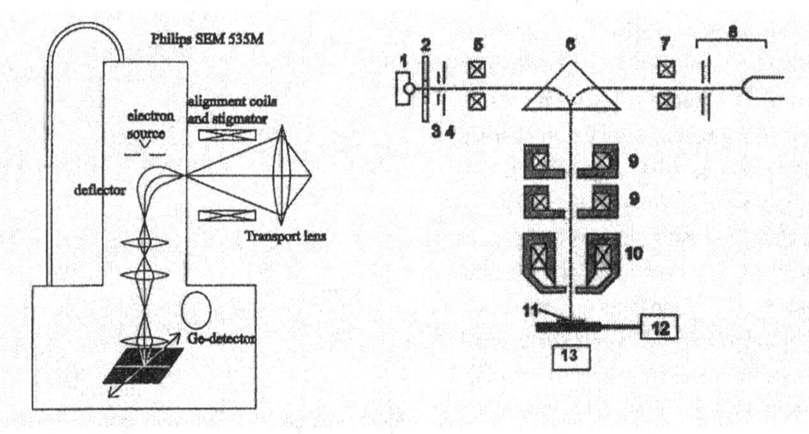


**Fig, 3.** Microbeam at TU Delft [5].

**Fig. 4.** Microprobe at Bonn [6].

The scanning, pulsed microbeam constructed at Munich [8-11] is shown in figure 5. Positrons of up to 30 keV energy in a microbeam of diameter ~ 1μm are scanned over an area 600μm square; a timing resolution of 200ps is achieved.

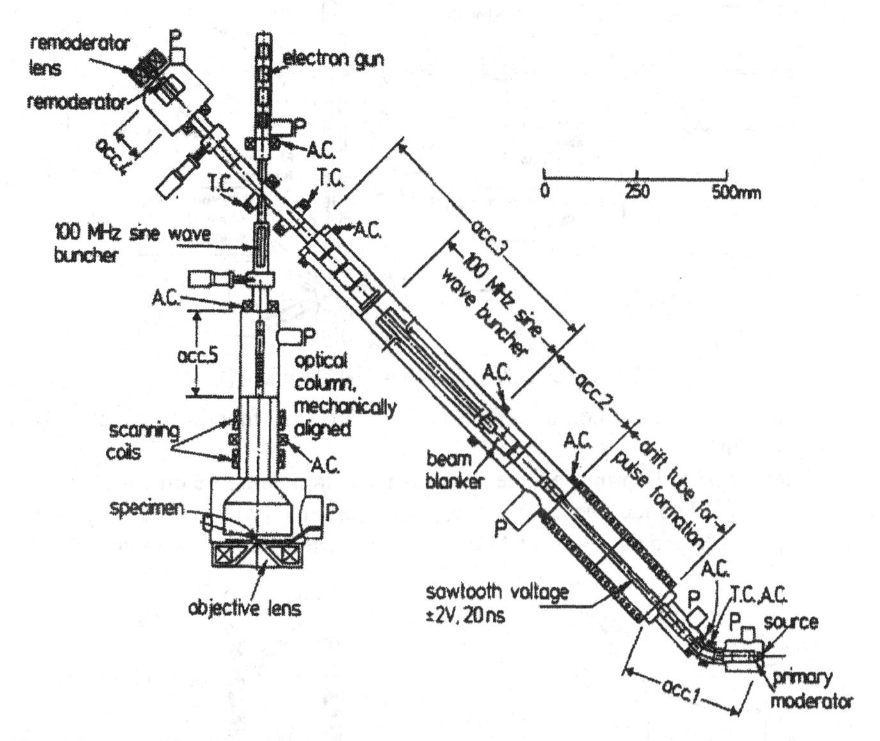


**Fig. 5.** Layout of the Scanning Positron Microscope at Munich: AC = alignment coils, TC = defelction coils, P = pumps [9].

At the Lawrence Livermore National Laboratory Howell *et al* are building a system with characteristics similar to those of the Munich microbeam [12]. High signal count rates are planned by using the intense LINAC-based slow positron source with a 20MHz positron pulsing rate and multiple radiation detectors.

## 2 Positron re-emission microscopes

The unique contrast mechanism for positron re-emission microscopy (PRM) lies in the sensitivity of the probability for positron re-emission into the vacuum to

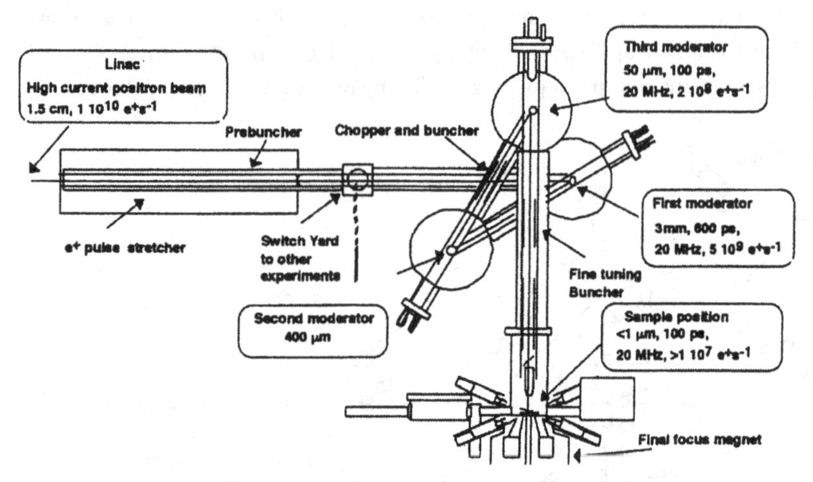


**Fig. 6.** Pulsed positron microbeam at Livermore [12].

surface conditions - i.e., the effective positron work function - which may be spatially variant.

Hulett *et al* [13] proposed the idea of a positron re-emission microscope in 1984 (figure 7). The idea became reality four years later with the development of microscopes at Brandeis [14] and Michigan [15]. The Michigan system (figure 8)

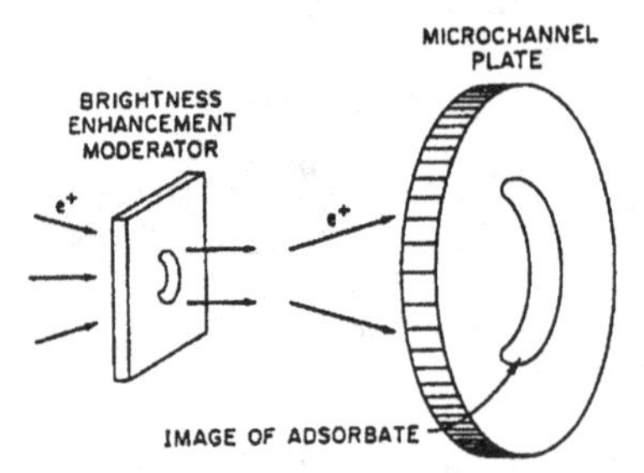


**Fig. 7.** Principle of the PRM in transmission mode, proposed by Hulett *et al* [13].

was reflection geometry; a 2.5mm diameter beam impinged upon the sample face at 50° and the re-emitted positrons were accelerated and focussed by a three-element objective lens and imaged by a projector lens on to a CEMA-phosphor screen detector with video camera. The magnification

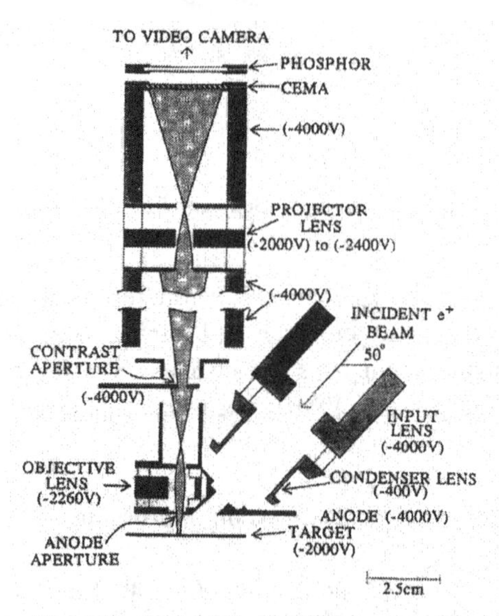


**Fig. 8.** First-generation reflection-geometry PRM at Michigan [15].

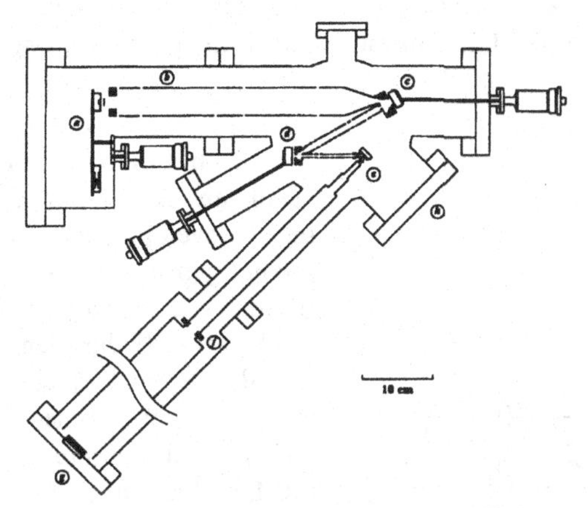

**Fig. 9.** The UEA positron microscope [17].

could be varied between x15 and x55 with a resolution of between 2.3 and 15 μm. At Brandeis a brightness-enhanced beam (figure 1) was implanted into a 150nm-thick Ni(100) foil; positrons which are thermalised and diffuse to the exit foil surface are ejected by the negative work function, accelerated, focussed and imaged. Because of the decoupling of incident and exit optics higher resolution is achievable with the transmiss-ion-geometry micro-scope; in ref. 14 images were obtain-ed with x1150 mag-nification and *c.* 300nm resolution. In later measure-ments these figures were improved to 4400x and 80nm. [16]

Goodyear and Coleman [17] have constructed a reflect-ion-geometry PRM for the study of thick samples (figure 9). The brightness-en-hanced beam, of ~ 10μm diameter,imp-inges on the sample at 45° through a micro-two-tube lens which is integrated with an immersion objective lens.

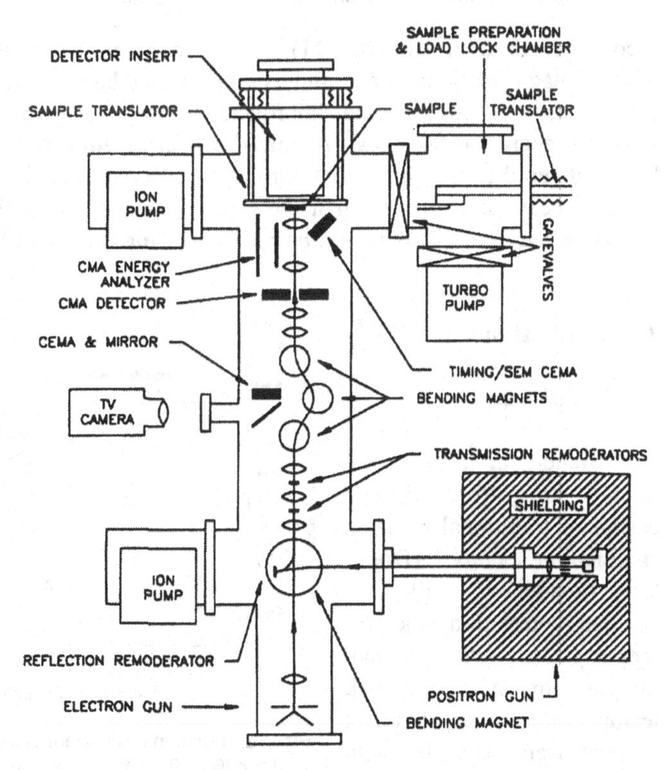


**Fig. 10.** Second-generation reflection-geometry PRM at Michigan [18].

Image acquisition is performed using a CEMA/resistive anode detector coupled to signal processing electronics and a digital framestore with image processing capability.

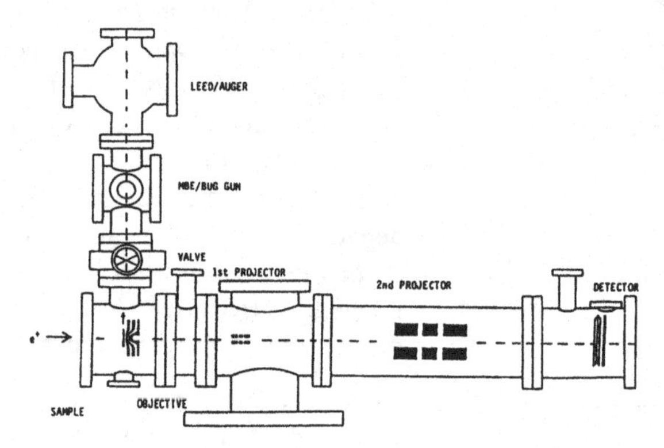


**Fig. 11.** Second-generation PRM at Brandeis [19].

Both Brandeis and Michigan have reported the development of second-generation microscope systems. At Michigan [18] the plans include normal incidence reflection-geometry optics and combined PRM and electron microscopy capabilities (fig. 10).

At Brandeis the second-generation PRM (fig. 11) has a second projector lens and can achieve 50000x magnification; a new immersion objective lens allows a 7.5 $MVm^{-1}$ field at the sample [19]. Several other improvements have been made to the optics and to sample preparation facilities. A 5μm-diameter positron beam at 4keV energy has been delivered to the sample. In order to achieve better than 10nm resolution an incident beam flux greater than the $\sim 10^6 s^{-1}$ currently available in the laboratory would be required. This would be possible at intense beam facilities (see Chapter 10).

## 3 Applications

### *3.1 Microbeams*

Brandes *et al* [4] point to the straightforward extension of traditional positron annihilation methods (Doppler broadening, lifetime, angular correlation spectroscopies) to very small samples, or to very small features on samples, by using a micro-beam. Coupled with variable implantation energy

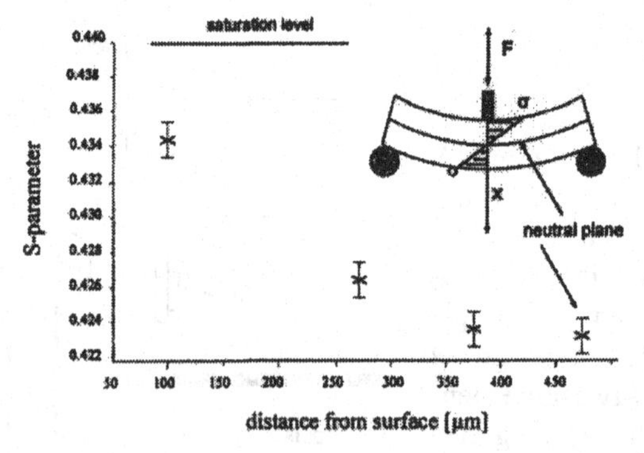


**Fig. 12.** Position-dependent *S* parameter measurement at Bonn [6,7]. The microbeam impinges at four different points across the cross-section of a bent copper plate.

capability a microbeam is able in principle to yield information on the three-dimensional distribution of sub-surface defects to depths of a few microns with a lateral resolution ~ 1μm and depth resolution of ~ 100nm; this may have particular relevance in the microelectronics industry. Greif *et al* [6,7] have measured the annihilation linewidth parameter $S$ as a function of position across a copper plate subjected to three-point bending fatigue (fig. 12). Practical application to defect identification and mapping would benefit from intense, pulsed positron beam generation, as proposed in refs. 11 and 12.

*3.2 Microscopes*

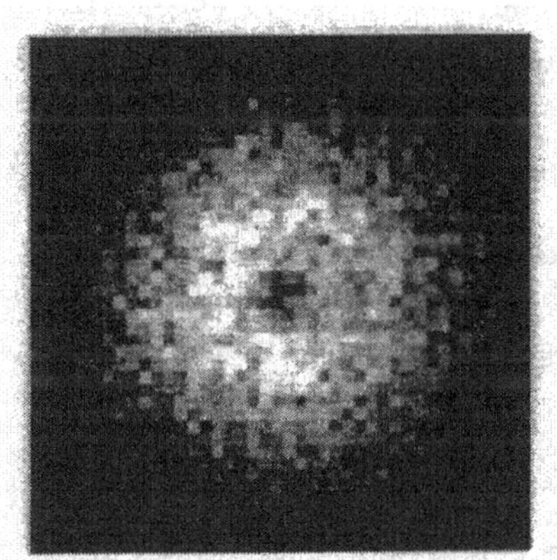

**Fig. 13.** Simulated PRM image of a near-surface monovacancy [20].

**Fig. 14.** PRM image of Ni(100) foil. Dark areas are attributed to high densities of disclocations [21].

Canter and Xie [20] state that "...It is important to emphasise that the reason for pursuing PRM studies is not just to have a novel way of seeing things that can already be seen with existing microscopies, but to be able see new things that cannot be seen in any other way." They show by modelling that single monovacancies can be imaged by a PRM with 1nm resolution, with the contrast depending on the depth of the trapping site below the surface of the sample (up to 2.5 nm) (fig. 13). Canter *et al* [21] have presented PRM images of regions of high concentrations of surface dislocations (fig. 14).

Hulett *et al* [22] discussed the used of the PRM as a 'chemical contrast microscope' - the re-emitted positron intensity at a point being determined, in the absence of defects, by the effective work-function, and hence chemical composition, of the surface at that point. The first images from the UEA PRM [23] demonstrate the contrast between re-emission from Ni and W, due both to chemical differences and the morphology of the sample surface.

Frieze *et al* [24] coined the phrase *Positron Tunnelling Microscopy* to describe the application of the PRM to the study of ultra-thin film growth on solid surfaces.

They pointed out that the re-emitted positron intensity depends on the physical and chemical nature of the film (of only a few atomic layers thickness) at each point.

Frieze *et al* [25] discuss various possible applications of positron re-emission microscopy. These include: (a) near-surface features which change the effective positron diffusion length and thus affect the probability of reaching and subsequently leaving the exit surface - e.g., open-volume and impurity-type defects, internal electric fields, and local changes in material composition : (b) surface catalysis - e.g. bimetallic systems for which the contrast in electron microscopy is small, systems in which high-atomic number substrates can swamp the signal in conventional microscopies, and the study of overlayer islanding and alloying: (c) non-destructive subsurface and surface defect mapping of semiconductor device structures (including defects at buried metal-semiconductor inter-faces): and (d) the prospect of imaging biological systems without the need for prior surface treatment such as staining, and with the low (~eV) energies of re-emitted positrons causing minimal damage to the systems under study.

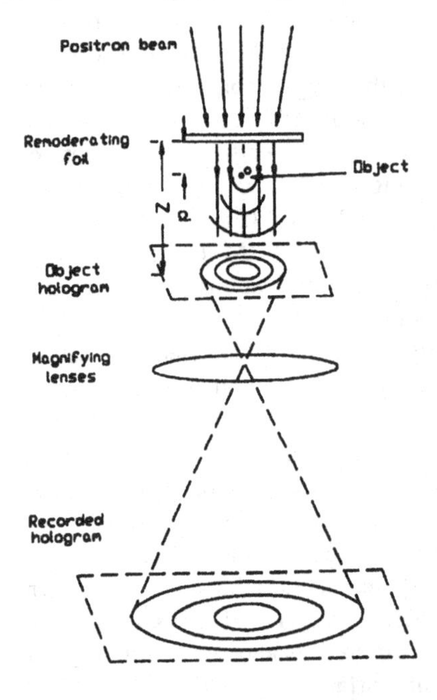


Fig. 15. Simulated positron holography experiment. Object: two atoms. *d* is the disatnce between object and moderator foil. *Z* is the distance between object hologram and moderator foil.

The resolution limit of a positron re-emission microscope is determined by the deBroglie wavelength of the positron, i.e., ~ 1nm. Information of small surface structures on a scale less than this limit is, however, achievable by holographic means. (Positron re-emitted by the surface work function are described by a plane wave state with a transverse coherence length of ~ 20nm.). Chen *et al* [26] have investigated - by computer simulation - the feasibility of performing positron re-emission holography with a PRM (fig. 15), and find that a $10^{10}$ positrons $s^{-1}$, 1μm beam would in general be required (i.e., the prospect is not out of the question). Tong *et al* have shown, also theoretically, that "...positron diffraction is better suited than electron diffraction for holographic reconstruction because of the positron's weak scattering and large damping in solids" [27].

## 4 References

1. A.P. Mills, Jr., 1980, *Appl. Phys.* **A23**, 189.
2. K.F. Canter, G.R. Brandes, T.N. Horsky and P.N. Lippel, 1987, *Atomic Physics with Positrons,* eds. J.W. Humberston and E.A.G. Armour (New York: Plenum), p. 153.
3. G.R. Brandes, K.F. Canter, T.N. Horsky, P.H. Lippel and A.P. Mills, Jr., 1988, *Rev. Sci. Instrum.* **59**, 228.
4. G.R. Brandes, K.F. Canter, T.N. Horsky, P.H. Lippel and A.P. Mills, Jr., 1989, *J.Phys.: Condens. Matter* **1** SA135.
5. L.J. Seijbel, R.F.J. Neelissen, P. Kruit, A. van Veen and H. Schut, 1995, *Appl. Surf. Sci.* **85**, 92.
6. H. Greif, M. Haaks, U. Holzworth, U. Männig, M. Tongbhoyai and K. Maier, 1997, *Mat. Sci. Forum* **255-7**, 641.
7. H. Greif, M. Haaks, U. Holzworth, U. Männig, M. Tongbhoyai, T. Wider, K. Maier, J. Bihr and B. Huber, 1997, *Appl. Phys. Lett.* **71**, 2115.
8. A. Zecca, R.S. Brusa, M.P. Duarte-Naia, G.P. Karwasz, J. Paridaens, A. Piazza, G. Kögel, P. Sperr, D.T. Britton, K. Uhlmann, P. Willutski and W. Triftshäuser, 1995, *Europhys. Lett.* **29**, 617.
9. G. Kögel and the SPM Group, 1997, *Appl. Surf. Sci.* **116**, 108.
10. A. David, G. Kögel, P. Sperr and W. Triftshäuser, 1997, *Mat. Sci,. Forum* **255-7**, 741.
11. W. Triftshäuser, G. Kögel, P. Sperr, D.T. Britton, K. Uhlmann and P. Willutski, 1997, *Nucl. Instrum. Methods* **B130**, 264.
12. R.H. Howell, W. Stoeffl, A. Kumar, P.A. Sterne, T.E. Cowan and J. Hartley, 1997, *Mat. Sci,. Forum* **255-7**, 644.
13. L.D. Hulett, Jr., J.M. Dale and S. Pendyala, 1984, *Mat. Sci. Forum* **2**, 133.
14. G.R. Brandes, K.F. Canter and A.P. Mills, Jr., 1988, *Phys. Rev. Lett.* **61**, 492.
15. J. Van House and A. Rich, 1988, *Phys. Rev. Lett.* **61**, 488.
16. G.R. Brandes, K.F. Canter and A.P. Mills, Jr., 1991, *Phys. Rev.* **B43**, 10103.
17. A. Goodyear and P.G, Coleman, 1995, *Appl. Surf. Sci.* **85**, 98.
18. D.W. Gidley. W.E. Frieze, T.L. Dull, G.B. DeMaggio, E.Y. Yu, H.C. Griffin, M. Skalsey, R.S. Vallery and B.D. Wissman, 1994, *Slow Positron Beam Techniques for Solids and Surfaces,* eds. E.H. Ottewitte and A.H. Weiss (New York: AIP Conf. Series 303) p. 391.
19. K.F. Canter, V. Dharmavaram, A.G. Smirnov, S.A. Wesley, K.H. Wong, R. Xie, G.R. Brandes and A.P. Mills, Jr., 1994, ibid, p. 385.
20. K.F. Canter and R. Xie, 1998, *Mat. Chem. Phys.* **52**, 221.

21. K.F. Canter, G. Amarendra, D. Vasumathi, S.A. Wesley, R. Xie, A.P. Mills, Jr., R.L. SAbatini and Y. Zhu, 1995, *Appl. Surf. Sci.* **85**, 339.
22. L.D. Hulett, Jr. and S. Pendyala, 1988, *Intense Positron Beams,* eds. E.H. Ottewitte and W. Kells (Singapore: World Scientific) p. 148.
23. P.G. Coleman, A. Goodyear and C.P. Burrows, 1997, *Appl. Surf. Sci.* **116**, 184.
24. W.E. Frieze, D. W. Gidley and B. D. Wissman, 1990, *Sol. State. Commun.* **74**, 1079.
25. W.E. Frieze, D. W. Gidley, A. Rich and J. Van House, 1990, *Nucl. Instrum. Methods* A**299**, 409.
26. X.M. Chen, K.F. Canter and A.P. Mills, Jr., 1991, *Appl. Phys.* A**53**, 203.
27. S.Y. Tong, H. Huang and X.Q. Guo, 1992, *Phys. Rev. Lett.* **69**, 3654.

CHAPTER 8

# MeV POSITRON BEAMS

H. STOLL
*Max-Planck-Institut für Metallforschung*
*Heisenbergstr. 1, D-70569 Stuttgart, Germany*

## 1 Introduction

In contrast to the large number of keV positron ($e^+$) beams (so-called "slow" positron beams) there are only a few relativistic positron beams with positron kinetic energies of a few MeV available. The Stuttgart beam was established early in 1987[1-3] followed by the MeV positron beam at the Brookhaven National Laboratory, USA[4]. An additional beam is in operation at the Lawrence Livermore National Laboratory[5]. Recently an MeV beam was built up in Gent, Belgium[6]. The setting-up of the Stuttgart beam is described in Sect. 2.

The incentive for setting up an MeV positron beam at Stuttgart was to take advantage of the $\beta^+\gamma$ coincidence technique for positron lifetime measurements (cf. Sect. 3). By letting positrons pass through a $\beta^+$ plastic scintillator before implantation into the sample a start signal with nearly 100% detection efficiency can be generated. For a good time resolution a thick start detector and a relativistic positron beam are necessary. Relativistic positrons also avoid jitter in the time of flight between the $e^+$ start detector and the sample. Compared with efficiencies of less than 10% for prompt $\gamma$ detectors in conventional $e^+$ lifetime measurements, the unity detection efficiency of the $\beta^+$ start detector permits better statistics and/or drastically reduced measuring time.

Time-resolved information on the evolution of positron states are obtained by measuring the individual positron lifetime (= positron age) together with the energy of one of the annihilation quanta (age-momentum correlation, AMOC) as described in Sect. 4.

Selected experiments which were performed with an MeV positron beam are described in Sect. 5.

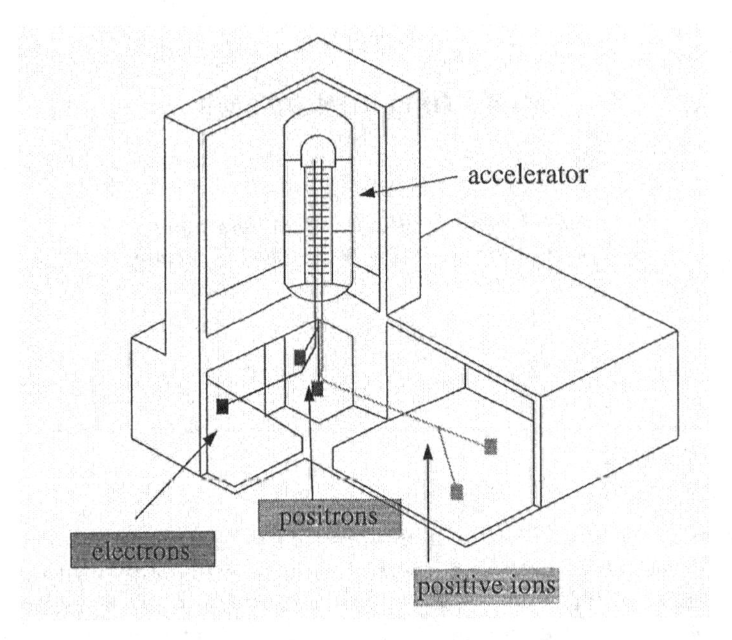


Figure 1. Perspective view of the Stuttgart accelerator building with the accelerator and the three beam line systems for positrons, positive ions, and electrons.

## 2 The Setting Up of an MeV Positron Beam

### *2.1 Accelerator*

Almost any accelerator capable of generating a few MV of positive voltage may be employed to produce an MeV positron beam. A 3 MV Dynamitron was built up at Brookhaven[4], a 3 MV Pelletron is used at Livermore[5] and a 3 MV Van De Graaff was converted to accelerate positrons at Gent[6]. The 6 MV Pelletron at Stuttgart not only accelerates positrons, but also positive ions and electrons[7] (Figure 1). For ion and electron operation the positron source in the accelerator terminal has to be pneumatically removed from the accelerator tube into a tungsten shielding box.

### *2.2 Positron Source*

The slow positron source located at the high-voltage terminal of the Stuttgart Pelletron is shown schematically in Figure 2. Positrons from a $^{22}$Na source are moderated by two moderators, by a monocrystalline tungsten foil in front of

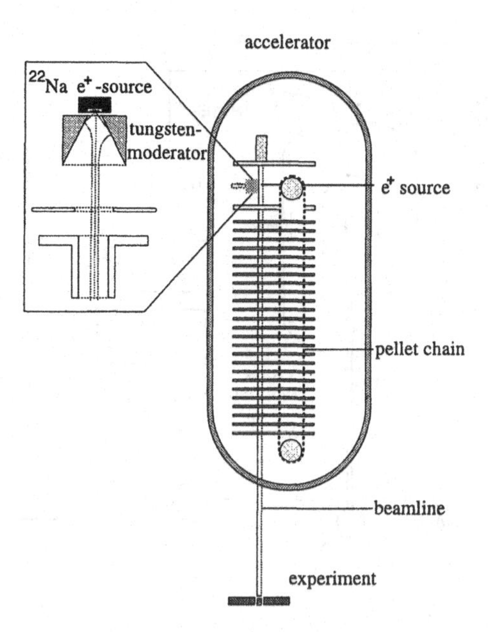


Figure 2. Schematic of the positron source located at the high-voltage terminal of the Stuttgart Pelletron.

the $^{22}$Na source in transmission mode and by a conical tungsten monocrystal in reflection mode (Figure 2). The moderated positrons are focused into the accelerator tube by means of a variable extraction voltage (0–30 kV) and two electrostatic lenses and accelerated to the desired energy. Most measurements at the Stuttgart relativistic positron beam were performed with a positron kinetic energy of 4 MeV. With the present $3.7 \cdot 10^{9}$ Bq (100 mC) $^{22}$Na source a positron beam flux of $1 \cdot 10^{6}$ $e^{+}/s$ is obtained at the target.

### *2.3 Positron Beam—Line*

The MeV positrons are directed to their target by means of magnetic steerers and lenses[2]. The shape and size of the positron beam may be monitored by scanning the beam spot over a small Si surface-barrier point detector positioned in the centre of the beam line. An easier way of determining the shape of the positron beam is to expose a film with it[8].

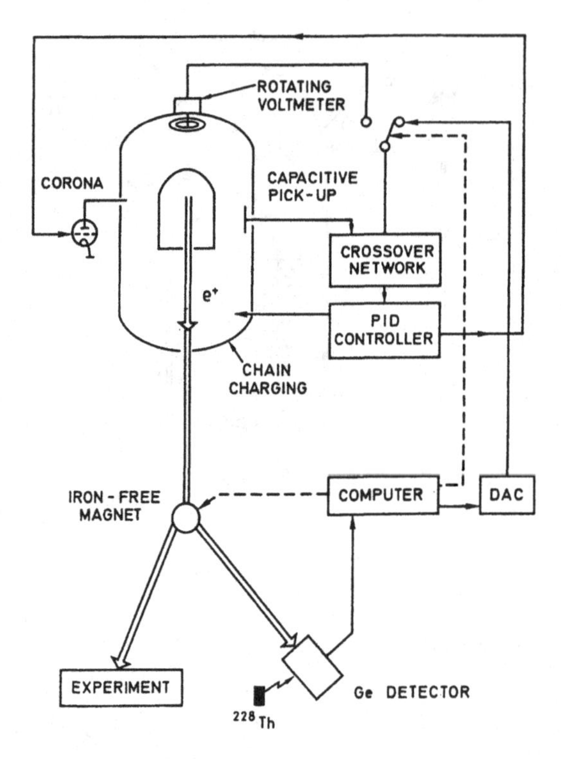


Figure 3. Block diagram of the circuit for a precise stabilization and calibration of the positron energy. The positron beam is periodically switched onto a windowless Ge detector for absolute energy measurement.

### *2.4 Energy Stabilization*

The voltage of the Stuttgart accelerator is measured by a rotating voltmeter (to detect voltage fluctuations from DC to 0.1 Hz) and a capacitive pickup capable to monitor voltage fluctuations between 0.1 Hz and about 100 Hz. The voltage is stabilized by a PID controller via corona needles and the charging of the Pelletron chains resulting in an energy stability of about $\Delta E/E \leq 10^{-3}$ (Figure 3).

In spite of the fact that for many experiments on MeV positron annihilation in condensed matter the exact positron kinetic energy is not important a much better energy stabilization and an absolute calibration of the positron kinetic energy can be obtained by switching the positron beam once a second for

0.1 s onto a windowless high-purity germanium detector by means of an iron-free deflecting magnet (cf. Figure 3). The germanium detector simultaneously monitors two $\gamma$ lines from a radioactive source. The $e^+$ and $\gamma$ spectra are analyzed using a computer. Every second the computer sends an analogous correction signal which, in addition to the high-frequency correction signal of the capacitive pick-up, serves as the input variable of a PID controller. The output of the PID module controls the corona current for fast adjustments and the charging of the Pelletron chains for low-frequency adjustments. The remaining voltage fluctuations of the accelerator are $\Delta E/E \leq 10^{-4}$ (FWHM). The absolute calibration of the positron energy with respect to the $\gamma$ calibration energies depends only on the linearity of the Ge detector system ($\Delta E_{\text{abs}} \leq 0.5\,\text{keV}$ at 2 MeV).

## 3 Beam-Based Positron Lifetime Measurements

### *3.1 $\beta^+\gamma$ Coincidence Technique*

The $\beta^+\gamma$ coincidence technique was introduced for the first time by Bell and Graham in 1953[9]. Positrons from a $^{22}$Na source were used. The broad energy distribution of the positrons available from a radioactive source causes a time of flight jitter between the $\beta^+$ start detector and the sample which could only be overcome by an energy filter. The drawback of this filter is the elimination of most of the produced positrons. Since there is no substantial advantage, e.g., in the count rate, of the $\beta^+\gamma$ technique over the $\gamma\gamma$ coincidence technique in source-based set-ups, the $\gamma\gamma$ coincidence measurements where the start signal of the positron lifetime measurement is generated by a prompt $\gamma$ quantum of the radioactive source (Figure 4) became the standard technique for positron lifetime spectroscopy for the following decades.

In 1978 Maier and Myllylä[10] took up the $\beta^+\gamma$ coincidence lifetime technique again. They concluded that this technique offers substantial advantages over the conventional $\gamma\gamma$ technique, the most striking one being the unity detection efficiency of the $\beta^+$ start detector compared to detection probabilities of less than 10% for the prompt $\gamma$ annihilation quanta used as start signal in the $\gamma\gamma$ coincidence technique. This should permit much shorter measuring times and/or better statistics. Maier and Myllylä also pointed out that the need for a positron-energy filter may be obviated by employing relativistic positrons whose speed is close to the velocity of light and therefore virtually unaffected by different initial energies or energy losses in the start detector.

Subsequent to a pilot $\beta^+\gamma$ lifetime experiment employing a $^{68}$Ge/$^{68}$Ga source[11] the work on the Stuttgart pelletron[1–3] confirmed fully the expecta-

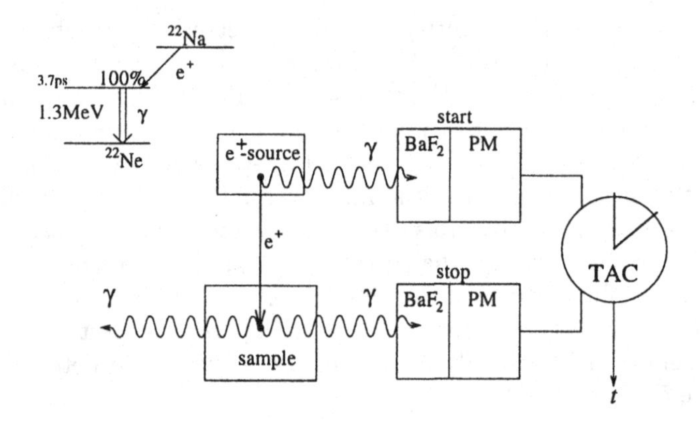


Figure 4. Schematic of conventional $\gamma\gamma$ coincidence positron lifetime spectrometer. The start signal for positron lifetime measurement is generated by a prompt $\gamma$ quantum emitted almost simultaneously with the positron by a radioactive source (e.g., $^{22}$Na). The stop signal for the lifetime measurement is triggered by one of the two annihilation quanta.

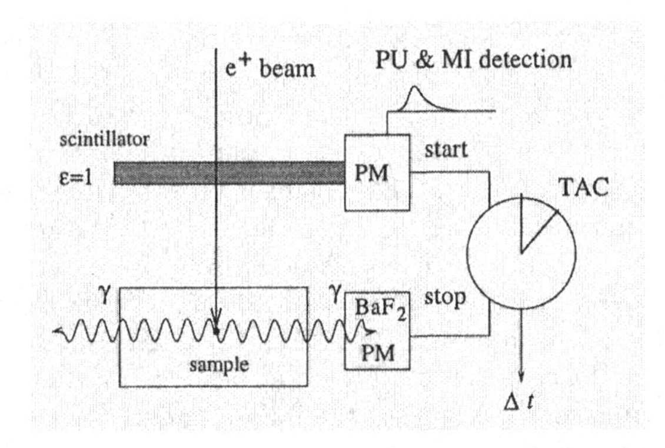


Figure 5. Schematic of a beam-based $\beta^+\gamma$ coincidence positron lifetime spectrometer. Positrons with kinetic energy of a few MeV pass a thin scintillator before implantation into the sample. In this way a start signal with virtual unity efficiency ($\varepsilon = 1$) is generated. The stop signal is generated by one of the annihilation quanta like in conventional $\gamma\gamma$ coincidence positron lifetime measurements. (PU = pile-up detection, MI = multiple impact detection, see text.)

tions of Maier and Myllylä.

The beam-based $\beta^+\gamma$ coincidence technique shown in Figure 5 as well as the beam-based $\beta^+\gamma\Delta E_\gamma$ age-momentum correlation technique described in Sect. 4 exhibit the following advantages over the conventionel (source-based)

$\gamma\gamma$ and $\gamma\gamma\Delta E_\gamma$ correlation technique:

1. The coincidence count rate is substantially increased which results in better statistics, or drastically reduced measuring times.

2. Background in the lifetime spectra caused by almost simultaneous annihilation of more than one positron in the sample can virtually be eliminated owing to the 100% detection efficiency of the start detector. These events are rejected by a pile-up (PU) detection circuit detecting overlapping start pulses and a multiple impact (MI) detection circuit (Figure 5) responding to pulses which are too close together in the time. Just this way, positron lifetime spectra can be measured with low background even at high count rates. This permits precise investigations of low-intensity long-lived components in the lifetime spectra, e.g., long-lived positron lifetime components in condensed rare gases (cf. Sect. 5.2.1).

3. The specimen preparation may be considerably simplified as MeV positrons are deeply implanted into the material. The positrons can easily be implanted into all kinds of solids or liquids, e.g., metal melts.

## 4 Age–Momentum Correlation

### *4.1 Beam-Based AMOC*

The two quantities which can be observed by annihilation of an individual positron in condensed matter are the positron age $\tau$, which is the time interval between implantation and annihilation of the positron, and the momentum $p$ of the annihilating positron-electron pair. Correlated measurements of these quantities (*A*ge-*Mo*mentum *C*orrelation, AMOC[a]) are an extremely powerful tool for the study of reactions involving positrons. It not only provides the information obtainable from the two constituent measurements but allows us to follow directly, in the time domain, changes in the $e^+e^-$ momentum distribution of a positron state (e.g., thermalization cf. Sect. 5.2.2) or transitions between different positron states (e.g., trapping of positrons or chemical reactions of positrons and positronium).

Since both 511 keV photons resulting from a $2\gamma$-annihilation event transmit equivalent information, one photon may be used to determine the age of the annihilation positron and the other one to deduce information on the momentum of the annihilating $e^+e^-$ pair, either by ACAR (Angular Correlation of Annihilation Radiation) measurement or by measurement of the

---

[a]The name age-momentum correlation appears to have been invented by MacKenzie and Mc Kee[12]. The name AMOC was introduced by Stoll, Wesołowski, Koch, Maier, Major, and Seeger[13] at the suggestion of P. Wesolowski.

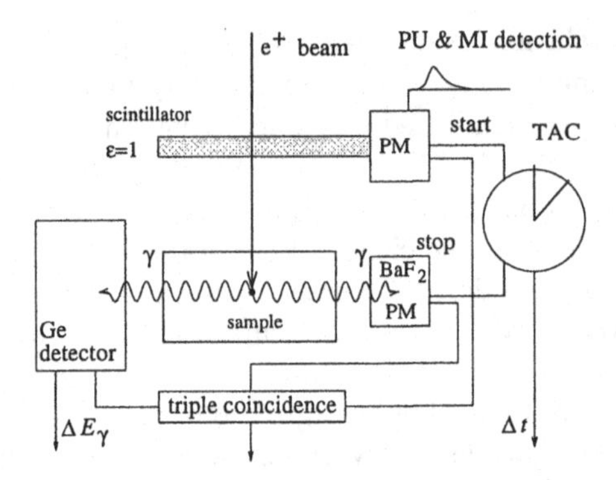


Figure 6. Schematic of the beam-based $\beta^+\gamma\Delta E_\gamma$ coincidence spectrometer. The triple coincidence triggers the data acquisition of the AMOC spectrum.

Doppler shift $\Delta E$. In correlated lifetime and ACAR measurements, however, the same $\gamma$ quanta have to be used for both, for the stop signal of the lifetime measurement and for measuring the non-collinearity of the $2\gamma$ emission in the laboratory system. This results in a poor time resolution of the combined ACAR/Stop detector. The same is true if only one single annihilation quantum is used in a combined $\gamma$ energy and stop detector. Thus, except for measurements where time resolution is not important, it is the best to use both annihilation quanta by separate positron lifetime and Doppler broadening equipment for correlated measurements of positron age and momentum.

### *4.2 The AMOC Set up and Data Acquisition*

The experimental set-up[13,14] for beam-based AMOC measurements is shown in Figure 6. An intrinsic Ge detector is added to the positron lifetime spectrometer (Figure 5) to measure the energy of the second annihilation quantum. A triple coincidence between start, stop, and energy signal triggers the data acquisition of the AMOC spectrum. Simultaneously the one-dimensional positron lifetime and Doppler broadening spectra are measured. A schematic of the electronic circuit is given in Figure 7 showing the timing-filter and spectroscopy amplifiers for the energy signal, the energy selection for the start and stop signal, and the pile-up and multiple impact rejection. The output of the triple coincidence triggers the two-dimensional AMOC data acquisition, otherwise the one-dimensional positron lifetime and Doppler broadening spectra are recorded by the multi-parameter data acquisition system.

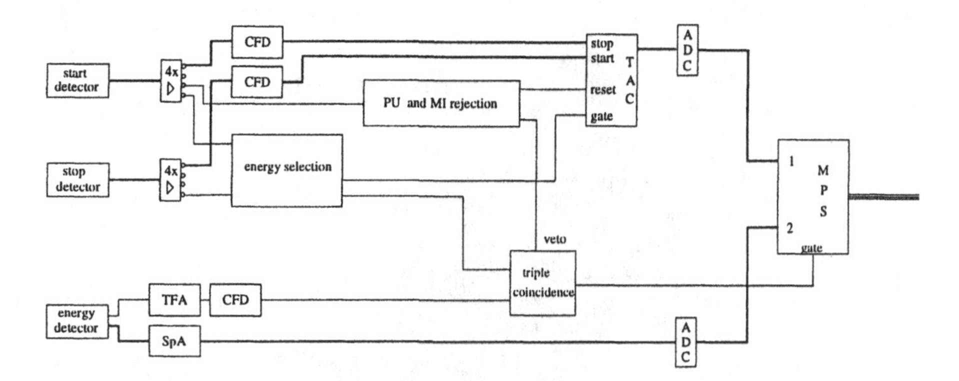


Figure 7. Schematic of the electronic circuit for combined AMOC, positron lifetime, and Doppler broadening measurements. Active power splitters are used behind the start and stop detectors. (CFD = constant fraction discriminator, ADC = analog-to-digital converter, TAC = time-to-amplitude converter, TFA = timing filter amplifier, SpA = spectroscopy amplifier, MPS = multiparameter data acquisition system.)

### *4.3 AMOC Relief, Lineshape Function, and "Tsukuba Plot"*

As an example, Figure 8 shows the so-called AMOC relief of fused quartz. The number of triple-coincidence counts is plotted on a logarithmic scale versus the positron age and the energy of one of the annihilation quanta. Sections of constant energy represent lifetime spectra at different momenta of the annihilation positron-electron pairs, sections of constant positron age represent energy spectra at different positron ages. The well-known positron lifetime spectrum or the Doppler broadening spectrum may be obtained by summation over all channels with constant energy or constant positron age, respectively (see Figure 9). For visualization of the AMOC data a time dependent $S$-parameter, the so-called lineshape function $S^t(t)$ (Figure 10a) has been found to be particularly useful especially to demonstrate changes in the population of different positron states as a function of positron age (cf. Sect. 5). Plots of the mean lifetime $\overline{\tau}$ as a function of $E_\gamma$ (Figure 10b) are another simple visualization possibility ("Tsukuba Plot")[15].

### *4.4 Two-Dimensional Data Analysis*

Full use of the two-dimensional AMOC data is made by fitting a two-dimensional model function to the AMOC relief directly. The fitting consists in developing models for the processes under investigation, solving the appropriate system of rate equations with suitable initial conditions, calculat-

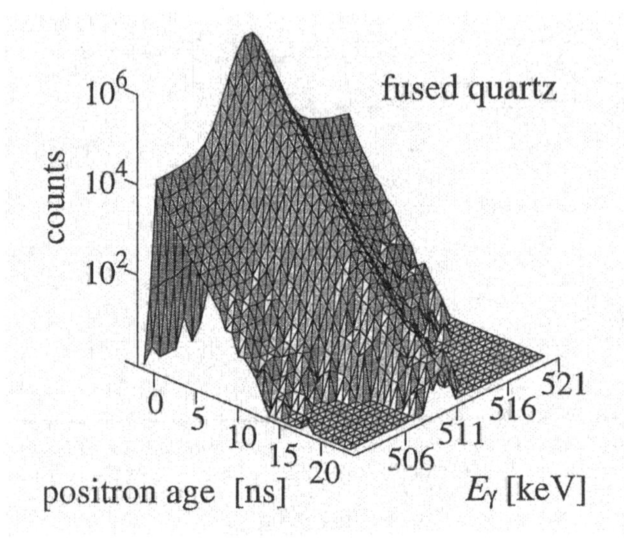


Figure 8. AMOC relief of fused quartz at room temperature.

ing the AMOC relief, and convoluting it with the time and energy resolution function of the set-up[16,17]. In this way fits to the measured AMOC reliefs can be made without prior data reduction. Parameters derivable from such two-dimensional data analysis are: (i) the annihilation rates and the Doppler broadening linewidths of all positron states involved, (ii) the transition rates between distinct positron states, and (iii) the fraction of positrons forming positronium. Additional quantities obtained from the data analysis are (iv) the time resolution of the AMOC spectrometer and (v) the background of the Doppler-broadening measurements.

## 5 Selected Experiments

### *5.1 Positron Single Quantum Annihilation and Channelling Experiments*

MeV positron beams made new experiments feasible which hardly could be performed with other techniques. Single quantum annihilation (SQA) of positrons with shell-bound atomic electrons could be studied in a number

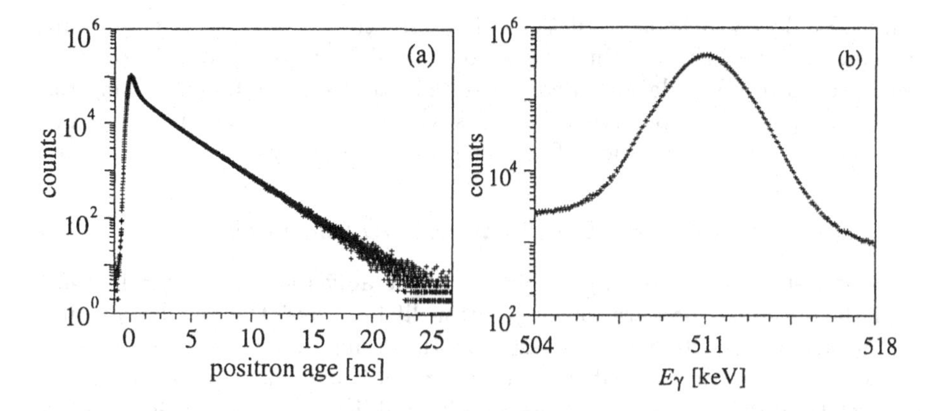


Figure 9. Positron lifetime spectrum (a) and Doppler broadening of the annihilation photon (b) in fused quartz derived from Figure 8.

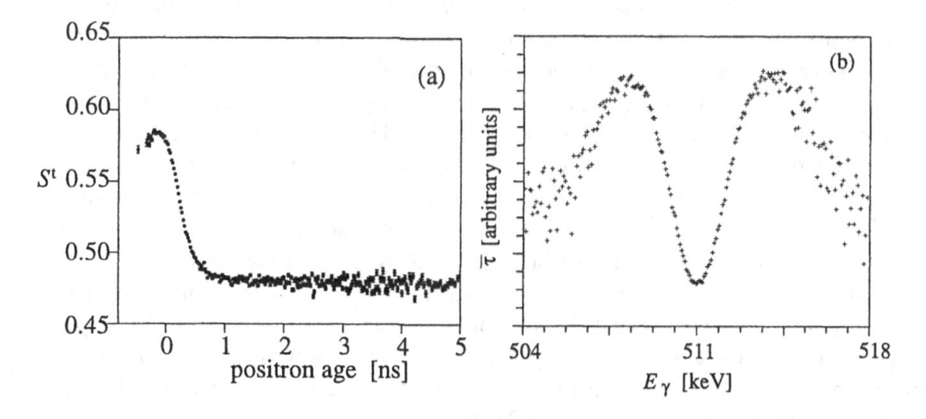


Figure 10. Lineshape function $S^t(t)$ (a) and mean lifetime $\overline{\tau}$ vs. $E_\gamma$ (b) ("Tsukuba Plot") of fused quartz derived from Figure 8.

of elements having atomic numbers between 49 and 90[18,19]. Spectra peaks corresponding to the K, L, and M shells could be resolved and observed distinctly for the first time. The dependence of the differential cross sections on the atomic number $Z$ and on the kinetic positron energy (energy range 1.02 MeV to 2.24 MeV) was measured.

High resolution angular studies of 1 MeV positrons through a single crystal of Si clearly revealed a detailed fine structure due to strong quantum channelling effects[20−22]. Building up of a MeV beam of polarized positrons was proposed. Calculations were presented showing that by combining the channelling effect for polarized positrons with the two-photon annihilation in flight the spatial distribution of spin densities in thin crystals can be probed.

### *5.2 Beam-Based Positron Lifetime and AMOC Experiments*

A relativistic positron beam simplifies considerably the experimental set-up and the specimen preparation of positron lifetime and Doppler broadening experiments, in particular in experiments involving liquids and non-metallic materials, for which the sealed-source technique[23] cannot be used. Examples are positron lifetime measurements on solid and liquid Ge and on Si up to the melting point[24,25] as well as on molten metals including undercooled metal melts[17].

In the field of positronium chemistry AMOC combines the ortho-positronium sensitive lifetime measurement with the Doppler broadening measurement which is suitable for the observation of the para-positronium state with its characteristic narrow momentum distribution. In addition AMOC allows time-domain observations of the occupations and transitions of different positron states tagged by their characteristic Doppler broadening. Chemical reactions of positronium "atoms" such as oxidation, complex formation as well as spin conversion and inhibition of positronium formation have been studied by beam-based AMOC[26−28]. The lifetimes of bound states between positrons ($e^+$) and halide ions could also be measured by beam-based AMOC[28].

AMOC measurements were performed in natural and synthetic diamonds[29−31] where the trapping of positrons at defects and the dependence of positron trapping on light and temperature was studied. In metals the range of defect concentrations which can be measured by positron lifetime experiments can be extended by AMOC measurements.

Two examples of the capability of AMOC experiments will be treated in detail: Positronium formation in condensed rare gases (Sect. 5.2.1) and thermalization of positronium (Sect. 5.2.2).

#### *5.2.1 Positronium States in Condensed Rare Gases*

The extremely low background of beam-based positron-lifetime measurements (cf. Sect. 3) facilitates the study of long-lived components of low intensities. Positron-lifetime and AMOC spectra were measured on Ne, Ar, and Kr in the liquid and in the solid states[17]. The o-Ps lifetimes of 2.4 ns in solid Ar

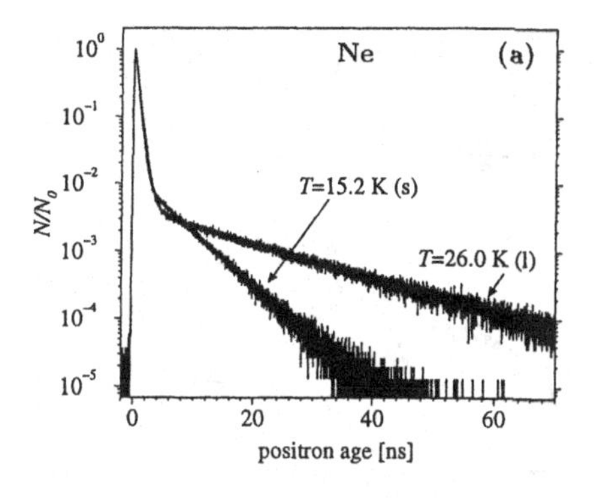


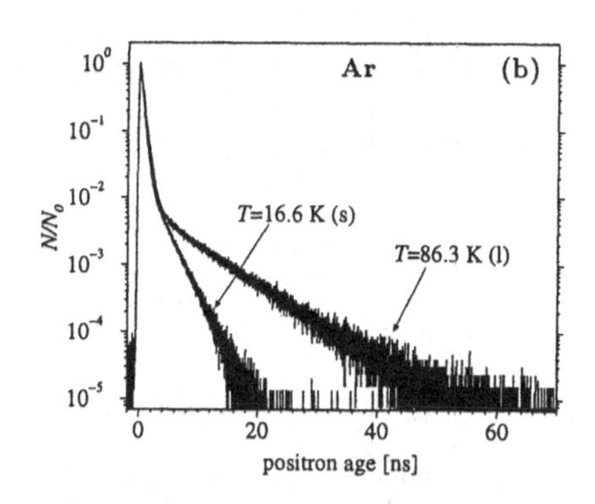


Figure 11. Positron lifetime spectra obtained on Ne (a) and Ar (b) in the solid and liquid phase. Lifetime spectra on Kr (not displayed) are very similar to that of Ar.

at 16K (Figure 11b) and of 2.1ns in solid Kr at 50K are accounted for pick-off annihilation of o-Ps. The much longer o-Ps lifetimes in the liquids are explained by o-Ps annihilation in long-lived self-localized states, the so-called positronium bubbles[32]. The fact that a long o-Ps lifetime component has been found in solid Ne (5.3ns at 15K, Figure 11a) suggests strongly that here positronium bubbles are formed, too.

The lineshape functions of Ar and Kr in the liquid state showed a surprising maximum at positron ages of about 3ns (Figure 12d and 12f). Analysis of the AMOC data obtained on liquid Ar and Kr revealed for the first time that the positronium bubbles are formed from an additional delocalized, metastable o-Ps state[17,33]. The temperature dependence of the lifetime of o-Ps in solid Ne, Ar, and Kr and of the two o-Ps states in the liquids are shown in Figures 13 and 14. Measurements on Ar and Kr were performed at a pressure of $1.2 \cdot 10^5$ Pa, Ne was measured at $1.0 \cdot 10^5$ Pa. The lifetime of the metastable o-Ps states in the liquids were found to be about the same as the o-Ps lifetimes in the solids in the vicinity of the melting point (cf. Figure 14).

The transition rate to the longer-lived and apparently more stable o-Ps bubble states in liquid Ar and Kr is about $3 \cdot 10^8 \mathrm{s}^{-1}$ [17]. A lower limit of the height of the energy barrier between the two different o-Ps states of about $10^{-1}$ eV is estimated by assuming that the barrier is overcome by overbarrier jumps with an attempt frequency of $10^{14} \mathrm{s}^{-1}$.

A bump in the lineshape function of liquid Ne may also be possibly visible at positron ages of about 4ns to 6ns (Figure 12b) indicating that a metastable Ps state may be formed in Ne, too.

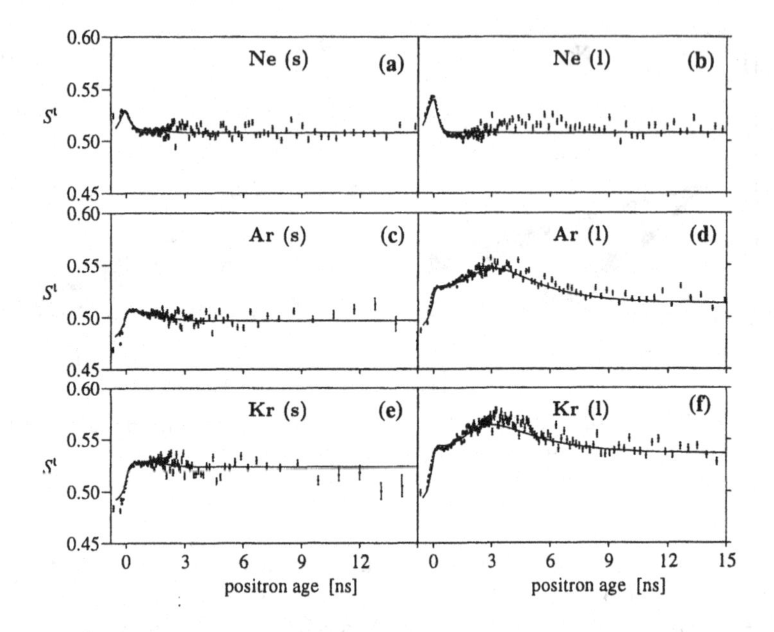


Figure 12. Lineshape functions $S^t(t)$ calculated from AMOC data and model functions (solid lines) for condensed rare gases: Solid Ne at 15.2 K (a), liquid Ne at 26.0 K (b), solid Ar at 83.3 K (c), liquid Ar at 86.3 K (d), solid Kr at 50.0 K (e), and liquid Kr at 120.0 K (f).

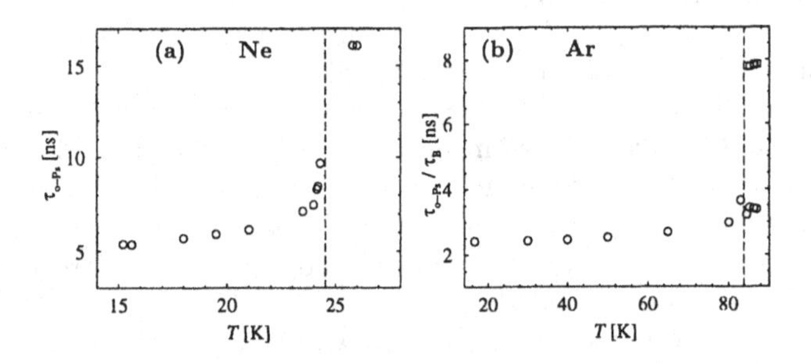


Figure 13. o-Ps lifetime $\tau_{\text{o-Ps}}$ in Ne (a) and Ar (b) as a function of temperature. In liquid Ar, the lifetime of delocalized o-Ps $\tau_{\text{o-Ps}}$ is indicated by ○ whereas □ denotes the lifetime $\tau_B$ of the bubble state. The dashed line indicates the melting temperature $T_m$. The statistical errors are smaller than the size of the symbols.

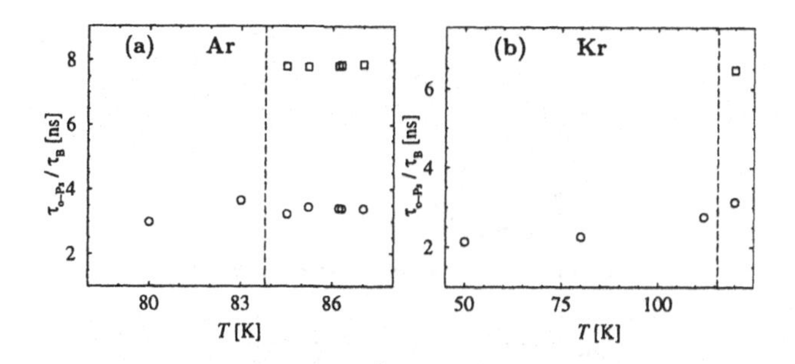


Figure 14. o-Ps lifetime in Ar (a) in the vicinity of $T_m$ (from figure 13 (b)) and in Kr as a function of temperature. The lifetime of delocalized o-Ps $\tau_{o\text{-}Ps}$ is indicated by ○ whereas □ denotes the lifetime $\tau_B$ of the bubble state in liquid Ar and Kr. The statistical errors are smaller than the size of the symbols.

### *5.2.2 Positronium Thermalization*

Beam-based AMOC allowed us to investigate Ps thermalization in condensed matter. This possibility rests on the following well-established phenomena:

(i) In condensed matter o-Ps predominantly annihilates by a pick-off process. The overlap between the $e^+$ wavefunction and that of the foreign electron results in a much shorter o-Ps lifetime ($\tau_{o\text{-}Ps}$: 1 ns to 4 ns) compared with the intrinsic annihilation of o-Ps ($\tau_{o\text{-}Ps} \approx 142$ ns) but it is still much longer than the lifetime of the intrinsic annhilation of p-Ps ($\tau_{p\text{-}Ps} \approx 125$ ps).

(ii) The momentum of thermalized Ps is very small compared with the momenta of the $e^-$ participating in the pick-off annihilation.

(iii) Non-thermalized p-Ps annihilates with the same annihilation rate as thermalized p-Ps but with a higher centre-of-mass momentum and therefore larger Doppler broadening of the annihilation line.

Items (i) to (iii) are summarized in Table 1. It is evident that correlated measurements of the positron age and the Doppler broadening allow us to

| | thermalized p-Ps | non-thermalized p-Ps | o-Ps pick-off annihilation | "free $e^+$" |
|---|---|---|---|---|
| positron age | young | young | old | intermediate |
| Doppler shift $\Delta E_\gamma$ | small | large | large | large |
| $S$-parameter | large | small | small | small |

Table 1. Annihilation characteristics of different positron states as studied in an AMOC experiment. The time spent by a positron in the sample is denoted as positron age.

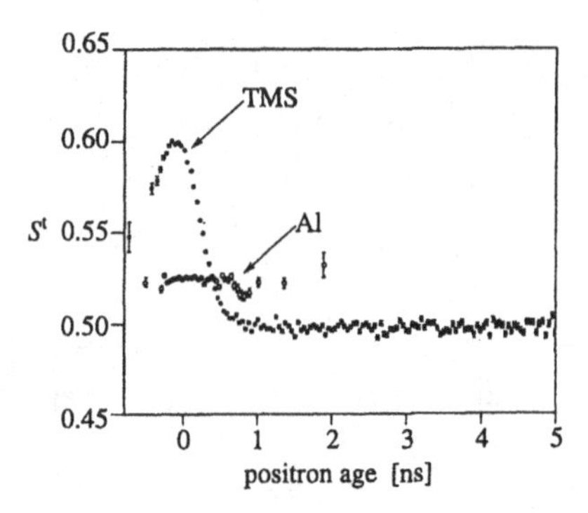


Figure 15. Lineshape functions $S^{t}(t)$ of TMS (tetramethylsilane) and Al at room temperature.

distinguish between the various annihilation modes, whereas separate measurements of Doppler broadening or of positron age distribution (positron lifetime measurements) do not allow unique decisions between the various possibilities listed in Table 1.

For all Ps-forming solids and liquids investigated so far[17,34,35], a decrease in the $S^{t}(t)$ values has been found at young positron ages as shown for TMS (tetramethylsilane) in Figure 15. This juvenile broadening has never been observed in materials for which there is no evidence for Ps formation (e.g., see Al results, Figure 15). When positrons form Ps, the creation of electron-hole pairs as an effective mechanism of losing kinetic energy ceases to operate. In materials with optical phonon branches the dominate slowing-down process for Ps is thought to be the transfer of kinetic energy to the lattice via collision with optical phonons[36,37].

The AMOC data of the juvenile broadening have been analysed in terms of a two state model[17] which approximates the Ps energy loss via interaction with phonons by allowing for transitions between epi-thermal Ps and thermalized Ps[b]. The room temperature thermalization times $t_{\text{th}}$ and the initial kinetic energies $E_0$ of Ps obtained by this procedure for materials with optical phonon branches are given in Table 2. The values obtained for $t_{\text{th}}$ have a higher reliability than those for $E_0$, which appear to be more model-dependent.

The hypothesis of Ps thermalization via optical phonons predicts that in positronium forming materials without optical phonons much longer Ps thermalization times should be found. This may be tested by AMOC measure-

[b]A better analysis of AMOC data on Ps thermalisation is suggested by Seeger[38].

ments in solid rare gases, crystallizing in the face centred cubic (fcc) structure which, being a Bravais lattice, does not have optical phonon branches. In Ar and, even more pronounced, in Kr (Figure 16), the lineshape functions $S^t(t)$ show indeed a clear shift of the juvenile Doppler broadening to higher positron ages. The narrowing annihilation of thermalized p-Ps giving a high $S^t$ value has vanished completely showing that most of the p-Ps annihilates from an epi-thermal state. This indicates Ps thermalization times in the order of the p-Ps lifetime.

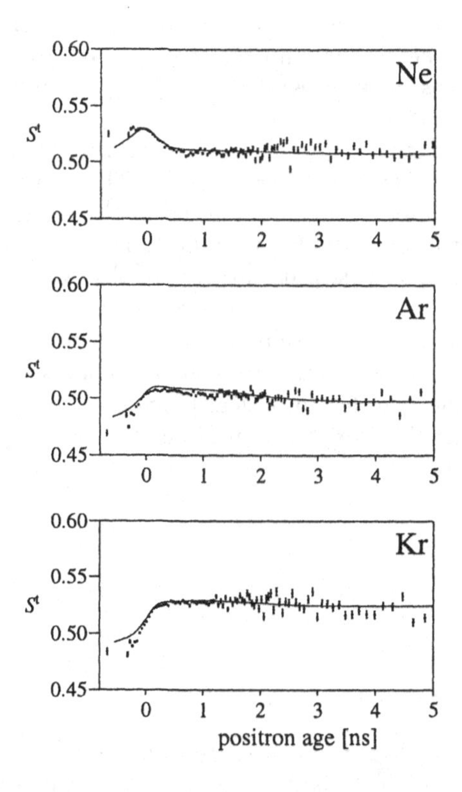


Figure 16. Lineshape functions $S^t(t)$ of solid rare gases (data points with error bars). The solid lines have been calculated from the AMOC relief data and are not fits to the $S^t(t)$ data points.

| | $E_0$ [eV] | $t_{th}$ [ps] |
|---|---|---|
| methanol | 4.3 ± 0.3 | 10 ± 3 |
| ethanol | 3.3 ± 0.3 | 14 ± 3 |
| benzene | 4.0 ± 0.2 | 25 ± 3 |
| water | 5.9 ± 0.3 | 36 ± 4 |
| TMS | 4.2 ± 0.2 | 24 ± 3 |
| polystyrene | 4.9 ± 0.3 | 24 ± 3 |
| fused quartz | 4.1 ± 0.2 | 21 ± 3 |

Table 2. Positronium thermalization times $t_{th}$ and initial positronium kinetic energies $E_0$ for various liquids and solids at room temperature (TMS = tetramethylsilane).

The analysis based on the two-state model mentioned above (solid lines in Figure 16) yields thermalization times $t_{th}$ of 125 ps to 250 ps for solid Ar, and of 400 ps to 600 ps for solid Kr[17,35]. They do not differ very much from those for the liquid states.

The lineshape function $S^t(t)$ of solid Neon (also presented in Figure 16) shows a maximum similar to TMS (Figure 15). The thermalization times $t_{th}$ according to the two state model analysis are in the range of 20 ps to 40 ps[17] and thus are similar to $t_{th}$ in materials with optical phonons (Table 2).

The rather long Ps thermalization times $t_{th}$ observed in Ar and Kr are in agreement with the idea that Ps cannot lose energy by generating electron-hole pairs and, since in the rare gases studied there are no optical phonon branches, the Ps energy can only be transferred to acoustic phonons which is significantly less effective than the transfer to optical phonons. The process responsible for the shorter thermalization times in Ne is not yet fully understood and requires further study.

## Acknowledgements

I would like to thank all of those who were involved in the Stuttgart MeV positron beam. This paper would not have been possible without the work of all of them. In particular I would like to thank Prof. A. Seeger for his continuos support, Prof. H. D. Carstanjen and Dr. J. Major for many fruitful discussions, and Dr. P. Bandžuch for his help in preparing the manuscript.

## References


1. W. Bauer, K. Maier, J. Major, H.-E. Schaefer, A. Seeger, H. D. Carstanjen, W. Decker, J. Diehl, and H. Stoll, *Appl. Phys.* B **43**, 261 (1987).
2. W. Bauer, J. Briggmann, H. D. Carstanjen, W. Decker, J. Diehl, K. Maier, J. Major, H.-E. Schaefer, A. Seeger, H. Stoll, and R. Würschum: The Stuttgart Positron Beam, its Performance and Recent Experiments, in: Positron Annihilation, eds. L. Dorikens-Vanpraet, M. Dorikens, and D. Segers (World Scientific, Singapore, 1989) p. 579.
3. W. Bauer, J. Briggmann, H. D. Carstanjen, S. H. Connell, W. Decker, J. Diehl, K. Maier, J. Major, H.-E. Schaefer, A. Seeger, H. Stoll, and E. Widmann, *Nucl. Instr. and Meth.* B **50**, 300 (1990).
4. P. Asoka-Kumar, J. S. Greenberg, S. D. Henderson, H. Huomo, M. S. Lubell, K. G. Lynn, R. Mayer, S. McCorkle, J. McDonough, J. C. Palathingal, B. F. Philips, A. Vehanen, M. Weber, and X. Y. Wu, *Nucl. Instr. and Meth.* A **337**, 3 (1993).
5. R. H. Howell, P. Asoka-Kumar, W. Stoeffl, J. Hartley, and P. Sterne, in Proc. Int. Workshop on Advanced Techniques of Positron Beam Generation and Control, RIKEN, Wako, Japan, 1998, p. 93.
6. B. Van Waeynberge, C. Dauwe, D. Segers, H. Stoll, and J. Major, submitted to Radiation Physics and Chemistry (Proceedings of the 6th Int. Workshop on Positron and Positronium Chemistry, Tsukuba, Japan, 1999).
7. H. D. Carstanjen, W. Decker, and H. Stoll, *Z. Metallkd.* **84**, 368 (1993).
8. R. H. Howell, personal communication.
9. R. E. Bell and R. L. Graham, *Phys. Rev.* **90**, 830 (1953).
10. K. Maier and R. Myllylä: Positrion Lifetime Spectrometer with $\beta^+\gamma$-Coincidences, in: Positron Annihilation, eds. R. R. Hasiguti and K. Fujiwara (Japan Institute of Metals, Sendai, 1979) p. 829.
11. W. Weiler, H.-E. Schaefer, and K. Maier, in: Positron Annihilation, eds. P. G. Coleman, S. C. Sharma, and L. M. Diana (North-Holland, Amsterdam, 1982) p. 865.
12. I. K. MacKenzie and B. T. A. McKee, *Appl. Phys.* **10**, 245 (1976).
13. H. Stoll, P. Wesołowski, M. Koch, K. Maier, J. Major, and A. Seeger, *Materials Science Forum* **105–110**, 1989 (1992).
14. H. Stoll, M. Koch, K. Maier, and J. Major, *Nucl. Instr. and Meth.* B **56–57**, 582 (1991).
15. Y. Kishimote and S. Tanigawa, in: Positron Annihilation, eds. P. G. Coleman, S. C. Sharma, and L. M. Diana (North-Holland, Amsterdam, 1982) p. 168, p. 404, p. 790, p. 815.

16. H. Schneider, A. Seeger, A. Siegle, H. Stoll, P. Castellaz, and J. Major, *Appl. Surf. Sci.* **116**, 146 (1997).
17. A. Siegle: Positronenzerstrahlung in kondensierter Materie — eine Untersuchung mit der Methode der Lebensalter–Impuls–Korrelation, Dr. rer. nat. Thesis, Universität Stuttgart, (Cuvillier, Göttingen, ISBN 3-89712-129-8, 1998).
18. J. C. Palathingal, P. Asoka-Kumar, K. G. Lynn, Y. Posada, and X. Y. Wu, *Phys. Rev. Lett.* **67**, 3491 (1991).
19. J. C. Palathingal, P. Asoka-Kumar, K. G. Lynn, and X. Y. Wu, *Phys. Rev.* A **51**, 2122 (1995).
20. J. C. Palathingal, J. P. Peng, K. G. Lynn, X. Y. Wu, and P. J. Schultz, *Materials Science Forum* **175–178**, 193 (1995).
21. R. Haakenaasen, Lene Vestergaard Hau, J. A. Golovchenko, J. C. Palathingal, J. P. Peng, P. Asoka-Kumar, and K. G. Lynn, *Phys. Rev. Lett.* **75**, 1650 (1995).
22. Lene Vestergaard Hau, J. A. Golovchenko, R. Haakenaasen, A. W. Hunt, J. P. Peng, P. Asoka-Kumar, K. G. Lynn, M. Weinert, and J. C. Palathingal, *Nucl. Instr and Meth.* B **119**, 30 (1996).
23. D. Herlach and K. Maier, *Appl. Phys.* **11**, 197 (1976).
24. R. Würschum, W. Bauer, K. Maier, J. Major, A. Seeger, H. Stoll, H. D. Carstanjen, W. Decker, J. Diehl, and H.-E. Schaefer: Positron Studies of Vacancies in Semiconductors, in: Positron Annihilation, eds. L. Dorikens-Vanpraet, M. Dorikens, and D. Segers (World Scientific, Singapore, 1989) p.671.
25. R. Würschum, W. Bauer, K. Maier, A. Seeger, and H.-E. Schaefer, *J. Phys.: Condensed Matter* **1**, Supplement A, 33 (1989).
26. H. Schneider, A. Seeger, A. Siegle, H. Stoll, I. Billard, M. Koch, U. Lauff, and J. Major, *J. Physique IV* **3**, C4 69 (1993).
27. P. Castellaz, J. Major, C. Mujica, H. Schneider, A. Seeger, A. Siegle, H. Stoll, and I. Billard, *J. Radioanal. Nucl. Chem.* **210**, 457 (1996).
28. P. Castellaz: Untersuchung von chemischen Reaktionen des Positrons und des Positroniums in Flüssigkeiten mit Hilfe der Lebensalter–Impuls–Korrelation (AMOC), Dr. rer. nat. Thesis, Universität Stuttgart, (Cuvillier, Göttingen, ISBN 3–8958–934–2, 1997).
29. H. Stoll, M. Koch, U. Lauff, K. Maier, J. Major, A. Seeger, P. Wesołowski, I. Billard, J. Ch. Abbé, G. Duplâtre, S. H. Connell, J. P. F. Sellschop, E. Sideras-Haddad, K. Bharuth-Ram, and H. Haricharun: Beam–Based Age–Momentum Correlation Studies of Positronium Spin–Conversion in Paramagnetic Solutions and of Positron Trapping at Defects in Diamonds, in: Proceedings of the Fifth International Workshop on Slow–

Positron Beam Techniques for Solids and Surfaces, eds. E. H. Ottewitte and A. Weiss (American Institute of Physics Conference Proceedings 303, New York, 1994) p. 179.

30. R. W. N. Nilen, U. Lauff, S. H. Connell, H. Stoll, A. Siegle, H. Schneider, P. Castellaz, J. Kraft, K. Bharuth-Ram, J. P. F. Sellschop, and A. Seeger, *Appl. Surf. Sci.* **116**, 198 (1997).
31. U. Lauff, R. W. N. Nilen, S. H. Connell, H. Stoll, K. Bharuth-Ram, A. Siegle, H. Schneider, P. Harmat, P. Wesołowski, J. P. F. Sellschop, and A. Seeger, *Appl. Surf. Sci.* **116**, 268 (1997).
32. R. A. Ferrell, *Phys, Rev.* **108**, 167 (1957).
33. A. Seeger: Evidence for the Existence of Two Kinds of Orthopositronium in Condensed Rare Gases and for Irreversible Transitions between them, in: Radiation Physics and Chemistry (Proc. of the 6th Int. Workshop on Positron and Positronium Chemistry, Tsukuba, Japan, 1999) in press.
34. H. Stoll, M. Koch, U. Lauff, K. Maier, J. Major, H. Schneider, A. Seeger, and A. Siegle, *Appl. Surf. Sci.* **85**, 17 (1995).
35. A. Seeger, *Materials Science Forum* **255-257**, 1 (1997).
36. A. Seeger, *Appl. Surf. Sci.* **85**, 8 (1995).
37. A. Seeger, *J. Phys.: Condens. Matter* **10**, 10461 (1998).
38. A. Seeger: Slowing-Down of Positronium: Analysis of the Age-Momentum Correlation, in: Radiation Physics and Chemistry (Proc. of the 6th Int. Workshop on Positron and Positronium Chemistry, Tsukuba, Japan, 1999) in press.

CHAPTER 9

# SPIN–POLARIZED POSITRON BEAMS IN CONDENSED–MATTER STUDIES

J. MAJOR
*Max–Planck–Institut für Metallforschung*
*Heisenbergstr. 1, D–70569 Stuttgart, Germany*
and
*Universität Stuttgart, Institut für Theoretische und Angewandte Physik*
*Pfaffenwaldring 57, D–70569 Stuttgart, Germany*

## 1 Introduction

The electron ($e^-$) and its antiparticle, the positron ($e^+$), are the lightest electrically charged members of the family of leptons, and consequently, they are stable particles. Our world, however, in particular the interior of condensed matter, consists of electrons with non-vanishing density nearly everywhere, and therefore the fate of a positron shortly after its implantation into the sample is the annihilation with an electron. The majority of experiments that are carried out with positrons, as witnessed by a long line of review articles, conference proceedings, and books (here we can cite only some of them[1–10], for a more complete list of references see, e.g., the accompanying papers in the present volume), make use of the fact that the most direct information about the electronic surroundings of the positron and about the positron itself may be obtained by detecting the time between positron implantation and annihilation or the energies or the flight directions of the annihilation photons. The corresponding techniques, the so-called positron lifetime, the Doppler broadening, and the angular correlation of the annihilation radiation (ACAR) methods, represent the basic measuring principles of positron annihilation experiments. Since, apart from special cases, positrons once implanted into matter will not come out again, as a matter of fact, there are no other quantities in connection with positron–matter interaction to be measured apart from those listed above[a].

Soon after the discovery of the positron, the existence of positronium (Ps), the bound state of a positron and an electron that resembles the hydrogen

[a] To the knowledge of the author, the polarization states of the annihilation photons have not been detected apart from the basic experiments performed in the early years (see e.g., reference[11]).

atom and the muonium (the bound state of a positive muon ($\mu^+$) and an electron) was proposed by Mohorovičić[12], and more than a decade later Deutsch[13] proved its existence experimentally. The spin quantum numbers in Ps may very strongly influence the results of positron experiments, in particular two–photon annihilation from the $1^3S_1$ positronium states (orthopositronium, o-Ps) is forbidden as a consequence of charge–conjugation symmetry[14]. Therefore, the fastest intrinsic o-Ps annihilation, i.e. the annihilation of the positron with the electron of the Ps itself, is three–photon annihilation, a relatively slow process compared to intrinsic two–photon p-Ps annihilation that is allowed in the $1^1S_0$ positronium state (parapositronium, p-Ps). The mean lifetime, the inverse of the annihilation rate in the case of a single annihilation channel, of the isolated, so-called vacuum p-Ps state is

$$\tau^0_{\mathrm{p-Ps}} = 125.1 \text{ ps} \tag{1}$$

and that of the isolated o-Ps state is

$$\tau^0_{\mathrm{o-Ps}} = 142.0 \text{ ns}. \tag{2}$$

The difference of three orders of magnitude between the so-called intrinsic lifetimes of the singlet (p-Ps) and the triplet (o-Ps) positronium states is probably the most striking signature of the spin quantum numbers.

The main field of positron and positronium research is the study of interactions of these species with the matter into which the positron is implanted. Due to the omnipresence of electrons, the long intrinsic lifetime of $\tau^0_{\mathrm{o-Ps}} = 142$ ns (Eq. (2)) of the triplet positronium state is never observed in practice. The process that most efficiently shortens the o-Ps lifetime is the so-called pick-off process, the annihilation of the positron in the positronium with an electron from the surroundings that is not bound in the positronium. 'Foreign' electrons with opposite spin compared to that of the electron of the Ps lead to the fast two–photon annihilation process in the case of o-Ps. Although this two–photon annihilation process is much faster than the three–photon process, the mean positron lifetime in this positron state, typically a few nanoseconds, is much longer than the mean lifetimes in other, e.g., free–positron or p-Ps states. However, in usual condensed matter this pick-off process exhibits the longest mean lifetime among all annihilation channels.

This long–lifetime annihilation component is an unambiguous signature of positronium in positron experiments. The identification of the presence of positronium, however, is not always a simple task by means of a traditional, radioactive–source–based positron–lifetime set-up. The main difficulty is usually the low intensity of the long–lived component that results in a conflict with the random background. This limitation may be overcome by applying an MeV positron–beam–based set-up for lifetime or age–momentum–correlation (AMOC) studies, which are described in an accompanying paper[15] in the present volume.

It is obvious that if we control the spin polarization of the positrons taking part in experiments whose results is influenced by the positron polarization, the gain in physical understanding may be enormous. Such a situation has been demonstrated in the previous example.

Since in condensed matter the majority of positrons possess thermal energy at the time of annihilation, the information that we obtain by observing the parameters of the annihilation photons describe mainly the state of the annihilation partner, the electron. E.g., from lifetime measurements we obtain the electron density at the positron site, from the Doppler and ACAR experiments we obtain the momentum distribution of the electrons taking part in the annihilation. If we perform experiments with spin–polarized positrons, we may obtain much broader information on the spin–dependent interaction of the positron with its surroundings, e.g., in magnetically ordered matter such as ferromagnets or antiferromagnets or in matter in which positronium can be formed.

The aim of the present work is to give an introduction to the use of spin–polarized positrons in condensed–matter experiments. In the next chapter the construction principles of a polarized–positron set-up are discussed. Such beams usually work with radioactive positron sources owing to the parity–non-conserving property of the $\beta$ decay, which results in longitudinally polarized positrons. In Sects. 3, 4, and 6, two practical methods for the determination of the spin polarization of positrons implanted into condensed matter are introduced. If the electrons in the sample are spin polarized, the so-called residual spin polarization of the positrons, i.e. the spin polarization of the positrons at the time of annihilation, can be determined. From such experiments the depolarization of the positrons during their diffusive motion in the sample can be deduced, the so-called depolarization function. If the electrons of the sample are not polarized but positronium may be formed in the sample, the polarization of the implanted positrons can be determined by the method of magnetic quenching of positronium hyperfine interaction. Such experiments may provide us with information on the various static and dynamic parameters of the positron or positronium species. In Sect. 5 positron–spin relaxation experiments on ferromagnetic $\alpha$ iron, in Sect. 7 results of magnetic–quenching experiments performed on fused quartz, natural diamond, silicon carbide, $C_{60}$ fullerite, and pyrolytic graphite are described. In Sect. 8, in addition to the summary, the expected future of the spin–polarized positron–beam methods is briefly discussed.

## 2 The principles of the spin–polarized positron set-up

The availability of spin–polarized positron beams is a somewhat peculiar question. The positron, as a nuclear probe, has the advantage that its detection is very easy compared to the case of other — non-nuclear — experimental techniques, e.g., of electron paramagnetic resonance (EPR) or nuclear magnetic resonance (NMR) experiments. For this reason, the intensities of the common positron sources, e.g., the directly applied radioactive sources or the beams of slow (eV to keV) or slightly relativistic (MeV) positrons, may be much lower compared to the intensities of other beams, e.g., beams of electrons or various ions that possess a typical current in the $\mu$A range. We can compare these currents with the number of positrons in positron experiments by converting, say, 1 $\mu$A to positron–beam intensity: 1 $\mu\mathrm{A} \sim 6 \cdot 10^{12}$ $\mathrm{e}^+/\mathrm{s}$, which is equivalent to the full intensity of a $6 \cdot 10^{12}$ Bq = 170 Ci radioactive source. An electron– or ion–beam current of 1 $\mu$A can be considered as the lowest limit, whereas on the other hand, a positron beam with an intensity of the order of $10^{10}$ $\mathrm{e}^+/\mathrm{s}$ or 1 Ci seems to be the ultimate goal for the positron community in the next decade. Consequently, beam polarization methods that are based on the extraction of particles with the requested spin direction and therefore accompanied by a high loss of beam intensity, are probably not applicable in positron research. Additionally, it is believed that conventional polarization filters, the prototype of which is a Stern–Gerlach magnet, do not work with free electrons or positrons[b].

Other methods suffer mainly from very low efficiencies[19] or from other technical drawbacks, e.g., self–polarization of a positron beam in a magnetic storage ring due to synchrotron radiation is an extremely slow process.

Fortunately, for positrons emitted from radioactive sources, a correlation exists between their spin and momentum at the time of emission. This phenomenon is a consequence of the weak interaction that mediates the $\beta$ decay of the radioactive nuclides. In such processes, the invariance under spatial reflection, that is the conservation of parity, is violated and, consequently, the emitted $\beta$ particle is spin polarized along its momentum direction, with other words it possesses a non-vanishing helicity. Non-conservation of parity in $\beta$-decay processes was predicted by Lee and Yang[20] in 1956 and was soon verified by Wu and co-workers[21] on the $\beta^-$ decay of $^{60}$Co. As a consequence of non-conservation of parity, positrons and electrons obtained from $\beta$ decays

[b]This was claimed by Bohr (cf. reference[16]) after a suggestion of Brillouin that the contrary may be true. Recently, the idea of Brillouin was discussed repeatedly and it was concluded that Stern–Gerlach–type experiments are feasible as electron or positron polarization filters or analysers[17,18]. The experimental situation is still unclear.

are longitudinally spin polarized in their directions of motion, independently of the polarization state of the emitter nuclides. On this basis, we can easily obtain ensembles of spin-polarized positrons by selecting the appropriate momentum without applying any spin filter. The efficiency of this selection lies in the order of unity, the corresponding intensity loss is affordable.

At the time of their forming in a $\beta^+$ decay, positrons possess a helicity of

$$\mathcal{H} = \frac{v}{c}, \tag{3}$$

where $v$ denotes the velocity of the positron and $c$ is the speed of light in vacuum. The helicity is defined as an ensemble average (denoted by $\langle ... \rangle$) according to

$$\mathcal{H} = \left\langle \frac{\boldsymbol{s} \cdot \boldsymbol{p}}{s \cdot p} \right\rangle, \tag{4}$$

where $\boldsymbol{s}$ is the spin vector and $\boldsymbol{p}$ is the momentum of the positron. If all positrons from the radioactive source take part in the experiment, the spin polarization of the positron ensemble, defined as

$$\mathcal{P} = \left\langle \frac{\boldsymbol{s} \cdot \boldsymbol{e}}{s} \right\rangle, \tag{5}$$

where $\boldsymbol{e}$ is the reference unit vector, vanishes for any $\boldsymbol{e}$. If certain positrons are excluded from the experiment, e.g., by the means of a collimator system, we may obtain a polarized ensemble of positrons. The direction of $\boldsymbol{e}$ for which $\mathcal{P}$ (cf. Eq. (5)) exhibits its maximum magnitude is called the direction of polarization, and the maximum magnitude itself is the polarization of the positron ensemble. In general, $\mathcal{P}$ can be identified as the polarization component of the positron ensemble in the $\boldsymbol{e}$ direction. The most common case is that we use only those positrons that are emitted from the source inside a cone with an opening angle of $2\alpha$. The positrons that reach the sample have a spin polarization of

$$\mathcal{P}_{\mathrm{c}} = \frac{\mathcal{H}}{2} \cdot (1 + \cos\alpha). \tag{6}$$

Since $\mathcal{H}$ depends only on the magnitude of $v$, radioactive positron sources with an emission velocity $v$ comparable to $c$ are the best candidates for the generation of a spin-polarized positron beam. In such cases the end-point energy of the $\beta^+$ decay must be around or higher then the rest energy of the positron, $mc^2 = 511$ keV ($m$ = positron mass). Table 1 lists some positron sources of practical interest together with the quantities that are significant

Table 1. List of some radioactive positron sources. Quantities that may be important for the construction of spin–polarized positron facilities are also included: $E_0$ is the end–point energy of the $\beta^+$ decay[22], $E_{\rm m}$ is the most likely $e^+$ kinetic energy[23] (which is approximately the average $e^+$ kinetic energy), $v/c$ is the helicity that corresponds to $E_{\rm m}$ (cf. Eq. (7)), $f$ is the relative positron yield of the decay[24], and $E_\gamma$ is the energy of prompt photons if they occur in practical fractions[22].

| Nuclide | Half–life | $E_0$ [MeV] | $E_{\rm m}$ [MeV] | $v/c$ | $f$ [%] | $E_\gamma$[MeV] |
|---|---|---|---|---|---|---|
| $^{11}$C | 20.4 min | 0.96 | 0.37 | 0.81 | 100 | – |
| $^{13}$N | 9.96 min | 1.20 | 0.50 | 0.86 | 100 | – |
| $^{14}$O | 70,6 s | 1.81 | | | 100 | 2.313 |
| $^{15}$O | 122 s | 1.73 | 0.65 | 0.90 | 100 | – |
| $^{17}$F | 64.5 s | 1.74 | | | 100 | – |
| $^{18}$F | 110 min | 0.633 | 0.21 | 0.71 | 97 | – |
| $^{19}$Ne | 17.3 s | 2.22 | 0.52 | 0.87 | 99 | – |
| $^{22}$Na | 2.6 y | 0.546 | 0.20 | 0.70 | 90 | 1.275 |
| $^{26}$Al | $7.4 \cdot 10^5$ y | 1.17 | | | 82 | 1.809 |
| $^{27}$Si | 4.2 s | 3.79 | | | 100 | – |
| $^{44}$Ti/$^{44}$Sc | 49 y | 1.47 | 0.45 | 0.85 | 95 | 1.157 |
| $^{48}$V | 16 d | 0.695 | 0.30 | 0.78 | 50 | 1.312, 0.984 |
| $^{57}$Ni | 36 h | 0.864 | | | 34 | 1.378 |
| $^{58}$Co | 70.8 d | 0.475 | 0.19 | 0.68 | 15 | 0.811 |
| $^{64}$Cu | 12.7 h | 0.653 | 0.27 | 0.76 | 19 | – |
| $^{68}$Ge/$^{68}$Ga | 271 d | 1.90 | 0.99 | 0.94 | 90 | – |
| $^{72}$Se/$^{72}$As | 8.4 d | 2.5 | | | 52 | 0.834 |
| $^{72}$Se/$^{72}$As | 8.4 d | 3.34 | 1.36 | 0.96 | 13 | – |
| $^{89}$Zr | 78.4 h | 0.90 | | | 22 | – |
| $\mu^+$ | 2.20 $\mu$s | 52.3 | | 1.00 | 100 | – |

for the construction of spin–polarized positron facilities. The helicities corresponding to the magnitudes of the most likely $e^+$ kinetic energies $E_{\rm m}$ were computed from the equation[25]

$$\frac{v}{c} = \sqrt{1 - \frac{1}{\left(1 + \frac{E}{mc^2}\right)^2}}\,, \tag{7}$$

where $E$ is the positron kinetic energy immediately after the $\beta^+$ decay.

The magnitudes of $E_{\rm m}$ are not known for all positron sources. However, if we plot the helicity $v/c$ listed in Table 1 vs. the end–point energy $E_0$, which is known for nearly all positron sources, we obtain an empirical plot (Figure 1) that may be useful for positron sources with unknown $E_{\rm m}$ values.

As we have seen above, the spin polarization of the positron beam may be influenced by the collimator system that excludes, roughly speaking, the positrons with momenta corresponding to 'wrong' polarization directions (cf. Eq. (6)). Obviously, the magnitude of the beam polarization and the intensity

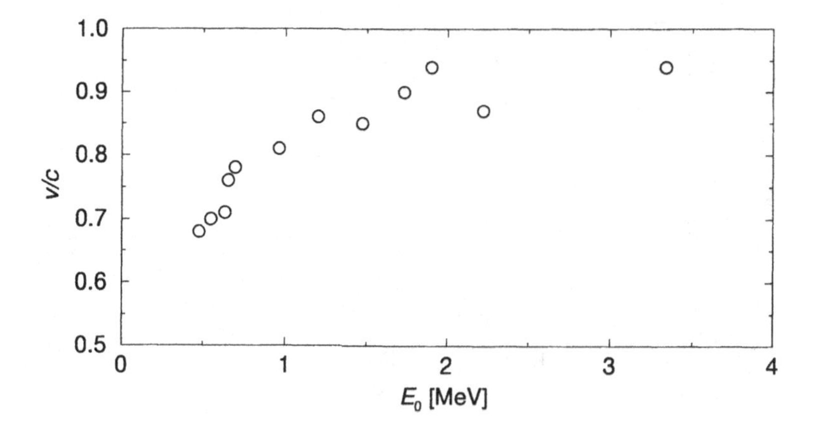


Figure 1. Helicity $v/c$ corresponding to the most likely $e^+$ kinetic energy vs. end–point positron kinetic energy $E_0$ for various radioactive positron sources (cf. Table 1).

of the beam are in conflict with each other during the construction of the collimator system. The physical effect to be determined is generally proportional to the beam polarization[c], and can be expressed in terms of mean values or variances of the detected spectra. Since the latter can be determined with relative statistical uncertainties that are proportional to $n^{-1/2}$ ($n$ = number of counts in the corresponding parts of the spectra), within a given duration of experiment we can determine the effect with the highest precision if the product of the beam polarization and the square root of the beam intensity is highest. We can name the square of this quantity,

$$q = \mathcal{P}^2 \cdot I, \tag{8}$$

the quality factor of the polarized positron source. The quantity $q$ can be used for the design of polarized–positron facilities (for a more complete discussion of the figure-of-merit for polarized sources and beams see reference[19]).

In the case of the simple conical geometry mentioned above, the polarization of the positrons that are implanted into the sample is proportional to $1 + \cos\alpha$ (cf. Eq. (6)), the intensity of the beam is proportional to the solid angle from which the positrons can reach the sample, i.e. proportional to $1 - \cos\alpha$. The quality factor, which is a function of the half opening angle

[c]In experiments with polarized particles the information is usually obtained from the relative difference or 'asymmetry' of the counting rates at opposing polarizations[19].

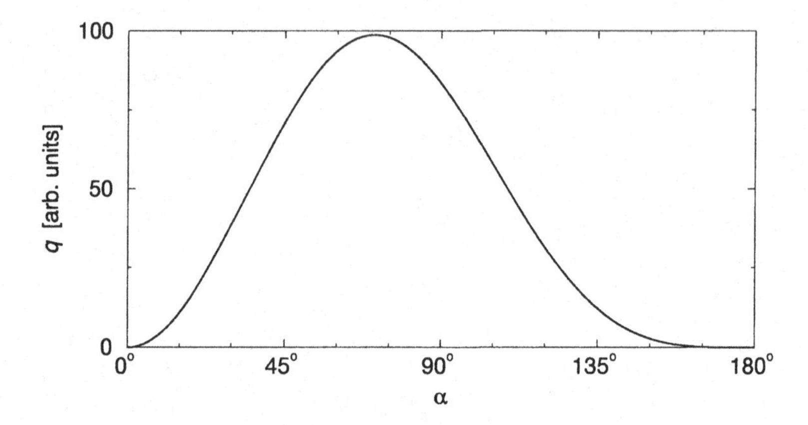


Figure 2. The quality factor $q = \mathcal{P}_c^2 \cdot I$ of a polarized positron beam as function of the half opening angle $\alpha$ of the cone in which the directions of positron momentum are uniformly distributed (arbitrary units, cf. Eq. (9)).

$\alpha$ of the cone, can be written as

$$q = \mathcal{P}_c^2 \cdot I \propto (1 + \cos\alpha)^2 \cdot (1 - \cos\alpha). \tag{9}$$

This function shows a maximum at $\alpha = 70.5°$ which is not very sharp (see Figure 2). In the case of practical constructions, any angle $\alpha$ between approximately 45° and 100° may be acceptable.

The acceptance angle $2\alpha$ is mainly defined by the geometry of the radioactive positron source and the moderator. Additionally, if a magnetic field is applied in this area, as is often the case, the acceptance angle $2\alpha$, and hereby also the quality factor $q$, may depend on the magnitude of the magnetic field.

In the practical design of facilities for polarized positron beams we have to deal with additional considerations. We mention here only those with major influence. The mean helicity of positrons arriving at the sample will be influenced by the absorption of the slowest positrons in the source itself or in the windows that are placed between the positron source and the sample. Although the cost of this helicity increase is a loss of intensity, the usage of low-$Z$ absorbers between the positron source and the sample is common practice[26] for improving the quality factor $q$. In contrast, commercial positron sources are usually optimized for positron yield by high-$Z$ backing material, typically tungsten, beneath the active source layer. This construction enhances the back–scattering effect that results in a higher positron yield but reduces the polarization and causes a reduction of the quality factor by introducing positrons with reversed spin. The latter can be avoided or at least minimized by the application of low-$Z$ backing material, e.g., beryllium.

It is generally accepted that after their implantation into the sample, positrons are slowed down to thermal velocities within a short time compared to their lifetime in matter, independently of their speed at the time of implantation. The slowing–down time is so short that within this time not only the probability of annihilation is small, but also, considering the magnitudes of the static magnetic fields that may be present in the sample, the relaxation of the positron–spin polarization may be neglected. However, if the positrons are relativistic, i.e. if their kinetic energies are at least of the order of $mc^2$, during the slowing–down process they may experience much higher magnetic fields that originate from the relativistic transformation of the strong electrostatic fields present in matter. In contrast to the case of purely magnetic fields in which, independently of the velocity of the positrons, the angle between the particle momentum and the spin, i.e. the helicity, is a constant of motion[d], the relativistically induced magnetic fields together with the electrostatic fields present may result in a strong depolarization of the positron spin.

This phenomenon has been studied experimentally[26,28–30] and theoretically[31–39] as well. Due to the theory of Bouchiat and Lévi-Leblond[33], during the positron slowing–down process in $(CH_2)_n$, as long as $E_{\mathrm{kin}} \ll mc^2$ holds for the initial positron kinetic energy, the fractional polarization loss $\delta$ of the positron spin is very small. In the range of $0.5 \cdot mc^2 \lesssim E_{\mathrm{kin}} \lesssim mc^2$, $\delta$ is a strongly increasing function of $E_{\mathrm{kin}}$, and for $E_{\mathrm{kin}} \gtrsim 5 \cdot mc^2$, $\delta$ is virtually independent of $E_{\mathrm{kin}}$ (for $E_{\mathrm{kin}}/mc^2 = 0.5, 1, 2, 5, 50$, $\delta = 0.06, 0.17, 0.30, 0.47, 0.64$, respectively). These results are in good agreement with the experimental results for typical positron kinetic energies in $\beta$ decays of radioactive nuclides[26], however, in the case of very large kinetic energies there are discrepancies between theory and experiment. In the case of the decay of the positive muon $\mu^+$, which is a $\beta$–decay process[40] with $E_0 = 52.3$ MeV, the experimentally observed[28–30] spin depolarization of $\delta \approx 0.8$ in the first stage of the slowing–down process (during which the $e^+$ kinetic energy decreases approximately from 50 MeV to 10 MeV) is in contradiction with the results of the calculations[33], $\delta \approx 0.2$. Although this contradiction is still unsolved, from the experimental point of view we can say that positrons from the decay of $\mu^+$, notwithstanding their very high initial polarization, are certainly not

---

[d]This is true only with limitations: the particle must not possess anomalous magnetic moment, the magnetic field must not change rapidly within distances comparable to the de Broglie or to the Compton wavelength of the positron, and the magnetic field must not be too strong[27].

feasible for polarized–positron studies in condensed matter due to their nearly complete depolarization during the first part of the slowing–down process.

One of the results of the theory of Bouchiat and Lévi-Leblond[33] which seems to hold for $e^+$ energies typical in decays of radioactive nuclides, is the determination of the $\overline{Z}$-dependence of depolarization, where $\overline{Z}$ is the mean atomic number of the material in which the positrons are stopped. Due to this theory, the depolarization is approximately proportional to $\exp(\overline{Z(Z+1)}/\overline{Z})$, the bars denote mean values. By this scaling factor we can obtain depolarization values for heavier materials. As a general result, it can be said that positrons emitted from any nuclide possess finite helicities that may be useful for experiments with polarized positrons. However, $\beta^+$ decays with most likely positron kinetic energies of $E_{\mathrm{m}} > mc^2$ are the best selections for sources of polarized positrons[e].

In addition to the mechanisms of positron–spin depolarization during slowing–down mentioned above, a possibility was reported that in some molecular substances strong spin relaxation occurs prior to positronium formation[41,42]. The experiments that revealed this depolarization belong to the class discussed in Sect. 6.

We can either implant the collimated polarized positrons as obtained from the $\beta^+$ decay immediately into the sample, as demonstrated in Sect. 5, or we can manipulate their energies and/or focus them prior to implantation. The discovery that positrons thermalized in solid moderators and subsequently emitted from the moderator surface preserve a substantial part of their polarization at the time of the $\beta$ decay[43] opened the way to the construction of variable–energy spin–polarized positron beams. Such beams may be ideally suited for studies of various surface phenomena, such as surface magnetism[44,45]. Nevertheless, no other slow polarized positron beam facilities, apart from the prototype beam[45] have been realized so far. These days, however, several facilities for monoenergetic polarized positrons are reportedly under construction[46–48]. Hopefully, the systematic exploitation of the possibilities which arise with the polarized eV–keV–energy positron–beam technique will start soon.

Undoubtedly, the use of positron beams is in most respects superior to the traditional integrated sample–source set-ups, such as the sandwich or sealed–source constructions, which are common in experiments with unpolarized positrons. However, there are at least two points in which the beam set-up is disadvantageous. First, the intensity of the positron beam may be

---

[e] Positrons obtained from pair production, a popular method for very high intensity positron beams, are not polarized since pair production is mediated by the parity–conserving electromagnetic interaction.

significantly lower than that of the source itself; second, positrons reflected from the sample may annihilate in parts of the equipment that surround the sample, and this may result in a high background or even in false signals. The first disadvantage is usually balanced by the advantage of the beam method, whereas the second one may be a critical problem in the design of experiments[f]. A possible way of reducing of the effects of reflected positrons, a combined veto-detector and shielding arrangement, is applied in the set-up[51] shown in Figure 8.

## 3 Determination of the positron-spin polarization

Although all information about the fate of the positron in the sample is provided by the annihilation photons only, we can additionally make use of the fact that positrons, as implanted into the sample from a source of polarized positrons, are polarized even at the end of the slowing-down process. If the positron experiences a generally time-dependent magnetic field, its magnetic moment and its spin may be depolarized at the time of annihilation. The knowledge of this depolarization may provide us with important details of the interaction of the positron in matter, e.g., about the diffusive motion of positrons in a ferromagnetic metal (cf. Sect. 5).

Methods similar to those used in EPR or NMR experiments are not applicable either. The reason for this is the limited strength of the positron source ($10^6$ Bq to $10^{10}$ Bq depending on the type of experiments) and the short positron lifetime in condensed matter. This results in the average number of positrons simultaneously present in the sample between $10^{-3}$ and 1, obviously by many orders of magnitude smaller than what is necessary for experiments such as EPR or NMR.

Owing to the fact that positrons annihilate in matter obeying the exponential annihilation law, methods that are analogous to the determination of the spin polarization of muons at the time of their decays may be applied for the determination of the time-dependent spin polarization of positrons. However, the decay process of $\mu^+$ or $\mu^-$ is mediated by the parity non-conserving weak interaction and, consequently, the emission direction of the decay product positrons or electrons is asymmetrical with respect to the muon spin. The various experimental techniques are known as muon spin rotation, relaxation, or resonance ($\mu$SR) experiments[52-55]. Since positron annihilation is mediated by the parity-conserving electromagnetic interaction, the rather simple and elegant method used in muon experiments cannot be applied. In contrast

[f] Reflectivity data for positrons[49] or electrons[50] are rare in the literature.

to $\mu$SR, the emission directions of the annihilation photons are not correlated with the spin direction of the positron.

The traditional methods for the determination of spin polarization of ensembles or beams of electrons or positrons at kinetic energies up to a few MeV are scattering on electrons (Møller or Bhabha scattering) and on nuclei (Mott scattering)[56]. Since the efficiencies (fractions of scattered to incident particles) are of the order of $10^{-5}$ or $10^{-4}$, the accuracies of such measurements are generally rather low. Positron polarimetry by these methods is, generally, even less sensitive than electron polarimetry[57]. The scattering positron–polarimetry methods possess a common drawback, since the positrons must be in a known and reproducible state for such experiments, e.g., in the form of a monoenergetic beam. In most positron experiments that are performed on condensed matter, this condition cannot be fulfilled, positrons can even be more or less localized in matter.

A practical way to experimentally investigate the polarization of positrons in matter is to employ the sample itself as spin analyser. A common attribute of such methods, which, instead of scattering effects, are based on condensed–matter or atomic–physical effects, is that the magnitude of the spin polarization cannot be obtained in such a simple way as in the case of $\mu$SR, EPR, or NMR. Although the data acquisition itself may be rather simple, the data evaluation process is usually rather complex. Such positron experiments, in contrast to scattering experiments, may provide us with information about the time dependence of the positron–spin polarization in matter, often referred to as spin–depolarization function, in particular in the case of time–differential experiments.

Two basic methods are known for the determination of the spin polarization of positrons in condensed matter:

(a) *Specimen with spin–polarized electrons.* The electrons in the sample are at least partially spin polarized and, additionally, the momentum distribution of the electrons is spin dependent. It is supposed that positronium is not formed in the sample, which is always the case in metals. Such systems are generally the ferromagnetically or ferrimagnetically ordered materials. In the case of a magnetically saturated $\alpha$-iron sample, for instance, the 3d electrons, which are responsible for the macroscopical magnetization of the sample, are polarized according to the magnetization direction, and their momentum distribution is relatively broad. The 4s conduction electrons, which are nearly free with a narrower momentum distribution, possess a much lower polarization in the direction which opposes that of the 3d electrons. Consequently, if we observe the Doppler broadening or the ACAR on such a sample, in principle we can determine the component of the positron–spin polarization parallel to the magnetization direction of the sample. Although only a small fraction of the electrons are spin polarized, such experiments are feasible since the spin–polarized electrons are 'tagged' by their broad momentum distribution. Usually, analogous effects in positron lifetime experiments do not exist due to the absence of electron–polarization dependent lifetime (i.e. electron–density) tagging.

Since positrons annihilate predominantly with electrons of opposite spins even if no positronium is formed (two–photon process), in case of polarized positrons the Doppler broadening or the ACAR linewidths are different at the two different magnetic–field directions, i.e. if the applied magnetic field is parallel or antiparallel to the spin–polarization direction of the positrons. Although the measured quantity, the change in the Doppler broadened or the ACAR linewidth, is rather small (e.g., in the case of the Doppler broadening measurements it is definitively smaller than the instrumental resolution broadening), the experiment is rather simple. Since we can reverse the magnetic field at any time, the experiment can be performed as a true differential measurement eliminating the effects of long–time instabilities, such as thermal drift.

Such method was first used for the determination of the positron–spin polarization by Hanna and Preston[58,59] and by Lovas[60]. Positrons from a $^{64}$Cu radioactive source were implanted into magnetized iron, and a change in the ACAR lineshape was found as a result of the reversal of the sample magnetization. For a review of the early experiments see reference[61].

Akahane and Berko[62] reported the first such experiment with detection of the Doppler broadening. This method, in spite of its low momentum resolution, is superior to the ACAR method in experiments for the determination of positron–spin polarization at least in three respects. (i) The detection count rate is much higher in the case of Doppler measurements, and statistically significant results for the polarization can be obtained in a shorter time. (ii) The experimental set-up is less complicated in Doppler broadening experiments. An important aspect is that the commonly used instrument for the determination of the energy of the annihilation photons, the intrinsic Ge detector, is nearly insensitive to the presence of external magnetic fields, at least for time–independent magnetic fields and in differential measurements, i.e. in experiments in which results obtained with the same magnitude of the applied magnetic field are directly compared. (iii) In Doppler broadening experiments only one of the annihilation photons must be detected. The second photon, which in the case of two–photon annihilation is emitted collinear with respect to the first photon, can be used by the means of a second, highly efficient detector for background suppression or by the means of a fast detector for the determination of the annihilation time. The latter choice may provide us with age–momentum correlation (AMOC) data[g]. The method (a) for the determination of the positron–spin polarization in the sample provides us with *the averaged residual spin polarization of the positrons*, i.e. with the mean polarization at the time of annihilation. By the means of appropriate models, we may obtain the spin–depolarization function even from the time–integral data if the experiments were conducted at multiple values of certain parameters, in particular the magnetic–field magnitude and/or the temperature. However, the resulting time dependence may be strongly model–dependent. Model–independent depolarization functions may be directly extracted from results of AMOC experiments with polarized positrons performed in external magnetic fields. Although such measurements are not merely lifetime experiments, they would be very useful despite the limited time resolution attainable in the presence of strong magnetic fields.

As it was mentioned above, in general electron–polarization induced positron–lifetime tagging is absent. In a special case, nevertheless, such a measurement proved to be feasible. Recently, the spin polarization of electrons in F-centres in additively coloured KCl in large applied magnetic fields has been determined by Maier and co-workers[63,64]. Since the annihilation rate of the spin–polarized positrons trapped in paramagnetic centres depend not only on the electron density in the trapping sites but also on the spin polarization of the paramagnetic electrons, both Doppler-broadening and lifetime experiments revealed a temperature–dependent effect.

(b) *Specimen in which positronium can be formed.* In samples in which positronium can be formed, the annihilation characteristics may strongly be influenced by the spin polarization of the positrons (cf. Sect. 1). This gives us a way to determine *the spin polarization of the implanted positrons* and, additionally, the influence of the surroundings on the various positron states in condensed matter.

In applied magnetic fields the $m = 0$ orthopositronium state ($m$ is the so-called the magnetic quantum number) and the parapositronium state are mixed. In other words, in the presence of an applied magnetic field the p-Ps and the o-Ps ($m = 0$) states are not energy eigenstates, nevertheless from these states the stationary states in an external magnetic field can be combined. The annihilation characteristics of these 'mixed' states can be described as a combination of the annihilation characteristics of the zero–field states according to their weights. These weights, however, depend on the magnitude

[g] Hitherto, to the knowledge of the author, AMOC with polarized positrons has not yet been realized.

and the direction of the applied magnetic field and on the polarization of the implanted positrons. The fraction of the pick-off annihilation, which leads to a broader momentum distribution of the annihilating electron–positron pairs or in a smaller annihilation rate, decreases with increasing magnetic field. This so-called magnetic quenching — the hyperfine interaction of the electron and the positron in the positronium is quenched — depends also on the initial polarization of the positrons, however, so weakly that only differential measurements are feasible, i.e. experiments in which the results obtained at both magnetic–field directions, parallel and antiparallel to the spin–polarization component to be determined, are directly compared. For a detailed description of the magnetic–field induced mixing of the positronium states see Sect. 6.

Such experiments, however, are rather complicated, since the sensitivity of the detectors to magnetic fields is not easy to overthrow (in this respect the type (a) and type (b) methods do not differ from each other). Intrinsic Ge detectors only suffer from reduced resolution even at magnetic fields as high as 1 T. On the other hand, photomultiplier tubes, which are indispensable in lifetime spectrometry, may completely stop working even at small magnetic fields such as 0.001 T or 0.01 T.

Magnetic shielding is only possible with strong limitations, in magnetic fields larger than 0.1 T it is essentially impossible. Accordingly, in such experiments it is necessary to remove the photomultipliers from the region of large magnetic fields, which can only be done on the cost of the time resolution (by application of lightguides) or of the count rate (by application of larger distances between the sample and the scintillator)[h]. ACAR experiments are not influenced directly by the magnetic field present at the sample site. However, most ACAR facilities apply strong magnetic field for focusing the positrons onto the sample, and a significant decrease of the magnitude of the applied magnetic field, which is also necessary in positronium studies, may reduce the coincidence count rate to an unacceptably low level.

Consequently, the various positron annihilation channels, e.g., the intrinsic two–photon annihilation, the annihilation of free positrons (those not bound in Ps), or the pick-off annihilation, are tagged by the momentum distribution of the annihilating positron–electron pairs as well as by the annihilation rates in experiments performed in applied magnetic fields.

The first experiment, in which the momentum of the annihilating electron–positron pairs was tagged, was performed by Page and Heinberg[65]. In their technique, positrons from a radioactive source were stopped in a

[h]These days photomultipliers of limited magnetic–field sensitivity and of relatively good time resolution are available, however, their usage is not yet disseminated in lifetime experiments.

gas and the magnetic field was applied parallel or antiparallel to the mean flight direction of the positrons. Since in the case of positronium formation in gases, pick-off annihilation does not occur, they chose experimental conditions in which the Ps atoms slowed down so slowly that from their average momenta at the time of annihilation, as it was revealed by ACAR measurements, showed an asymmetry for magnetic–field reversal. From this asymmetry, which reflects the effect of the initial spin polarization, they could determine the (positive) sign of the helicity of the positrons emitted from the $^{22}$Na positron source.

Greenberger and co-workers[66] revealed the effects of magnetic field and positron polarization in ACAR experiments on quartz single crystals, Herlach and Heinrich[67] on coloured KCl. In a more recent work, similar measurements performed on positronium formed in porous silica glass were reported[68].

An alternative method, proposed by Telegdi[69] and Lundby[70], uses a timing coincidence technique instead of the ACAR method. This method makes use of the fact that the various positronium states are also tagged by their different decay rates. The main advantage of this technique is the much higher coincidence count rate compared to those typical in ACAR experiments. This made the time–resolved positronium magnetic quenching method the most accurate positron polarimeter for beam–based atomic physics experiments[57,71–73]. In the high–precision versions of this method, positronium is usually formed at the surface of the particles of ceramic powders and can be regarded as unperturbed (vacuum) positronium and consequently, because of its long lifetime, a limited experimental time resolution is sufficient. In condensed–matter experiments, however, the situation is more complicated. The pick-off process is, generally, rather fast, and an excellent time resolution may be a prerequisite for at least a partial separation of the various positron and positronium species and their states. For this purpose external magnetic fields typically as large as 1 T or 2 T may be required, which at present makes the construction of a lifetime spectrometer with the necessary time resolution rather difficult[42,72].

In contrast to experiments of type (a), in type (b) measurements — magnetic quenching of the positronium hyperfine interaction — time–resolved experiments (AMOC) cannot show directly the time–dependent longitudinal spin polarization of the positrons. For model–independent observation of the relaxation processes, e.g., slowing-down or spin exchange, however, AMOC experiments with polarized positrons may be substantial.

It is common to type (a) and type (b) experiments that *the differential effect is absent in zero applied magnetic field.* In the limit of *large magnetic fields*, the effect of the magnitude of the magnetic field disappears. In experiments of type (a), *the differential effect will be saturated* since the electron spin polarization saturates. In experiments of type (b), the electron and the positron spins are decoupled from each other (Paschen–Back effect), the magnetic mixing is saturated, and the o-Ps and p-Ps occupation numbers will not depend on the direction of the magnetic field; consequently *the differential effect is absent* in the large–field limit.

## 4 Positron–spin relaxation ($e^+$SR) experiments

Positrons, when implanted into condensed matter, generally slow down to thermal or nearly thermal kinetic energies within a short time compared to their mean lifetimes in the sample. During slowing–down, only a part of their polarization will be lost, as discussed in Sect. 2. Unlike in muon experiments, the precession of the positron spin in magnetic fields present at the positron site cannot directly be observed, in spin–polarized positron experiments only the polarization component parallel to the external magnetic field can be determined (cf. Sect. 3). In these so-called longitudinal experiments, the initial spin direction is usually parallel to the direction of the applied magnetic field. (In transverse experiments, the initial spin possesses a component perpendicular to the direction of the applied magnetic field, and the perpendicular components of the spin are detected[74].)

In analogy to the acronym $\mu$SR for '**mu**on **s**pin **r**otation or **r**esonance or **r**elaxation', this technique has been dubbed $e^+$SR (='**p**ositron **s**pin **r**elaxation')[75]. Owing to the effect of random local magnetic fields, the initial polarization of the positrons may relax with the time the positrons spend in the sample, as e.g., in ferromagnets (cf. Sect. 5). This phenomenon, which we can trace by means of an experiment of type (a), is clearly a relaxation process. If, in the other typical case, the positron is bound to an electron and forms positronium, the positron magnetic moment experiences the magnetic field due to the magnetic moment of the electron and precesses with the so-called hyperfine frequency. Since in experiments of type (b) we can only detect the longitudinal polarization component of the positron spin, the hyperfine precession can be considered as a mechanism for the loss of mean longitudinal positron–spin polarization (cf. Sect. 6). The name **p**ositron **s**pin **r**elaxation ($e^+$SR) is justified in this case as well. The results of dynamical processes, e.g., spin flip or transitions between species with different hyperfine couplings, are also detected in a similar way as in the case of type (a) experiments.

Similar experiments belong to the basic methods in muon studies[52–55]. The mean muon lifetime $\tau_\mu = 2.2$ $\mu$s is, however, much longer than the mean positron lifetimes in condensed matter, but this is balanced by the much larger gyromagnetic ratio of the positrons ($\gamma_e = 1.761 \cdot 10^{11}$ s$^{-1}$T$^{-1}$) compared to that of the muons ($\gamma_\mu = 8.516 \cdot 10^8$ s$^{-1}$T$^{-1}$). In order to rotate the positron spin by an angle of $\pi$ in $t = 0.1$ ns, which corresponds to the shortest mean positron lifetime in condensed matter, a magnetic field

$$B = \pi/(\gamma_e t) = 0.17 \text{ T} \tag{10}$$

perpendicular to the positron spin is required. This is of the order of the dipolar fields present in ferromagnetic metals such as $\alpha$-Fe, or the hyperfine fields acting on the positron bound in a positronium. We can conclude that e$^+$SR studies with both type (a) and type (b) experiments are feasible.

## 5 e$^+$SR experiments on $\alpha$ iron

In magnetic fields the Zeeman energy levels of the positron are separated by

$$\Delta E_Z = \hbar \gamma_e B^{||}_{e+} \,, \tag{11}$$

where $B^{||}_{e+}$ denotes the longitudinal magnetic field (parallel to the initial spin direction) felt by the positron magnetic moments ($h = 2\pi\hbar$ is the Planck constant). In magnetic fields that are accessible in the laboratory, the Zeeman energy splitting is usually small compared to thermal energies. E.g., at $B^{||}_{e+} =$ 1 T the energy difference is $\Delta E_Z = 1.86 \cdot 10^{-23}$ J $= k_B \cdot 1.35$ K ($k_B$ is the Boltzmann constant). Consequently, the thermal equilibrium of the spin ensemble at not too low temperatures approximately corresponds to a random spin distribution.

In longitudinal external magnetic fields the initial positron–spin distribution of a longitudinally polarized spin ensemble corresponds to a metastable equilibrium independent of whether the polarization is parallel or antiparallel to the field. Transitions between the Zeeman levels can be generated by random magnetic fields with nonvanishing $B^{\perp}_{e+}$ components perpendicular to the longitudinal direction if this perpendicular component possesses a Fourier component with a circular frequency approximately

$$\omega_{e+} = \gamma_e B^{||}_{e+} \,. \tag{12}$$

These transitions lead to the equilibrium spin distribution, i.e. they tend to depolarize the positron spins. Such time–dependent transverse magnetic fields may arise from the local magnetic fields that exist in ferromagnets or ferrimagnets. The time dependence of the field is induced by the diffusive motion of the positron after its slowing down in its thermalized state. It is obvious that the angular frequency $\omega_{e^+}$ (Eq. (12)) that is required for the depolarization can be adjusted by variation of the magnitude of the external magnetic field or of the speed of positron diffusion, e.g., by altering the temperature.

In contrast to $\mu$SR, however, the classes of materials that may be investigated are severely restricted. On one hand, the sample itself must be capable of playing the role of the polarimeter, i.e. it must fulfill the requirements of a type (a) experiment (cf. Sect. 3). On the other hand, the magnetic field experienced by the positron must fulfill the requirements considered above. For instance, in ferromagnetic Ni the first condition is satisfied but not the second one. The interstices, the most likely sites for the positrons, possess cubic point symmetry, i.e. the dipolar magnetic fields in the interstices are parallel to the direction of magnetization[76]. This direction corresponds to the longitudinal axis in a type (a) experiment, and consequently, transverse magnetic fields in Ni that may depolarize the positron spin do not exist.

The relaxation rate of the positron polarization depends on the 'magnetic history' of the positrons, i.e. on the time dependence of the magnetic fields experienced by the positrons during the time spent in the sample. The prototype of such experiments, in which the residual positron–spin polarization was determined in a type (a) measurement, was performed on $\alpha$ iron[77,78]. In $\alpha$ iron, due to its bcc structure, interstices possess tetragonal symmetry. Consequently, in these sites local magnetic–field components perpendicular to the magnetization may appear. In particular, the strongest depolarization effect of the dipolar field arises, when the sample is magnetized along the $\langle 111 \rangle$ axis[76]. The $\langle 111 \rangle$ direction is the axis of hard magnetization in $\alpha$-Fe. If the sample is magnetized along the axis of easy magnetization, the $\langle 100 \rangle$ axis, the depolarization effect is absent[76].

In the study discussed in the preceding paragraph, the magnetic history of the positrons resulted from the diffusive motion of the positrons in the sample that was magnetized in the $\langle 111 \rangle$ crystallographic axis[76]. As a result of this magnetization direction, the dipolar magnetic fields $\boldsymbol{B}_{\rm dip}$ at the interstices are perpendicular to the direction of magnetization, i.e. to the longitudinal axis. An appropriate measure of the rate of change of the magnetic field experienced by the positrons is the correlation time $\tau_c$ of the magnetic field. Its inverse is proportional to the positron scattering rate if the positron states are described by delocalized 'Bloch states' (band model), or, if the positron states are described by localized 'particles' that hop between different interstitial sites, it is proportional to the positron jump frequency. If the correlation function of the time–dependent part of the magnetic field acting on the positron magnetic moments is proportional to $\exp(-t/\tau_c)$, the positron–spin longitudinal relaxation rate $\Gamma_1$, which is usually denoted by $1/T_1$ in NMR or EPR work, is given by (cf. reference[76])

$$\Gamma_1 = (\gamma_e B_{\rm dip})^2 \frac{\tau_c}{1+\tau_c^2\omega_{e+}^2} \, . \tag{13}$$

According to Eq. (13), $\Gamma_1$ shows a maximum at constant $\omega_{e+}$ at the value of $\tau_c$ for which

$$\omega_{e+}\tau_c = 1 \tag{14}$$

holds. In bcc structures the positron diffusivity $D^+$ is related to the correlation time $\tau_c$ by

$$D^+ = \frac{a_0^2}{36\tau_c} \tag{15}$$

for jumps between octahedral interstices ($a_0$ is the side of the elementary cube; for tetrahedral interstices the factor 36 has to replaced by 72). However, if the band model is the appropriate model,

$$D^+ = \frac{k_B T \tau_c}{m^+} \tag{16}$$

holds ($T$ is the temperature and $m^+$ the positron effective mass)[78].

The Stuttgart $e^+$SR set-up (Figure 3) uses polarized positrons from a $^{68}Ge/^{68}Ga$ radioactive source which has been deposited electrolytically on a Ni foil (initial activity: 2 mCi $= 7.4 \cdot 10^7$ Bq). The source holder is made of tungsten with a light metal backing. The annihilation photons, which arise from positrons annihilating inside the source, in the surroundings of the source, or in the backing metal, are effectively shielded by the tungsten body of the source holder, and the light–metal backing minimizes the polarization loss due to back–scattered positrons (cf. Sect. 2). The positrons enter into the vacuum chamber through a window made of Ti foil with a thickness of 25 $\mu$m. The vacuum chamber and all its components consist of Cu or other non-ferromagnetic metals in order to confine the analyser effect entirely to the sample.

The source holder and the vacuum chamber with the cryostat cold finger and heater (sample holder) and the sample are located between the pole pieces of an iron–core electromagnet in such a way that a substantial fraction of the emitted positrons are focussed onto the sample. With this set-up magnetic fields up to $B = 2.6$ T can be achieved. The focusing effect of the magnetic field and consequently the intensity of the positron 'beam' and its polarization depend on the strength of the magnetic field. This dependence, which can be calculated from the energy distribution of the emitted positrons and from

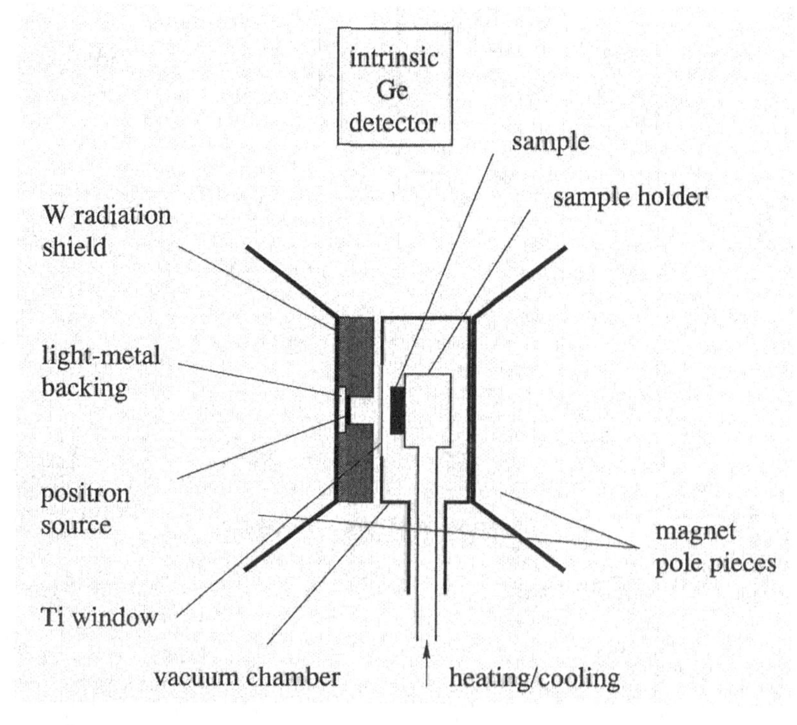


Figure 3. The Stuttgart polarized–positron ($e^+$SR) set-up for the study of positron diffusion in $\alpha$ iron[77]. The holder of the $^{68}$Ge/$^{68}$Ga positron source is made of tungsten for an efficient shielding of the annihilation photons that arise from positrons annihilating outside the sample. The Doppler–broadened lineshape of the annihilation photons is determined by the intrinsic Ge detector.

the geometry of the collimator (tungsten radiation shield), must be taken into consideration during data evaluation.

The monocrystalline iron sample (approximately $8 \times 8 \times 1$ mm$^3$) attached to the sample holder is orientated with the $\langle 111 \rangle$ axis parallel to the magnetic field and the positron beam polarization, i.e. with the longitudinal direction. The temperature of the sample, which is placed in vacuum, can be controlled between 10 K and 750 K by means of liquid helium or nitrogen and a resistor heating coil situated near to the sample. The Doppler–broadened lineshape of the annihilation photons is determined by an intrinsic Ge detector. The performance of the Ge detector is only slightly influenced by the stray magnetic field of the magnet. The slightly shifted zero position and the somewhat broadened resolution function have been shown to be virtually independent

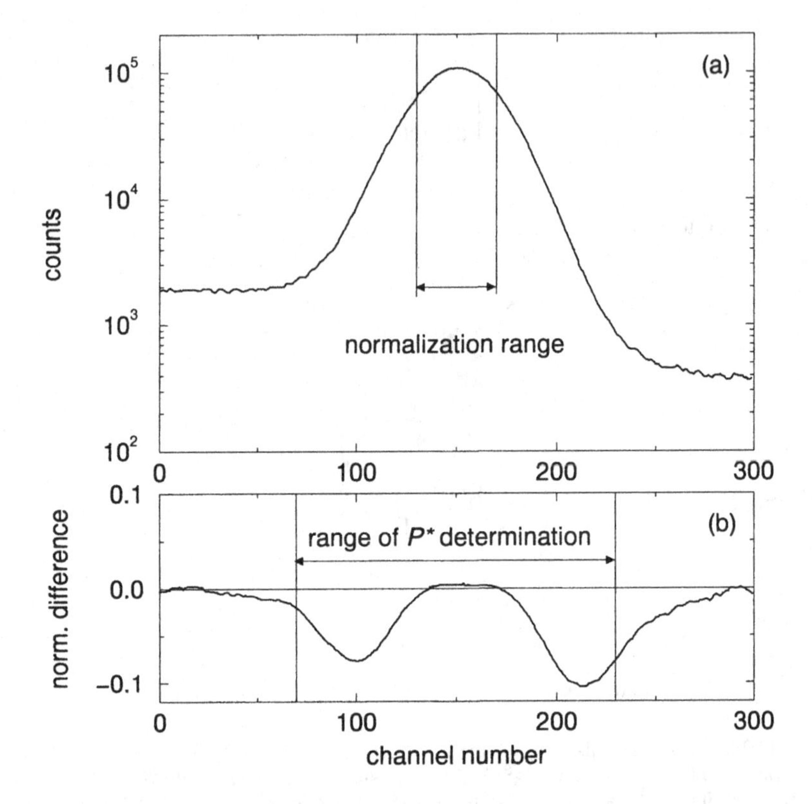


Figure 4. (a) Typical Doppler–broadened 511 keV annihilation line (one energy channel corresponds to 62 eV) measured on $\alpha$-Fe at room temperature. (b) Corresponding normalized difference spectrum as determined from results obtained at opposite sample magnetizations. The ranges used for the normalization of the lines and for determining $P^*$ are indicated.

of the polarity of the magnetic field. The spectrometer is stabilized against energy–scale drifts by a computer which also controls all other parameters of the measurement, e.g., the magnetic field and the sample temperature.

The spectrometer was tested by employing paramagnetic metallic samples. As expected, no effect of the field reversal on the Doppler broadening of the annihilation photons was detected.

To a good approximation, the relative difference between the photon lineshapes at the two magnetic–field polarities and hence at opposite magnetizations of the Fe sample is proportional to the residual spin polarization of the positrons. Since in this experiment the dependence of the polarization on the positron age was not directly measured, information on $\Gamma_1$ must be deduced from the dependence of the residual spin polarization on temperature and applied magnetic field.

During each measuring run a certain number of Doppler spectra (typically ten, each with a counting time of 500 s) are recorded, corrected for long–time shifts, and added. After the magnetic–field reversal this procedure is repeated. The added spectra taken at two different directions of magnetization are normalized for equal integrals in the peak region and subtracted or added channel by channel. The difference obtained in this way is divided by half the sum of counts in each channel. A typical normalized difference spectrum derived by the above procedure is shown in Figure 4 together with one of the Doppler–broadened lines from which it was deduced. In the normalized difference spectrum the mean value per channel is determined in the range indicated in Figure 4b, and this differential lineshape parameter, denoted by $P^*$, is the measure of the residual positron–spin polarization.

Measurements on three $\alpha$ iron single crystals (two of them doped with 200 at.-ppm and 50 at.-ppm nitrogen, respectively) showed no change of the residual positron–spin polarization over the temperature range from 10 K to 100 K. The experimental results above 100 K are shown in Figure 5, where the quantity $P^*_{\mathrm{corr}}$, which is proportional to the residual positron–spin polarization, has been plotted[77,78]. Since the demagnetizing factors of the samples are different, the lowest magnetic fields at which the samples are still uniformly magnetized at low temperatures vary from specimen to specimen.

The sample doped with 200 at.-ppm nitrogen exhibits a minimum in the temperature dependence of the residual positron–spin polarization, which is shifted towards higher temperatures with increasing magnetic fields (cf. Figure 5a). The results obtained on the other specimens at higher magnetic fields are compatible with this behaviour. At low magnetic fields no clear systematics of the temperature dependence of $P^*_{\mathrm{corr}}$ can be discerned.

From the fact that a maximum in the depolarization function has been observed at an external magnetic field of e.g., $B_{\mathrm{demag}} = 1.92$ T ($T = 370$ K), as can be seen in Figure 5a, we can estimate the magnitude of the correlation time $\tau_c$. From the general relationship for the magnetic field at the site of the positron

$$\boldsymbol{B}_{\mathrm{e+}} = \boldsymbol{B}_{\mathrm{appl}} + \boldsymbol{B}_{\mathrm{demag}} + \boldsymbol{B}_{\mathrm{Lorentz}} + \boldsymbol{B}_{\mathrm{dip}} + \boldsymbol{B}_{\mathrm{Fermi}}\,, \tag{17}$$

where $\boldsymbol{B}_{\mathrm{appl}}$ is the applied magnetic field, $\boldsymbol{B}_{\mathrm{demag}}$ the demagnetizing field, $\boldsymbol{B}_{\mathrm{Lorentz}}$ the Lorentz field, and $\boldsymbol{B}_{\mathrm{Fermi}}$ the Fermi contact field[76]. Since in iron magnetized along the $\langle 111 \rangle$ axis, $\boldsymbol{B}_{\mathrm{dip}}$ is perpendicular to the magnetization

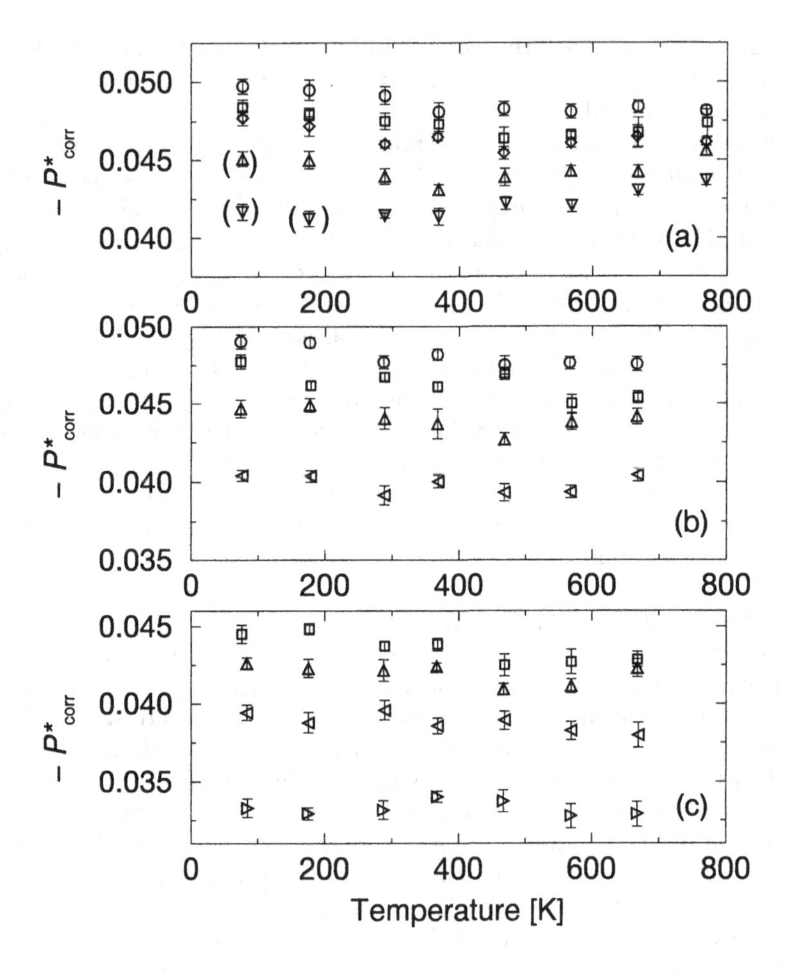


Figure 5. Results of $e^+$SR experiments on $\langle 111 \rangle$-orientated, differently doped ((a): 200 at.-ppm N, (b): 50 at.-ppm N, (c): undoped) $\alpha$ iron samples at external magnetic fields of 2.61 T ($\circ$), 2.24 T ($\Box$), 2.08 T ($\Diamond$), 1.92 T ($\triangle$), 1.75 T ($\triangledown$), 1.64 T ($\triangleleft$), and 1.33 T ($\triangleright$)[78]. The differential annihilation lineshape parameter $P^*$ (for its definition see Figure 4) is corrected in order to compensate for the temperature dependence of the polarimeter analyzing power. The resulting quantity, $P^*_{\rm corr} = P^* M_{\rm sp}(0)/M_{\rm sample}(T)$ is proportional to the residual spin polarization of the positrons. $M_{\rm sp}(0)$ denotes the spontaneous magnetization of iron at 0 K and $M_{\rm sample}(T)$ is the sample magnetization at the temperature of the experiment. Data points that are obtained from measurements with low magnetic fields at which the samples cannot be magnetically saturated, are marked by brackets.

direction and the other terms in Eq. (17) are longitudinal,

$$B_{e+}^{\parallel} = B_{\text{appl}} + B_{\text{demag}} + B_{\text{Lorentz}} + B_{\text{Fermi}}\,, \tag{18}$$

$$B_{e+}^{\perp} = B_{\text{dip}} \tag{19}$$

holds. In addition to the known magnitudes of $B_{\text{demag}} = -1.66$ T (calculated with a demagnetizing factor of the sample of 0.8 from the magnetization of iron at $T = 370$ K of 2.08 T/$\mu_0$, where $\mu_0 = 4\pi \cdot 10^{-7}$ Vs/Am) and $B_{\text{Lorentz}} = 0.69$ T, the value of the Fermi field is required, too. This quantity, which is associated with the spin polarization of conduction electrons at the positron site, is not known from direct observations. For positive muons in iron the low–temperature value $B_{\text{Fermi}} = -1.13$ T was obtained[79]. Since, owing to the smaller mass of the positron compared with that of the muon, the wave functions of the positron are definitively more spread out than the muon wave functions, the modulus of $B_{\text{Fermi}}$ must be substantially smaller than for the positive muon. The average magnetization of the conduction electrons in $\alpha$ iron is negative, i.e. opposite to the magnetization, consequently $B_{\text{Fermi}} < 0$ must be fulfilled for positrons, too. If we suppose that $B_{\text{Fermi}}$ in $\alpha$ iron for positrons amounts to a value between $-0.3$ T and $-0.7$ T[77], we can calculate the correlation time $\tau_c$ from Eqs. (12), (14), and (18). From the experiment discussed above, we obtain

$$\tau_c(370\,\text{K}) = (6.4 \pm 1.4) \cdot 10^{-12}\,\text{s}\,. \tag{20}$$

This value cannot be explained by the spin relaxation of thermalized positrons in Bloch states[80] mainly because the spin–relaxation rates predicted by the theory are much smaller than those observed, and the magnitude of the positron diffusivity that can be obtained from Eq. (16) is much higher than generally accepted.

In contrast to this, the particle model of self–localized positrons accounts at least qualitatively for the order of magnitude and for the temperature dependence of the observed relaxation rate[78]. For the positron diffusivity, however, Eq. (15) gives us a value that is several orders of magnitudes smaller than expected. A way out of this contradiction is the assumption that in the temperature range investigated, most positrons are in Bloch states and hence do not contribute to the spin relaxation. A small fraction of the positrons, however, is thermally excited into a metastable self–localized state, and these are responsible for the observed relaxation.

In the weakly doped (Figure 5b) and in the undoped (Figure 5c) samples, relaxation maxima are not present as clearly as in the 200 at.-ppm nitrogen–doped sample (Figure 5a). The influence of positron scattering by foreign atoms on the relaxation behaviour is not yet understood and further experiments are necessary. In particular, time-resolved experiments may provide us with direct knowledge of the positron-spin relaxation function, and the interpretation of the results would become more model-independent.

## 6 Magnetic quenching of the positronium hyperfine interaction

As discussed in Sect. 3, the existence of positronium gives us a way of studying positron-spin relaxation. Since the hyperfine coupling in positronium[i] results in a loss of the longitudinal spin polarization of the positrons and, by means of the longitudinal magnetic fields, we may quench the hyperfine interaction, the $e^+$SR experiment is ideally suited for the study of the positronium species in condensed matter. From such experiments we may obtain detailed information on the character and the fate of the various positronium states that are perturbed by the surrounding matter. This information can be in the form of parameters such as occupation numbers or hyperfine coupling constants or, even in the case of time-integral measurements, in the form of dynamic parameters such as annihilation or transition rates.

The first positronium magnetic-quenching experiments were performed by Deutsch and Dulit[81], by the means of unpolarized positrons. They determined the ratio of the number of three-photon annihilation events to that of two-photon events as a function of the applied magnetic field, and deduced the magnitude of the hyperfine energy splitting of positronium for the first time. It is straightforward to combine this method with positron polarimetry of type (b) (cf. Sect. 3) by making use of an ensemble of polarized positrons (cf. Sect. 2). The main advantages of this combination are the same as those of the experiments presented in Sect. 5, i.e. positronium magnetic-quenching experiments can be performed as differential Doppler-broadening experiments in longitudinal magnetic fields with alternating field direction.

Under usual experimental conditions the positronium wave function can be separated into a spatial and a spin part[82]. The spatial part is not affected by external magnetic fields but is influenced by the surrounding matter. Since in the context of the present work only the electron density $|\Psi_e(0)|^2$ at the site of the positron is of any importance, the modification of the spatial part of the positronium wave function can be described by the so-called density or

---

[i]In positronium the fine structure and the hyperfine structure are of the same order since, in contrast to hydrogen, the magnetic moments of the two constituents are of equal magnitude.

contact parameter[83,84] $\kappa$ defined as

$$\kappa = \frac{|\Psi_e(0)|^2}{|\Psi_e^0(0)|^2} . \tag{21}$$

$\Psi_e^0$ denotes the wave function of the electron in unperturbed (vacuum) positronium, $\Psi_e$ is the wave function of the electron in positronium in matter. The quantity $\kappa$ is proportional to the intrinsic annihilation rate of positronium and, for the 1S state in the case of isotropic hyperfine interaction, to the singlet–triplet energy splitting in zero applied magnetic field[82,84].

The spin part of the positronium Hamiltonian, the so-called hyperfine Hamiltonian[85], can be written as

$$\mathcal{H}_{\mathrm{hf}} = \frac{1}{2}\,\hbar\gamma_e\,B_{\mathrm{appl}}\,\sigma^z_{e^+} - \frac{1}{2}\,\hbar\gamma_e\,B_{\mathrm{appl}}\,\sigma^z_{e^-} + \frac{1}{4}\,\Delta E_{o-p}\,\boldsymbol{\sigma}_{e^+}\,\boldsymbol{\sigma}_{e^-}\,, \tag{22}$$

where

$$\Delta E_{o-p} = \kappa \cdot 0.8412\,\mathrm{meV} \tag{23}$$

is the hyperfine energy splitting, i.e. the energy difference between the $^1S_0$ and $^3S_1$ positronium states at $B_{\mathrm{appl}} = 0$, and $\boldsymbol{\sigma}_{e^-}$ and $\boldsymbol{\sigma}_{e^+}$ are the Pauli spin matrices for the electron and the positron. The external magnetic field is taken parallel to the z axis. In (22) and for the rest of the present work we restrict the hyperfine Hamiltonian to the case of isotropic hyperfine interaction.

The eigenstates of Eq. (22) with a principal quantum number $n = 1$ can be obtained in the form of linear combinations of the well–known $n = 1$ eigenstates of the Hamiltonian (22) for $B_{\mathrm{appl}} = 0$. The corresponding $B_{\mathrm{appl}} = 0$ wave functions are

$$|0,0\rangle = \frac{1}{\sqrt{2}}(|\uparrow\downarrow\rangle - |\downarrow\uparrow\rangle) \tag{24}$$

for the $^1S_0$ singlet state and

$$|1,0\rangle = \frac{1}{\sqrt{2}}(|\uparrow\downarrow\rangle + |\downarrow\uparrow\rangle)\,, \tag{25}$$

$$|1,1\rangle = |\uparrow\uparrow\rangle\,, \tag{26}$$

$$|1,-1\rangle = |\downarrow\downarrow\rangle \tag{27}$$

for the $^3S_1$ triplet states. On the left–hand sides of Eqs. (24-27) the first number refers to the total spin, the second one to the magnetic quantum number $m$ of the two–spin system; the symbols on the right–hand sides denote the spin states of the positron (first arrow) and that of the electron (second arrow).

The eigenstates of the Hamiltonian (22) at $B_{\text{appl}} \neq 0$ are described by the wave functions

$$|1\rangle = \sqrt{\operatorname{Sech}\xi}\left(-\operatorname{Sinh}\frac{\xi}{2}\cdot|0,0\rangle + \operatorname{Cosh}\frac{\xi}{2}\cdot|1,0\rangle\right), \tag{28}$$

$$|2\rangle = |1,1\rangle\,, \tag{29}$$

$$|3\rangle = |1,-1\rangle\,, \tag{30}$$

$$|4\rangle = \sqrt{\operatorname{Sech}\xi}\left(\ \operatorname{Cosh}\frac{\xi}{2}\cdot|0,0\rangle + \operatorname{Sinh}\frac{\xi}{2}\cdot|1,0\rangle\right). \tag{31}$$

The external magnetic field in the wave functions (28) and (31) is included by the dimensionless quantity $\xi$, which is a non-linear combination[j] of $B_{\text{appl}}$ and of the hyperfine energy splitting $\Delta E_{\text{o-p}}$ of positronium in zero external magnetic field,

$$\xi = \operatorname{Arsinh}\frac{B_{\text{appl}}}{B_0} \tag{32}$$

with the hyperfine magnetic field in positronium

$$B_0 = \frac{\Delta E_{\text{o-p}}}{2\hbar\gamma_{\text{e}}} = \kappa\cdot 3.6\,\text{T}\,. \tag{33}$$

The quantity $\xi$, which is used instead of the external magnetic field $B_{\text{appl}}$, possesses several advantages[82]. At small fields $\xi$ and $B_{\text{appl}}$ are proportional to each other, in larger magnetic fields $\xi$ depends logarithmically on $B_{\text{appl}}$; hence in plots against $\xi$ the full $B_{\text{appl}}$ range can be covered, in contrast to plots in which the magnetic field is linear.

The eigenstates $|2\rangle$ and $|3\rangle$ (Eqs. (29 and 30)) are identical with the $m = \pm 1$ triplet (o-Ps) energy eigenstates at $B_{\text{appl}} = 0$ (Eqs. (26, 27)). We may further denote them as $m = \pm 1$ orthopositronium. The stationary states $|1\rangle$ and $|4\rangle$ are mixtures of the parapositronium and $m = 0$ orthopositronium states at $B_{\text{appl}} = 0$ (Eqs. (24, 25)), and we may denote them as plesiopara-positronium (pp-Ps) for $|4\rangle$ and as meiktopositronium (m-Ps) for $|1\rangle$, as proposed in a recent publication[82]. In the limiting case of $B_{\text{appl}} = 0$, the pp-Ps state reduces to p-Ps and the m-Ps state reduces to o-Ps. The generic name for the states, which reduce to o-Ps for $B_{\text{appl}} = 0$, i.e. for the m-Ps and both o-Ps states ($|1\rangle$, $|2\rangle$, and $|3\rangle$), is leipopositronium[k] (l-Ps).

---

[j] Here we recite some limiting values of the hyperbolic functions used in the present work: $\operatorname{Sinh}0 = 0$, $\operatorname{Sinh}(+\infty) = +\infty$; $\operatorname{Cosh}0 = 1$, $\operatorname{Cosh}(+\infty) = +\infty$; $\operatorname{Tanh}0 = 0$, $\operatorname{Tanh}(+\infty) = 1$ ($\operatorname{Sech}x \equiv 1/\operatorname{Cosh}x$).

[k] The translations of the ancient Greek words used in the names of the positronium energy eigenstates in applied magnetic fields are *plesios* = near, almost; *meiktos* = mixed; *leipo* = to leave behind[82].

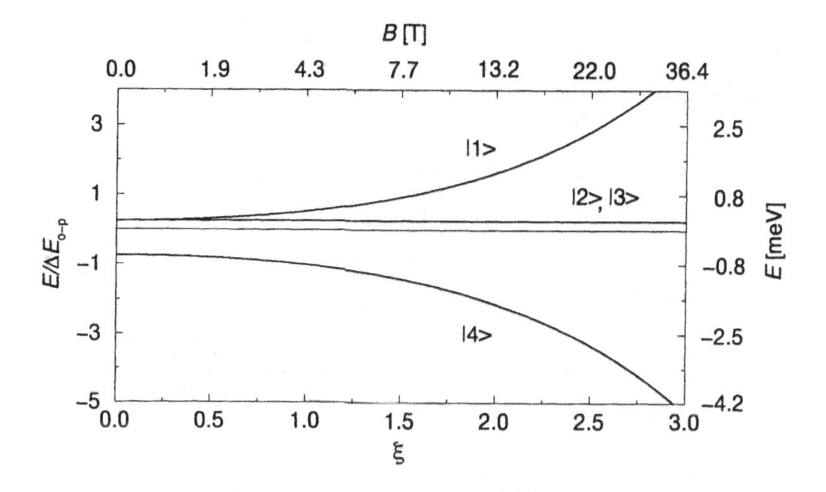


Figure 6. The eigenvalues of the positronium spin Hamiltonian as a function of the applied magnetic field $B_{\text{appl}}$ (Breit–Rabi diagram) for the 1S states. The energy $E$ is measured in units of the energy difference between o-Ps and p-Ps, $\Delta E_{\text{o-p}}$; the scales on the right and on the top correspond to $\kappa = 1$. The dimensionless variable $\xi$, which is defined in Eq. (32), is proportional to $B_{\text{appl}}$ for small magnetic fields, for large fields it depends logarithmically on $B_{\text{appl}}$. The corresponding energy eigenstates are denoted[82] by meiktopositronium (m-Ps, $|1\rangle$, $m = 0$), orthopositronium ($|2\rangle$, $m = 1$ and $|3\rangle$, $m = -1$), and plesioparapositronium (pp-Ps, $|4\rangle$, $m = 0$). The generic name for the states $|1\rangle$, $|2\rangle$, and $|3\rangle$ is leipopositronium (l-Ps).

The corresponding energy eigenvalues of the Hamiltonian (22) in applied magnetic fields are[82]

$$E_1 = -\frac{1}{4}\Delta E_{\text{o-p}}(1 - 2\,\text{Cosh}\,\xi)\,, \tag{34}$$

$$E_2 = \frac{1}{4}\Delta E_{\text{o-p}}\,, \tag{35}$$

$$E_3 = \frac{1}{4}\Delta E_{\text{o-p}}\,, \tag{36}$$

$$E_4 = -\frac{1}{4}\Delta E_{\text{o-p}}(1 + 2\,\text{Cosh}\,\xi)\,. \tag{37}$$

The zero on the energy scale has been chosen such that the sum of the four energy values (cf. Eqs. (34-37)), which is independent of $B_{\text{appl}}$, is zero. Since positronium possesses no permanent magnetic moment, $E_2$ and $E_3$ are independent of $B_{\text{appl}}$, and $E_1$ and $E_4$ show a quadratic Zeeman effect. The eigenvalues of the positronium spin Hamiltonian as a function of the applied

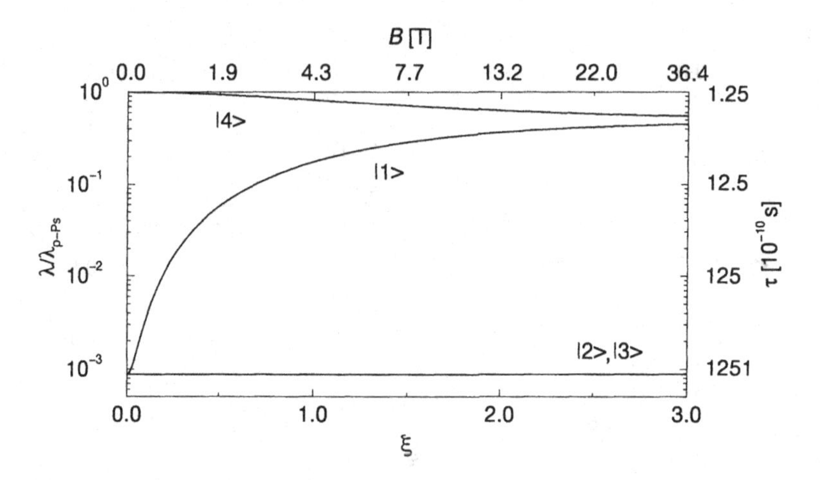


Figure 7. The intrinsic annihilation rates (Eqs. (28-31)) of the 1S positronium energy eigenstates as a function of the applied magnetic field $B_{\mathrm{appl}}$. The annihilation rate $\lambda$ is measured in units of the annihilation rate $\lambda_{\mathrm{p-Ps}}$ of p-Ps; the scales on the right and on the top correspond to $\kappa = 1$.

magnetic field are plotted in Figure 6, the so-called Breit–Rabi diagram.

Owing to the mixing ratio of p-Ps and o-Ps in plesioparapositronium and meiktopositronium, we may calculate the intrinsic annihilation rates of the $n = 1$ energy eigenstates of positronium in external magnetic fields. The annihilation rate of the state $|4\rangle$ (Eq. (31)) is

$$\begin{aligned} \lambda_{\mathrm{pp-Ps}} &= \left[\lambda_{\mathrm{p-Ps}} \operatorname{Cosh}^2 \frac{\xi}{2} + \lambda_{\mathrm{o-Ps}} \operatorname{Sinh}^2 \frac{\xi}{2}\right] \operatorname{Sech} \xi \\ &= \frac{1}{2}(\lambda_{\mathrm{p-Ps}} + \lambda_{\mathrm{o-Ps}}) + \frac{1}{2}(\lambda_{\mathrm{p-Ps}} - \lambda_{\mathrm{o-Ps}}) \operatorname{Sech} \xi \end{aligned} \tag{38}$$

and that of the state $|1\rangle$ (Eq. (28)) is

$$\begin{aligned} \lambda_{\mathrm{m-Ps}} &= \left[\lambda_{\mathrm{p-Ps}} \operatorname{Sinh}^2 \frac{\xi}{2} + \lambda_{\mathrm{o-Ps}} \operatorname{Cosh}^2 \frac{\xi}{2}\right] \operatorname{Sech} \xi \\ &= \frac{1}{2}(\lambda_{\mathrm{p-Ps}} + \lambda_{\mathrm{o-Ps}}) - \frac{1}{2}(\lambda_{\mathrm{p-Ps}} - \lambda_{\mathrm{o-Ps}}) \operatorname{Sech} \xi \,. \end{aligned} \tag{39}$$

The intrinsic annihilation rate $\lambda_{\mathrm{o-Ps}}$ of the states $|2\rangle$ and $|3\rangle$ (Eqs. (29, 30)) is not influenced by the magnetic quenching and remains constant (cf. Eq. (2))

according to

$$\lambda_{\text{o-Ps}} = \frac{\kappa}{\tau^0_{\text{o-Ps}}} \,. \tag{40}$$

The intrinsic annihilation rate of p-Ps is also scaled by the density parameter $\kappa$ (cf. Eq. (1)) according to

$$\lambda_{\text{p-Ps}} = \frac{\kappa}{\tau^0_{\text{p-Ps}}} \,. \tag{41}$$

The intrinsic annihilation rates (Eqs. (38-40)) of the positronium states $|1\rangle$, $|2\rangle$, $|3\rangle$, and $|4\rangle$ (Eqs. (28-31)) are plotted in Figure 7 as functions of the external magnetic field.

In these rates the influence of the surrounding matter is only partially included through the parameter $\kappa$, which takes the relaxation of the spatial part of the positronium wave function into consideration. A practical way of incorporating the additional influence of condensed matter, i.e. spin–flip effects or transitions between the different positron species (free positron or positronium) or between positronium states, is the method of rate equations for the population numbers. The general form of such a system of equations for the population numbers $n_0$ (free positrons), $n_1$, $n_2$, $n_3$, and $n_4$ (positronium states $|1\rangle$, $|2\rangle$, $|3\rangle$, and $|4\rangle$), is as follows[l]. The population numbers may be time–dependent, their time derivatives are denoted by dots above the symbols.

$$\begin{aligned}
\dot{n}_0 &= -\left[\lambda_{\text{f}} + k_{\text{Ps}}\right] n_0 + k_{\text{diss}}\,(n_1 + n_2 + n_3 + n_4)\,, && (42)\\
\dot{n}_1 &= -\left[\lambda_{\text{pp-Ps}} + \lambda_{\text{po}} + (1/4)\,k_{\text{ex}}\,(2 + \text{Sech}^2\,\xi) + k_{\text{diss}}\right] n_1 \\
&\quad + (1/4)\,k_{\text{Ps}} n_0 \\
&\quad + (1/4)\,k_{\text{ex}}\left[(\text{Sech}^2\,\xi)\,n_4 + (1 - \text{Tanh}\,\xi)\,n_2 + (1 + \text{Tanh}\,\xi)\,n_3\right], && (43)\\
\dot{n}_2 &= -\left[\lambda_{\text{o-Ps}} + \lambda_{\text{po}} + (1/2)\,k_{\text{ex}} + k_{\text{diss}}\right] n_2 \\
&\quad + (1/4)\,k_{\text{Ps}} n_0 \\
&\quad + (1/4)\,k_{\text{ex}}\left[(1 - \text{Tanh}\,\xi)\,n_1 + (1 + \text{Tanh}\,\xi)\,n_4\right], && (44)\\
\dot{n}_3 &= -\left[\lambda_{\text{o-Ps}} + \lambda_{\text{po}} + (1/2)\,k_{\text{ex}} + k_{\text{diss}}\right] n_3 \\
&\quad + (1/4)\,k_{\text{Ps}}\, n_0 \\
&\quad + (1/4)\,k_{\text{ex}}\left[(1 + \text{Tanh}\,\xi)\,n_1 + (1 - \text{Tanh}\,\xi)\,n_4\right], && (45)\\
\dot{n}_4 &= -\left[\lambda_{\text{m-Ps}} + \lambda_{\text{po}} + (1/4)\,k_{\text{ex}}(2 + \text{Sech}^2\,\xi) + k_{\text{diss}}\right] n_4 \\
&\quad + (1/4)\,k_{\text{Ps}} n_0 \\
&\quad + (1/4)\,k_{\text{ex}}\left[(\text{Sech}^2\,\xi)\,n_1 + (1 + \text{Tanh}\,\xi)\,n_2 + (1 - \text{Tanh}\,\xi)\,n_3\right]. && (46)
\end{aligned}$$

[l] See reference[82], also references[86–88].

Here $\lambda_f$ is the annihilation rate of free positrons, i.e. positrons that are not bound as positronium, $\lambda_{po}$ denotes the pick-off annihilation rate of positronium, $k_{Ps}$ is the rate of delayed positronium formation, $k_{diss}$ is the rate of positronium dissociation, and $k_{ex}$ denotes the spin–flip rate of the electron in the positronium due to Heisenberg spin exchange[87]. Although not strictly true if positronium diffusion plays a role[89], all rate coefficients in Eqs. (42-46) may be taken as age–independent.

Knowledge of the initial occupation numbers is necessary for the solution of the system of equations (42-46). These depend on the initial spin polarization $\mathcal{P}$ of the positrons. If the electrons are initially unpolarized, the following initial conditions for the occupation numbers hold[65,82]:

$$n_0(0) = 1 - r\,, \tag{47}$$
$$n_1(0) = (r/4)\;[1 - \mathcal{P}\,\mathrm{Tanh}\,\xi]\,, \tag{48}$$
$$n_2(0) = (r/4)\;[1 + \mathcal{P}]\,, \tag{49}$$
$$n_3(0) = (r/4)\;[1 - \mathcal{P}]\,, \tag{50}$$
$$n_4(0) = (r/4)\;[1 + \mathcal{P}\,\mathrm{Tanh}\,\xi]\,, \tag{51}$$

where $r$ denotes the fraction of implanted positrons that form positronium instantaneously and $\mathcal{P}$ is the polarization of the positron beam[m]. In equations (47-51) $\mathcal{P}$ is positive if the positron spin is orientated parallel to the direction of the magnetic field $B_{appl}$ and negative if the positron–spin orientation is the opposite[n].

An appropriate way of solving the set of differential equations (42-46) is the method of Laplace transforms. The Laplace transforms of the time–dependent population numbers are defined as

$$\mathcal{L}_j(s) = \int_0^\infty \mathrm{e}^{-st} n_j(t)\mathrm{d}t \qquad (j = 0,\,1,\,2,\,3,\,4)\,. \tag{52}$$

Owing to the transformation properties of the time–derivatives of the population numbers, i.e.

$$s\mathcal{L}_j(s) - n_j(0) = \int_0^\infty \mathrm{e}^{-st} \dot{n}_j(t)\mathrm{d}t \qquad (j = 0,\,1,\,2,\,3,\,4)\,, \tag{53}$$

---

[m] As a result of the focusing effect of the applied magnetic field, $\mathcal{P}$ may depend on $\xi$ (cf. Sect. 2).

[n] The spin and the magnetic moment of the positron are parallel to each other and the positrons emitted in a $\beta^+$ decay are spin polarized in the direction of their velocity.

the Laplace–transformed system of equations will be a linear system of equations for the Laplace–transformed population numbers $\mathcal{L}_j(s)$, which can easily be solved. In the case of time–resolved measurements, by performing inverse Laplace transforms we may obtain the set of time–dependent population numbers $n_j(t)$, from which we may construct a model that describes the outcome of the magnetic–quenching experiment.

The experiments that are reported in Sect. 7 integrate with respect to the positron age. Since for the population numbers

$$\lim_{t \to \infty} n_j(t) = 0 \qquad (j = 0,\, 1,\, 2,\, 3,\, 4) \tag{54}$$

holds, for the Laplace–transformed population numbers

$$\mathcal{L}_j(0) = \int_0^\infty n_j(t)\mathrm{d}t \qquad (j = 0,\, 1,\, 2,\, 3,\, 4)\,. \tag{55}$$

Consequently, for the evaluation of the results of the time–integral experiments it is not necessary to perform the inverse Laplace transforms.

## 7 $e^+$SR experiments on samples in which positronium may be formed

The Stuttgart $e^+$SR set-up (cf. Figure 3) with polarized positrons from a $^{68}$Ge/$^{68}$Ga radioactive source is also suited for performing $e^+$SR experiments on non-metallic materials in which positronium may be formed[51,77,90,91]. The experimental set-up was slightly modified (Figure 8), mainly in order to decrease the background due to annihilation events in the vicinity of the sample. The sample can fill the volume of the hollow scintillator (veto scintillator, sample diameter: 16 mm, sample height: 10 mm). Those positrons which escape from the sample generate a veto signal in the photomultiplier and for such events the data is discarded. The veto counter, which is made of plastic scintillator of a thickness of 1 mm, possesses an efficiency close to unity for positrons and essentially zero for the annihilation photons. A reference for software stabilization of the spectrometer is provided by monitoring the 497 keV photon line of a $^{103}$Ru radioactive $\gamma$ source. The Doppler-broadened lineshape of the annihilation photons is measured with an intrinsic Ge detector (Princeton Gamma-Tech IGC30, its diameter is 2", its length 2"; energy resolution: 1.75 keV FWHM, relative efficiency: 0.29 at 1.33 MeV).

The theory of magnetic quenching of the positronium hyperfine interaction, as described in Sect. 6, allows us to evaluate the results of the present $e^+$SR experiments. The beam polarization ($\mathcal{P}$), which is a prerequisite of

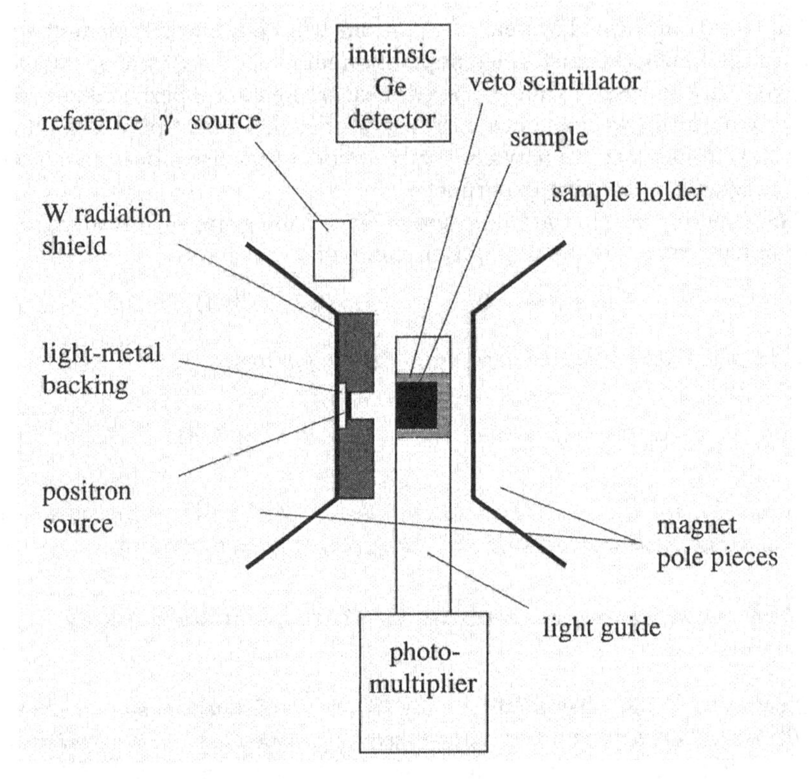


Figure 8. The modified Stuttgart polarized–positron ($e^+$SR) set-up (cf. Figure 3) for the study of positronium at room temperature[51]. The sample is surrounded by the thin plastic veto scintillator that is coupled to the photomultiplier by a long light guide. Veto signals are generated only by those positrons that do not annihilate in the sample.

such experiments, is included in equations (47-51), which describe the initial magnitudes of the occupation numbers of the various positron states.

The data acquisition in experiments on samples in which positronium may be formed is performed in a similar way as in the experiments on ferromagnetic iron samples described in Sect. 5. For the description of the magnetic quenching of the positronium hyperfine coupling, the same differential annihilation lineshape parameter $P^*$ is used as that defined in Sect. 5 (cf. Figure 4).

We report here $e^+$SR positronium magnetic quenching experiments that were performed on fused (Suprasil) quartz, type IIa natural diamond, polycrystalline silicon carbide (SiC, polytype 6H/15R), $C_{60}$ fullerite, and highly–

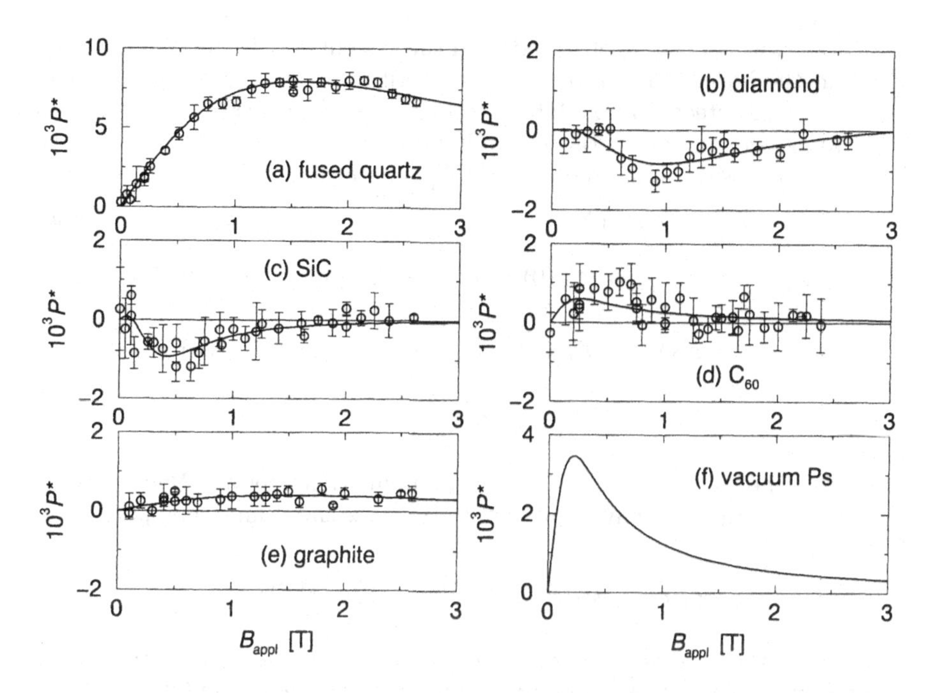


Figure 9. Results of room temperature e+SR positronium magnetic–quenching experiments on fused (Suprasil) quartz (a), type IIa natural diamond (b), polycrystalline silicon carbide (SiC, polytype 6H/15R) (c), $C_{60}$ fullerite (d), and highly–orientated pyrolytic graphite (HOPG) (e). The solid lines represent the results of the fit that is described in the text. The calculated quenching curve that corresponds to the unperturbed (vacuum) positronium is also shown (f). For the definition of the differential lineshape parameter $P^*$ see Figure 4.

orientated pyrolytic graphite (HOPG)[92]. The $P^* = P^*(B_{\text{appl}})$ quenching plots, which were obtained from room temperature measurements, are displayed in Figure 9. Figure 9f shows a calculated quenching curve corresponding to unperturbed (vacuum) positronium. The great variety of the different curve shapes in Figure 9 demonstrates that e+SR is a rather sensitive and informative method for the study of magnetic quenching of the positronium hyperfine interaction in condensed matter. The popular experimental method of magnetic quenching of the positron lifetime suffers from several practical difficulties (cf. Sect. 3), its ability to provide us with detailed information is in some respects limited, however it may be very useful for e.g., the determination of the positronium contact parameter $\kappa$[42,72].

Owing to the time–integrated population numbers of the various positronium states, which can be obtained from the system of rate equations (42-46) by the method of Laplace transforms as described in Sect. 6, we can calculate the number of detected annihilation photons from free–positron annihilation ($N_{\mathrm{f}}$), from two–photon intrinsic annihilation of positronium ($N_{\mathrm{p-Ps}}$), and from pick-off positronium annihilation ($N_{\mathrm{po}}$) according to

$$N_{\mathrm{f}} = N_0 \lambda_{\mathrm{f}} \mathcal{L}_0(0)\,, \tag{56}$$

$$N_{\mathrm{p-Ps}} = N_0 \lambda_{\mathrm{p-Ps}} \left[\mathcal{L}_1(0) \,\mathrm{Sinh}^2 \frac{\xi}{2} + \mathcal{L}_4(0) \,\mathrm{Cosh}^2 \frac{\xi}{2}\right] \mathrm{Sech}\,\xi\,, \tag{57}$$

$$N_{\mathrm{po}} = N_0 \lambda_{\mathrm{po}} \left[\left(\mathcal{L}_1(0) \,\mathrm{Cosh}^2 \frac{\xi}{2} + \mathcal{L}_4(0) \,\mathrm{Sinh}^2 \frac{\xi}{2}\right) \mathrm{Sech}\,\xi \right.$$
$$\left. + \mathcal{L}_2(0) + \mathcal{L}_3(0)\right], \tag{58}$$

where $N_0$ denotes the number of detected annihilation photons. By the means of these quantities, we may approximate the differential lineshape parameter $P^*$ as

$$P^*_{\mathrm{m}}(|\xi|) = K(|\xi|) \frac{N_{\mathrm{p-Ps}}(|\xi|) - N_{\mathrm{p-Ps}}(-|\xi|)}{N_0}\,. \tag{59}$$

$P^*_{\mathrm{m}}$ gives us the relative change of the number of annihilation events in the broad Doppler components due to magnetic–field reversal, and $K$ is a coefficient that depends on the experimental circumstances, in particular on the modulus of the external magnetic field. Inserting Eqs. (56-58) and the Laplace transforms of the occupation numbers as obtained from Eqs. (42-46) and from Eqs. (47-51) into Eq. (59), we may write

$$P^*_{\mathrm{m}}(|B_{\mathrm{appl}}|) = K^*(|B_{\mathrm{appl}}|) r \mathcal{P}(|B_{\mathrm{appl}}|) \frac{b_1 |B_{\mathrm{appl}}|}{c_0 + c_2 B^2_{\mathrm{appl}}}\,, \tag{60}$$

where $K^*$ is proportional to $K$ and $b_1$, $c_0$, and $c_2$ are known combinations of the unknown parameters such as $\kappa$, $\lambda_{\mathrm{f}}$, $\lambda_{\mathrm{po}}$, $k_{\mathrm{diss}}$, $k_{\mathrm{Ps}}$, and $k_{\mathrm{ex}}$ and some additional instrumental and other constants.

The 'model' differential lineshape parameter $P^*_{\mathrm{m}}(|B_{\mathrm{appl}}|)$ can be fitted to the experimental $P^*(|B_{\mathrm{appl}}|)$ results by the standard weighted least–squares fitting method by adjusting the magnitudes of the parameters incorporated in the model function.

The number of these parameters, which are unknown coefficients in the rate equations (42-46) (the density parameter $\kappa$, the annihilation rate $\lambda_{\mathrm{f}}$ of the free positrons, the pick-off annihilation rate $\lambda_{\mathrm{po}}$, the transition and spin–flip rates $k_{\mathrm{Ps}}$, $k_{\mathrm{diss}}$, $k_{\mathrm{ex}}$, the initial positronium formation probability $r$, and other factors), which depends on the experimental details, e.g., the magnetic–field dependent resolution factor of the Ge detector and the background, is too large for such a fitting procedure. This is still true if we consider that a part of the unknown parameters are less relevant for the fit (e.g., $\lambda_f$), or that some of them can be set to zero for a large class of specimens (e.g., $k_{Ps}$, $k_{diss}$, $k_{ex}$).

This is a common disadvantage of the so-called longitudinal–field quenching (LFQ) methods. In addition to $e^+$SR, the LFQ$\mu$SR method[93,94] also belongs to this class. The LFQ experiments generally suffer from the fact that all species and states of the probe (e.g., positron or muon), which are present in the sample, contribute to the measured quantity and this require a very complex data evaluation. This drawback is compensated by the advantage that the LFQ methods, like all longitudinal–field methods on spin systems, are particularly sensitive to dynamical processes. In transverse–field experiments, which are common in $\mu$SR[52–55], the spin coherence of the muon ensemble is a prerequisite for the precession signal. However, as we have seen in Sect. 3, no equivalent transverse counterparts exist to the $e^+$SR method.

It seems that evaluation of the positronium quenching curves (Figure 9) by fitting of Eq. (60) is not possible in many cases. This situation can be changed if we exploit the full information content of the results of $e^+$SR experiments. In particular, during data evaluation, any data compression such as data integration that provides the $P^*$ lineshape parameter, results in a loss of information. On the basis of the numbers of the annihilation photons that belong to different Doppler broadening (cf. Eqs. (56-58)), we can construct a two–dimensional 'model', the number of counts as function of the magnetic field and the energy measured by the Ge detector. For this two–dimensional model it is necessary to introduce some additional unknown parameters such as the magnetic–field dependent resolution width of the detector $\sigma_{res}$, the Doppler broadening due to the two–photon intrinsic annihilation of positronium $\sigma_{p-Ps}$, due to the pick-off annihilation $\sigma_{po}$, or due to the annihilation of the free positron $\sigma_f$. Here all Doppler lineshapes and the detector resolution function are approximated by Gaussian functions and the standard deviations $\sigma_i$ ($i$ = f, p-Ps, po) are the corresponding parameters.

The dependence of the two–dimensional lineshape on the energy $E$ of the annihilation photons, as measured by the Ge detector, and the applied magnetic field $B_{appl}$ can be approximated by the model function

$$\begin{aligned} N_m(E, B_{appl}) &= \sum_i [G_i(E, B_{appl}) + U_i(E, B_{appl})] \\ &\quad + G_{ref}(E, B_{appl}) + U_{ref}(E, B_{appl}) + U \\ &\qquad (i = \mathrm{f}, \mathrm{p-Ps}, \mathrm{po}), \end{aligned} \tag{61}$$

where the quantities $G_i$ describe the numbers of counts from the annihilation events from the various positron states, $U_i$ are the corresponding contributions to the background, $G_{\mathrm{ref}}$ and $U_{\mathrm{ref}}$ describe the number of counts and the corresponding background from the reference photon source, and $U$ is the random background[88].

For the contributions from the annihilation events we can write

$$G_i\left(k, B_{\mathrm{appl}}\right) = \frac{N_i}{\sqrt{2\pi\left(\sigma_i^2 + \sigma_{\mathrm{res}}^2\right)}} \exp\left[-\frac{\left(k - k_0\right)^2}{2\left(\sigma_i^2 + \sigma_{\mathrm{res}}^2\right)}\right] \tag{62}$$

and

$$U_i\left(k, B_{\mathrm{appl}}\right) = \frac{1}{2} H N_i \operatorname{erfc}\left[\frac{k - k_0}{\sqrt{2\left(\sigma_i^2 + \sigma_{\mathrm{res}}^2\right)}}\right]; \tag{63}$$

for those from the reference line

$$G_{\mathrm{ref}}\left(k, B_{\mathrm{appl}}\right) = \frac{N_0^{\mathrm{ref}}}{\sqrt{2\pi\sigma_{\mathrm{res}}^2}} \exp\left[-\frac{\left(k - k_0^{\mathrm{ref}}\right)^2}{2\sigma_{\mathrm{res}}^2}\right] \tag{64}$$

and

$$U_{\mathrm{ref}}\left(k, B_{\mathrm{appl}}\right) = \frac{1}{2} H N_0^{\mathrm{ref}} \operatorname{erfc}\left[\frac{k - k_0^{\mathrm{ref}}}{\sqrt{2\sigma_{\mathrm{res}}^2}}\right], \tag{65}$$

where we introduced the channel number $k$ of the data acquisition system instead of the energy $E$ and erfc denotes the complementary error function. $k_0$ is the channel number that corresponds to the photon energy of 511 keV, $k_0^{\mathrm{ref}}$ corresponds to 497 keV, the energy of the reference photons. For details and justification of Eqs. (61-65) see reference[88]. In Eqs. (61-65), in addition to $N_i$ and $N_0^{\mathrm{ref}}$, the parameters $\sigma_{\mathrm{res}}$, $k_0$, $k_0^{\mathrm{ref}}$, and $H$ may also depend on $B_{\mathrm{appl}}$, however these dependences can be determined during the fitting procedure together with the magnitudes of the physical parameters.

The set of physical and instrumental parameters that are comprised in the actual form of Eq. (61) were evaluated by fitting Eq. (61) to the results of the $e^+$SR measurements illustrated in Figure 9. As it was discussed above, the fit of the Eq. (60) cannot usually provide us with an unequivocal set of resulting parameters. The two–dimensional fit of Eq. (61), i.e. a fit with a summation over all energy channels $k$ and over all magnetic fields $B_{\mathrm{appl}}$ of both polarities, which always gives us a unique set of fit parameters, may result in a lengthy computing task. A combined data evaluation, the simultaneous fit of both Eqs. (60) and (61), however, results in a good compromise, by means of affordable computing time we obtain a reliable set of parameters[88] and, additionally, this combined fitting procedure is rather robust.

The most typical experimental result among those shown in Figure 9 is the one obtained on fused quartz (Figure 9a). Similar curves have been reported previously[51,82,88,91]. The results shown in Figures 9d and 9e also belong to the same class. On natural diamond (Figure 9b) and on SiC (Figure 9c), however, the results of the $e^+$SR experiments reveal a pathological feature: the sign of the quantity $P^*$ is opposite to that obtained in all other cases as well as that calculated for unperturbed positronium (Figure 9f). This may be interpreted as an indication that the intrinsic annihilation of parapositronium results in a Doppler line that is broader than the lines from the two other channels of annihilation, i.e. the annihilation of the free positrons and the pick-off annihilation of orthopositronium.

Although previously no positronium formation has been reported in SiC and only ACAR experiments revealed the presence of positronium in diamond[95] and, moreover, lifetime and AMOC experiments[96] performed on the same diamond sample as studied here (Figure 9b) failed to show any evidence of positronium, this unexpected behaviour of the observed $P^*(B_{\mathrm{appl}})$ curves was interpreted as a trace of positronium formation at epithermal kinetic energies[92]. Therefore, we have to modify the model functions Eqs. (60) and (61) accordingly. E.g., in the functions $G_i\,(k, B_{\mathrm{appl}})$ ($i$ = p-Ps, po) (defined in Eq. (62)) we must introduce a Doppler broadening for the annihilation of the epithermal positronium with an explicit time dependence and we have to integrate it with respect to time. According to the calculation of Seeger[97], the slowing-down of positronium is mainly governed by the scattering of optical phonons and the centre of mass momentum of positronium decreases linearly with time at least in the relevant part of the slowing-down process. Consequently, two additional parameters, the positronium kinetic energy $E_0$ at its formation, which corresponds to the initial Doppler broadening of the parapositronium annihilation, and the slowing-down time $\tau_{\mathrm{sd}}$ are necessary for the description of the measured time–integrated Doppler broadening. An additional complication is that during slowing-down, the modulus of the positronium momentum possesses a linear time dependence but its angular distribution is random, and therefore the corresponding time–integrated Doppler–broadening function cannot be written in a closed form, it must be evaluated numerically (for details see reference[92]).

As obtained from least–squares fits, the characteristic parameters of positronium in various materials are shown in Table 2. The fitting procedure was performed in all cases by the means of the modified Eqs. (60) and (61) since, even if it is not obvious from Figure 9, delayed slowing-down may also play a role in cases of $P^* > 0$. As a matter of fact, only in graphite and $C_{60}$ was the delayed slowing-down absent. The parameters of the positron

Table 2. Characteristic parameters of positronium in natural type IIa diamond, crystalline silicon carbide (SiC, 6H/15R polytype), fused quartz (Suprasil), highly orientated pyrolytic graphite (HOPG), and $C_{60}$ fullerite as obtained from least-squares fits of the modified Eqs. (60) and (61)[92] to the room-temperature $e^+$SR data (cf. Figure 9): fraction $r$ of positrons which form positronium, electron density parameter $\kappa$, pick-off lifetime $\tau_{po} = (\lambda_{po})^{-1}$, slowing-down time $\tau_{sd}$, kinetic energy $E_0$ of positronium at the time of Ps formation, Gaussian width $\sigma_{po}$ characterizing the momentum distribution of pick-off electrons, and Gaussian width $\sigma_{p-Ps}$ characterizing the momentum distribution of p-Ps. The errors are statistical $1\sigma$ errors. No error is given if the quantity was kept constant during the fitting.

| | diamond | SiC | fused quartz | graphite | $C_{60}$ |
|---|---|---|---|---|---|
| $100r$ | $7 \pm 1$ | $3.7 \pm 0.4$ | $67 \pm 4$ | $0.9 \pm 0.6$ | $3 \pm 1$ |
| $\kappa$ | $0.99 \pm 0.05$ | $0.15 \pm 0.02$ | $1.4 \pm 0.1$ | $0.7 \pm 0.1$ | $0.3 \pm 0.1$ |
| $\tau_{po}$ (ns) | $31 \pm 7$ | $20 \pm 8$ | $1.6 \pm 0.1$ | $0.58 \pm 0.03$ | $25 \pm 10$ |
| $\tau_{sd}$ (ns) | $12.6 \pm 0.6$ | $9 \pm 2$ | $0.04 \pm 0.01$ | | |
| $E_0$ (eV) | $8.03 \pm 0.06$ | $6.85 \pm 0.07$ | $3.08 \pm 0.08$ | | |
| $\sigma_{po}$ (keV) | $2.1 \pm 0.3$ | $1.8 \pm 0.2$ | $1.07 \pm 0.03$ | $3.1 \pm 1.3$ | $2.8 \pm 1.4$ |
| $\sigma_{p-Ps}$ (keV) | 0.1 | 0.1 | 0.1 | $0.4 \pm 0.2$ | $0.5 \pm 0.3$ |

slowing-down in fused quartz are in good agreement with those defined by the more direct – time-resolved – AMOC experiments (see the results of the recent analysis[98] of AMOC experiments on fused quartz[99]).

The results of the data evaluation (Table 2) show some very interesting properties. The small fractions $r$ of positrons which form positronium prove the extreme sensitivity of the $e^+$SR experiments. Presumably, the very small magnitudes of these fractions are responsible for the fact that previously no positronium was observed in SiC, graphite, and $C_{60}$. The contact parameter $\kappa$ is much smaller than unity in SiC and $C_{60}$, and in contrast, in fused quartz it is much larger than unity. The $\kappa < 1$ case represents positronium with enlarged spatial dimensions, $\kappa > 1$ corresponds to a spatially compressed positronium wave function. The pick-off lifetime $\tau_{po}$ is generally connected to the electron density in the surroundings of positronium. The extremely large $\tau_{po}$ magnitude in $C_{60}$ reveals that positronium is located in the interior of the $C_{60}$ molecule.

The slowing-down time $\tau_{sd}$ is very long in diamond and in SiC. Since these materials exhibit extreme mechanical hardness, the ineffectiveness of the positronium–phonon scattering may cause this phenomenon. However, the initial positronium kinetic energies $E_0$ obtained from this experiment are comparable to the binding energy of the unperturbed positronium of 6.8 eV. A rather interesting feature in graphite and in $C_{60}$ is the relatively strong – time-independent – Doppler broadening of the line of the intrinsic parapositronium annihilation.

## 8 Summary and future prospects of the spin–polarized positron–beam technique

It has been demonstrated in the present work how positrons emitted from radioactive sources, which are – as a 'Gift of Nature' – spin–polarized, can be utilized. One important point is that as a result of the small mass of the electron or the positron, all available radioactive positron sources provide useful spin polarization for $e^+$SR measurements. The other feature of the radioactive positron sources is that the majority of the positrons at the time of $\beta$ decay possesses a kinetic energy of the order of the rest energy of the electron or positron, 511 keV, and therefore in most cases the positrons are not substantially depolarized during slowing-down in condensed matter. Since the positron slowing-down, if no positronium formation occurs, is a rather fast process, $e^+$SR experiments are feasible for the study of the magnetic 'history' of the positron, i.e. the time dependence of the magnetic fields experienced by the magnetic moment of the positron during its life in the sample. Since the direct observation of the so-called residual spin polarization of the positrons, i.e. the spin polarization of the positrons at the time of annihilation is impossible (in contrast to $\mu$SR or NMR methods), the sample itself must also play the role of the spin analyser. This condition severely limits the candidates for $e^+$SR experiments, albeit, we still have a large number of possibilities for important and interesting measurements.

In experiments of type (a) (cf. Sect. 3), the specimen must possess spin–polarized electrons. In such experiments, for instance, the diffusive behaviour of the thermalized positrons in materials, in which strong local magnetic fields may depolarize the positron spin, can be studied, e.g., in $\alpha$ iron. These experiments are related to common spin–relaxation measurements, such as NMR or $\mu$SR.

In experiments of type (b) (cf. Sect. 3), positronium must be formed in the specimen. In such experiments the dependence of the magnetic quenching of the hyperfine interaction of the electron and the positron in the positronium in the applied longitudinal magnetic field on the positron spin polarization at the time of positronium formation are exploited. By the means of this kind of experiments, a very detailed description of the positronium state in condensed matter is possible even in the case of very small ($< 1\%$) fraction of positrons that form positronium. In the experiments reported in Sect. 7, positronium was found for the first time in SiC, graphite, and $C_{60}$[92]. In addition to the density parameter $\kappa$, these experiments usually provide us with values of the dynamical parameters of the positronium states, such as annihilation rates or rates of transitions between different positron and positronium states. The latter may still be true in the case of time–integrating data acquisition, e.g., in Doppler broadening experiments.

Positron–spin relaxation ($e^+$SR) experiments give us a very sensitive method for the study of the details of the interaction of positrons with condensed matter. Nevertheless, this experimental technique is still far from its full potential and we can suppose that $e^+$SR may soon undergo further progress. Since all $e^+$SR experiments require the presence of strong applied magnetic fields, probably the most straightforward way of realizing such measurements is the time–integral Doppler–broadening experiment as reported in the present work. The time–sensitive information that is provided by such a time–integral experiment, however, is somewhat limited, a time–differential set-up may be very useful for future detailed studies of the positronium states. The construction of such an instrument with limited, but appropriate, time resolution may today be doubtlessly feasible.

Another way that may significantly enhance the quality of $e^+$SR results is the application of the dual–detector Doppler–broadening spectrometer that was proposed earlier by Lynn and re-discovered recently[100–102]. This set-up may reduce the fraction of the background in the wings of the Doppler line, exactly in the region which provides the most important information in an $e^+$SR experiment. If the second detector is made of a dense, highly effective scintillator, e.g., of NaI or $BaF_2$, the loss in the count rate may be insignificant. In this way the present duration of the measurements, i.e. typically couples of weeks for a set of data as that displayed in Figure 9a, may be made shorter. This second detector may also play the role of the 'stop' counter in a time–resolved experimental set-up.

In addition to the $e^+$SR methods, other experiments may also exploit the existence of the polarized positron beams. An example for such a related effect was recently demonstrated by Lynn and co-workers[103]. In this experiment, collimated positrons transmitted through a thin single crystal of Si reveal a fine angular structure due to strong quantum channeling effects. It was also shown that basically the same experiment, i.e. the detection of the lattice–steering (channelling) pattern of a high–energy spin–polarized positron beam may provide us with information on the electron–spin polarization in the crystal studied.

Another example is the spin–polarized positron experiment of Maier and co-workers[63,64] on coloured KCl samples (cf. Sect. 3). Their work shows the feasibility of the spin–polarized positron technique, which is a nuclear method, in EPR experiments for detecting the electron spin polarization of small concentration of paramagnetic centres.

The $e^+$SR itself and other more-or-less related methods that have yet only been demonstrated may exploit future high–intensity spin–polarized positron–beam facilities of positron kinetic energies of the order of eV or keV (cf. Sect. 2). Such beams may make the construction of time–resolved $e^+$SR set-ups easier, and in any case allow for significantly faster data acquisition. The lowest– (variable–) energy versions of such beams may help to extend the long–known possibility of using slow positrons as surface probes[3] of spin–polarized positrons.

Beyond the technical challenges, high–intensity spin–polarized positron beams may open the way for detailed, systematic studies of various positronium systems in condensed matter, e.g., in the field of positronium chemistry, in which spin–polarized positron research may significantly extend our knowledge on the dynamical processes. Systematic measurements may also help us to answer questions that have remained unsolved for a long time, e.g., the question of the existence and properties of the anisotropic positronium (in analogy to the well-known anomalous muonium[53]), which was recently detected by $e^+$SR in crystalline quartz[104,105], or the question of the existence of the 'anomalous' positronium, in which the hyperfine splitting and the annihilation rates do not scale in the same way, as evidenced by magnetic–quenching positron–lifetime experiments, e.g., in naphthalene[106].

## Acknowledgements

The author is particularly indebted to Professor A. Seeger who initiated the Stuttgart polarized–positron activity several years ago and since then has continuously collaborated in the further development of the $e^+$SR experiment and its theory. The co-operation of earlier and present researchers at the Stuttgart $e^+$SR facility, F. Banhart, Th. Gessmann, P. Harmat, F. Jaggy, and O. Stritzke, is also acknowledged. The author wishes to thank J. Ehmann, A. Siegle, and H. Stoll for many fruitful discussions.

## References


1. Positrons in Solids, ed. P. Hautojärvi (Springer, Berlin, 1979).
2. Positron Solid–State Physics, Proc. Int. School of Physics E. Fermi, Varenna, Italy, 1981, eds. W. Brandt and A. Dupasquier (North Holland, Amsterdam, 1983).
3. P. J. Schultz and K. G. Lynn, *Rev. Mod. Phys.* **60**, 701 (1988).
4. Positron and Positronium Chemistry, eds. D. M. Schrader and Y. C. Jean (Elsevier, Amsterdam, 1988).
5. Positron Spectroscopy of Solids, Proc. Int. School of Physics E. Fermi, Varenna, Italy, 1993, eds. A. Dupasquier and A. P. Mills jr. (IOS, Amsterdam, 1995).
6. O. E. Mogensen: Positron Annihilation in Chemistry (Springer, Berlin, 1995).
7. Proceedings of the Seventh International Workshop on Slow–Positron Beam Techniques for Solids and Surfaces, eds. W. B. Waeber, M. Shi, and A. A. Manuel, *Applied Surface Science* **116** (1996).
8. Proceedings of the $5^{th}$ International Workshop on Positron and Positronium Chemistry, eds. Zs. Kajcsos, B. Lévai, and K. Süvegh, *J. Radioanalytical and Nuclear Chemistry, Articles* **210-211** (1996).
9. Cs. Szeles and K. G. Lynn, *Encycl. Appl. Phys.* **14**, 607 (1996).
10. Proceedings of the $11^{th}$ International Conference on Positron Annihilation, eds. Y. C. Jean, M. Eldrup, D. M. Schrader, and R. N. West, *Mat. Sci. Forum* **255-257** (1997).
11. C. S. Wu and I. Shaknov, *Phys. Rev.* **77**, 136 (1950).
12. St. Mohorovičić, *Astron. Nachr.* **253**, 94 (1934).
13. M. Deutsch, *Phys. Rev.* **82**, 455 (1951).
14. L. Wolfenstein and D. G. Ravenhall, *Phys. Rev.* **88**, 279 (1952).
15. H. Stoll, in this volume.
16. W. Pauli: Die Allgemeinen Prinzipien der Wellenmechanik, in: Handbuch der Physik, Vol. V., Part I, ed. S. Flügge, (Springer, Berlin, 1958) p. 165.
17. H. Dehmelt, *Z. Phys.* D **10**, 127 (1988).
18. H. Batelaan, T. J. Gay, and J. J. Schwendiman, *Phys. Rev. Lett.* **79**, 4517 (1997).
19. J. Kessler: Polarized Electrons, $2^{nd}$ Ed. (Springer, Berlin, 1985).
20. T. D. Lee and C. N. Yang, *Phys. Rev.* **104**, 254 (1956).
21. C. S. Wu, E. Ambler, R. W. Hayward, D. D. Hoppes, and R. P. Hudson, *Phys. Rev.* **105**, 1413 (1957).
22. Table of Isotopes, $8^{th}$ Ed., eds. R. B. Firestone and V. S. Shirley (Wiley, New York, 1996).

23. F. G. Houtermans, J. Geiss, H. Müller in Landolt–Börnstein, 6$^{th}$ Ed., Vol. I, Part 5, ed. K. H. Hellwege (Springer, Berlin, 1952).
24. Table of Isotopes, 7$^{th}$ Ed., eds. C. M. Lederer and V. S. Shirley (Wiley, New York, 1978).
25. L. D. Landau and E. M. Lifshits: Classical Theory of Fields, (Pergamon, Oxford, 1971).
26. J. Van House and P. W. Zitzewitz, *Phys. Rev.* A **29**, 96 (1984).
27. V. B. Berestetskiĭ, E. M. Lifshits, and L. P. Pitaevskiĭ: Relativistic Quantum Theory, Part 1, (Pergamon, Oxford, 1971).
28. L. A. Dick, L. Feuvrais, and M. Spighel, *Phys. Lett.* **7**, 150 (1963).
29. A. Buhler, N. Cabibbo, M. Fidecaro, T. Massam, Th. Muller, M. Schneegans, and A. Zichichi, *Phys. Lett.* **7**, 368 (1963).
30. S. Bloom, L. A. Dick, L. Feuvrais, G. R. Henry, P. C. Macq, and M. Spighel, *Phys. Lett.* **8**, 87 (1964).
31. G. W. Ford and C. J. Mullin, *Phys. Rev.* **108**, 477 (1957).
32. S. M. Neamtan, *Phys. Rev.* **110**, 173 (1958).
33. C. Bouchiat and J. M. Lévi–Leblond, *Nuovo Cimento* **33**, 193 (1964).
34. C. K. Iddings, G. L. Shaw, and Y. S. Tsai, *Phys. Rev.* **135**, B1388 (1964).
35. L. Braicovich, B. De Michelis, and A. Fasana, *Phys. Rev.* **164**, 1360 (1967).
36. L. Braicovich, *Nuovo Cimento* **55A**, 609 (1968).
37. V. L. Lyuboshits, *Sov. J. Nucl. Phys.* **31**, 509 (1980).
38. V. L. Lyuboshits, *Sov. J. Nucl. Phys.* **32**, 362 (1980).
39. V. G. Baryshevsky, *Nucl. Instrum. Methods* B **44**, 266 (1990).
40. E. Segrè: Nuclei and Particles, 2$^{nd}$ Ed., (Benjamin, Reading, Massachusetts, 1977).
41. A. Bisi, G. Consolati, F. Quasso, and L. Zappa, *Nuovo Cimento* D **11**, 635 (1989).
42. G. Consolati, *J. Radioanal. Nucl. Chem., Articles* **210**, 273 (1996).
43. P. W. Zitzewitz, J. C. Van House, A. Rich, and D. W. Gidley, *Phys. Rev. Lett.* **43**, 1281 (1979).
44. D. W. Gidley, A. R. Köymen, and T. Weston Capehart, *Phys. Rev. Lett.* **49**, 1779 (1982).
45. A. Rich, J. Van House, D. W. Gidley, R. Conti, and P. W. Zitzewitz, *Appl. Phys.* A **43**, 275 (1987).
46. T. Kumita, M. Chiba, R. Hamatsu, M. Hirose, H. Iijima, M. Irako, N. Kawasaki, Y. Kurihara, T. Matsumoto, N. Nakabushi, T. Omori, Y. Takeuchi, M. Washio, and J. Yang, *Appl. Surf. Sci.* **116**, 1 (1997).
47. T. Nakajyo, M. Tashiro, T. Koizumi, I. Kanazawa, F. Komori, and Y. Ito, *Appl. Surf. Sci.* **116**, 168 (1997).

48. M. Irako, M. Chiba, M. Fukusima, R. Hamatsu, M. Hirose, T. Hirose, H. Iijima, T. Kumita, N. C. Mazumdar, and M. Washio, *Materials Science Forum* **255-257**, 750 (1997).
49. P. G. Coleman, D. R. Cook, S. F. Schaffner, L. M. Diana, S. C. Sharma, and S. Y. Chuang, in: Positron Annihilation, eds. P. G. Coleman, S. C. Sharma, and L. M. Diana (North-Holland, Amsterdam, 1982) p. 174.
50. T. Tabata, R. Ito, and S. Okabe, *Nucl. Instrum. Methods* **94**, 509 (1971).
51. Th. Gessmann, P. Harmat, J. Major, and A. Seeger, *Appl. Surf. Sci.* **116**, 114 (1997).
52. Muons and Pions in Materials Research, eds. J. Chappert and R. I. Grynszpan, (North-Holland, Amsterdam, 1984).
53. A. Schenck: Muon Spin Rotation Spectroscopy (Adam Hilger, Bristol, 1985).
54. V. P. Smilga and Yu. M. Belousov: The Muon Method in Science (Nova Science, New York, 1994).
55. J. H. Brewer, *Encycl. Appl. Phys.* **11**, 23 (1994).
56. L. A. Page, *Rev. Mod. Phys.* **31**, 759 (1959).
57. A. Rich, *Rev. Mod. Phys.* **53**, 127 (1981).
58. S. S. Hanna and R. S. Preston, *Phys. Rev.* **106**, 1363 (1957).
59. S. S. Hanna and R. S. Preston, *Phys. Rev.* **109**, 716 (1958).
60. I. Lovas, *Nucl. Phys.* **17**, 279 (1960).
61. S. Berko: Positron Annihilation in Ferromagnetic Solids, in: Positron Annihilation, eds. A. T. Stewart and L. O. Roellig (Academic, New York, 1967) p. 61.
62. T. Akahane and S. Berko, in: Positron Annihilation, eds. P. G. Coleman, S. C. Sharma, and L. M. Diana (North-Holland, Amsterdam, 1982) p. 874.
63. U. Lauff, J. Major, A. Seeger, H. Stoll, A. Siegle, Ch. Deckers, H. Greif, K. Maier, and M. Tongbhoyai, *Phys. Lett.* A **182**, 165 (1993).
64. Ch. Deckers, J. Ehmann, H. Greif, F. Keuser, W. Knichel, U. Lauff, K. Maier, A. Siegle, and M. Tongbhoyai, *J. Physique IV* **5**, C1 81 (1995).
65. L. A. Page and M. Heinberg, *Phys. Rev.* **106**, 1220 (1957).
66. A. Greenberger, A. P. Mills, A. Thompson, and S. Berko, *Phys. Lett.* A **32**, 72 (1970).
67. D. Herlach and F. Heinrich, *Helv. Phys. Acta* **45**, 10 (1972).
68. Y. Nagashima and T. Hyodo, *Phys. Rev.* B **41**, 3937 (1990).
69. V. L. Telegdi, cited by L. Grodzins, *Prog. Nucl. Phys.* **7**, 163 (1959).
70. A. Lundby, *Prog. Elem. Part. Cosmic Ray Phys.* **5**, 1 (1960).
71. L. A. Dick, L. Feuvrais, L. Madansky and V. L. Telegdi, *Phys. Lett.* **3**,

326 (1963).

72. A. Bisi, A. Fiorentini, E. Gotti, and L. Zappa, *Phys. Rev.* **128**, 2195 (1962).
73. G. Gerber, D. Newmann, A. Rich, and E. Sweetman, *Phys. Rev.* D **15**, 1189 (1977).
74. A. Seeger: Positive Muons and Pions in Material Research: Perspectives and Future Developments, in: Muons and Pions in Materials Research, eds. J. Chappert and R. I. Grynszpan, (North-Holland, Amsterdam, 1984), p. 251.
75. A. Seeger, J. Major, and F. Jaggy, in: Positron Annihilation, eds. P. C. Jain, R. M. Singru, and K. P. Gopinathan, (World Scientific, Singapore, 1985) p. 137.
76. A. Seeger: Positive Muons as Light Isotopes of Hydrogen, in: Hydrogen in Metals, Vol. 1, eds. G. Alefeld and J. Völkl, (Springer, Berlin, 1978), p. 349.
77. A. Seeger, J. Major, and F. Banhart, *phys. stat. sol.* (a) **102**, 91 (1987).
78. F. Banhart, J. Major, and A. Seeger, in: Positron Annihilation, eds. L. Dorikens-Vanpraet, M. Dorikens, and D. Seegers (World Scientific, Singapore, 1989) p. 281.
79. L. Schimmele, A. Seeger, T. Stammler, W. Templ, C. Baines, T. Grund, M. Hampele, D. Herlach, M. Iwanowski, K. Maier, J. Major, and T. Pfiz, *Hyperfine Interactions* **85**, 337 (1994).
80. R. Blank, L. Schimmele, and A. Seeger, in: Positron Annihilation, eds. L. Dorikens-Vanpraet, M. Dorikens, and D. Seegers (World Scientific, Singapore, 1989) p. 278.
81. M. Deutsch and E. Dulit, *Phys. Rev.* **84**, 601 (1951).
82. J. Major, A. Seeger, J. Ehmann, and Th. Gessmann: Positronium in Condensed Matter Studied with Spin-Polarized Positrons, in: Atomic Physics Methods in Modern Research, eds. K. P. Jungmann, J. Kowalski, I. Reinhard, and F. Träger (Springer, Berlin, 1997) p. 381.
83. V. I. Goldanskii and E. P. Prokopev, *JETP Letters* **4**, 284 (1966).
84. A. Dupasquier: Positroniumlike Systems in Solids, in: Positron Solid State Physics, Proc. Int. School of Physics E. Fermi, Varenna, Italy, 1981, eds. W. Brandt and A. Dupasquier (North Holland, Amsterdam, 1983) p. 510.
85. G. Breit and I. I. Rabi, *Phys. Rev.* **38**, 2082 (1931).
86. A. P. Mills jr., *J. Chem. Phys.* **62**, 2646 (1975).
87. M. Senba, *Can. J. Phys.* **74**, 385 (1996).
88. Th. Gessmann, P. Harmat, J. Major, and A. Seeger, *Z. Phys. Chem.* **199**, 213 (1997).

89. R. Würschum and A. Seeger, *Z. Phys. Chem.* **192**, 47 (1995).
90. J. Major, A. Seeger, and O. Stritzke, *Materials Science Forum* **105-110**, 1939 (1992).
91. Th. Gessmann: Untersuchung von Positronium in kondensierter Materie mit Spinpolarisierten Positronen, Dr. rer. nat. Thesis, Universität Stuttgart, (Cuvillier, Göttingen, ISBN 3-89588-956-3, 1997).
92. Th. Gessmann, J. Major, and A. Seeger, *J. Phys.: Condens. Matter* **10**, 10493 (1998).
93. R. Scheuermann, L. Schimmele, A. Seeger, Th. Stammler, T. Grund, M. Hampele, D. Herlach, M. Iwanowski, J. Major, M. Notter, and Th. Pfiz, *Phil. Mag.* B **72**, 161 (1995).
94. R. Scheuermann, L. Schimmele, J. Schmidl, A. Seeger, Th. Stammler, E. E. Haller, D. Herlach, and J. Major, *Hyperfine Interactions* **105**, 357 (1997).
95. S. Fujii, Y. Nishibayashi, S. Shikata, A. Uedono, and S. Tanigawa, *J. Appl. Phys.* **78**, 1510 (1995).
96. R. W. N. Nilen, U. Lauff, S. H. Connell, H. Stoll, A. Siegle, H. Schneider, P. Castellaz, J. Kraft, K. Bharuth-Ram, J. P. F. Sellschop, and A. Seeger, *Appl. Surf. Sci.* **116**, 198 (1997).
97. A. Seeger, *Appl. Surf. Sci.* **85**, 8 (1995).
98. A. Siegle: Positronenzerstrahlung in kondensierter Materie – eine Untersuchung mit der Methode der Lebensalter–Impuls–Korrelation, Dr. rer. nat. Thesis, Universität Stuttgart, (Cuvillier, Göttingen, ISBN 3-89712-129-8, 1998).
99. H. Schneider, A. Seeger, A. Siegle, H. Stoll, P. Castellaz, and J. Major, *Appl. Surf. Sci.* **116**, 146 (1997).
100. K. G. Lynn, J. R. MacDonald, R. A. Boie, L. C. Feldmann, J- D. Gabbe, M. F. Robbins, E. Bonderup, and J. Golovchenko, *Phys. Rev. Lett.* **38**, 241 (1977).
101. V. J. Ghosh, M. Alatalo, P. Asoka-Kumar, K. G. Lynn, and A. C. Kruseman, *Appl. Surf. Sci.* **116**, 278 (1997).
102. A. C. Kruseman, H. Schut, A. van Veen, P. E. Mijnarends, M. Clement, and J. M. M. de Nijs, *Appl. Surf. Sci.* **116**, 192 (1997).
103. R. Haakenaasen, Lene Vestergaard Hau, J. A. Golovchenko, J. C. Palathingal, J. P. Peng, P. Asoka-Kumar, and K. G. Lynn, *Phys. Rev. Lett.* **75**, 1650 (1995).
104. A. Seeger, *Materials Science Forum* **255-257**, 1 (1997).
105. Th. Gessmann, J. Major, A. Seeger, and J. Ehmann, to be published.
106. A. Bisi, G. Consolati, and L. Zappa, *Hyperfine Interactions* **36**, 29 (1987).

CHAPTER 10

# THE FUTURE: INTENSE BEAMS

RICHARD H HOWELL
*Physics Department*
*Lawrence Livermore National Laboratory*
*Livermore CA, 94550, USA*
*E-mail: howell5@llnl.gov*

There is a critical need by all positron experiments for high intensity positron beams. Attempts to fill this need have been successful and beams with intensities as high as $10^{10}e^+s^{-1}$ are available. These beams are necessarily located in association with nuclear reactors or high power electron linacs. New facilities under development offer the promise of even higher positron currents.

## 1. The need for intensity

Many commonly used spectroscopies have gone through an evolution from a phase when the utility of the technique was clear but applications were limited by the strength of the source. When higher flux, brighter sources became available there was a revolution in the application of the technique. Witness the explosion of accessible problems in photo spectroscopies when bright photon sources became available at synchrotrons. Positron spectroscopies are making that transition. Low positron intensity is one of the most severe limits in positron beam spectroscopic methods. Since the development of the first monoenergetic, keV positron beams there has been a constant search for methods to produce positron beams with higher currents and with improved beam characteristics such as spot size and brightness. Achieving higher currents has have been the subject of focused conferences [Ottewitte 1987] and has lead to a continuing stream of proposals and workshops for new high current beam facilities. Two techniques to produce high current beams have had success and several facilities with high positron currents have been developed at reactors and electron linacs. Intense beam facilities tend to be planned to service several experiments as the floor plans in figures 1,3,4, and 5 show.

The need to use large accelerators or nuclear reactors with high capital costs to produce an intense beam of positrons defines a significant characteristics of intense beam facilities. Up to now this has limited attempts to produce intense beams to shared use of existing facilities and has lead to operational restrictions of successful beams. Some successful high current beams have ceased operation when the host accelerator or nuclear reactor ceased operation.

Low intensity positron beams from commercial radioactive sources are used for materials spectroscopy in most laboratories. The community using these low intensity positron beams to perform spectroscopy for materials analysis has made great progress in developing positron moderators to produce monochromatic beams and in developing bright positron beams with minimal loss of initial positron current. The best beams from commercial sources have currents $\leq 10^6$ $e^+s^{-1}$. To separate our discussion we will consider any beams greater than $10^7$ $e^+s^{-1}$ as intense.

Low intensity systems have been used with great skill to investigate surface and near surface features. These experiments have produced exciting data on surface structure and the depth distribution of defects at surfaces and buried interfaces. Low intensity beams have also been used extensively in atomic and fundamental physics experiments. However, only two positron annihilation spectroscopes have been routinely performed with lower intensity beams. They are electron momentum spectroscopy by Doppler broadening of the annihilation gamma rays and positron lifetimes with a pulsed beam. The value of other positron spectroscopies such as diffraction, Auger emission and microscopy has been proven by initial measurements with a low beam but limits in the data rate and beam quality set by low positron intensity exclude most interesting solid state and surface science problems.

There is a long list of experiments which require higher positron beam currents in order to effectively take data. These include electron momentum measurements by two dimensional angular correlation [Howell 1985], low energy positron diffraction [Rosenberg 1980], positron imaging microscopy [Brandes 1988], positron microprobe spectroscopy [Howell 1997], positron stimulated Auger spectroscopy [Weiss 1988], positron stimulated desorption spectroscopy [Kanazawa 1997] and experiments in atomic physics and fundamental positron interactions.

## 2 Methods of production

All positron beam systems can be broken down into three main components: a source of positrons at high energy, a moderating system that thermalizes the positrons and emits a mono-energetic eV energy positron flux and a transport system that accelerates and transports the positron beam. We see in chapter 2 that highly efficient systems to accelerate and transport positrons are readily available and there is little that can be done to increase positron beam currents through improvement of the transport systems.

This leaves two strategies to increase the current in a positron beam. First is to provide a stronger positron source and second is to develop a more efficient method to moderate the source positrons into a monoenergetic beam. Positron moderators have been extensively studied by the positron community for over two decades and relatively efficient moderating systems have been developed. The factors affecting the moderator efficiency are well understood and limitations in moderator efficiency are based on the physics of the positron interaction in the moderating medium. The issues surrounding the use of moderators in making positron beams, moderator performance and the development of better moderators are covered in chapter 2 of this book. We will concentrate on the positron source strength and coupling to commonly used moderators systems.

The one component of positron beams that has no physical limits is the initial positron source strength. Due to the matter-antimatter asymmetry positrons are not naturally occurring in our local environment. They must be made by some process that produces an anti-matter particle from interactions involving ordinary matter. The two processes that produce positrons are pair production from photons interacting with atomic nuclei and radioactive decay of nuclei off the beta stability curve.

### *2.1 Strong radioactive sources*

Some radioactivity occurs in our natural environment but all useful levels of positron emitters are manufactured by exposing some non-radioactive element to a nuclear particle that transforms it into a positron emitting species. This may be done by a beam of ions from an accelerator or by exposure to neutrons and photons in the core of a nuclear reactor. There is no physical limit on the amount of radioactivity that can be made in this manner and in principle arbitrarily high positron current could be obtained by increasing the positron source strength.

There are practical limits on the use of radioactive sources. Since we wish to harvest the positrons emitted by the radioactive decay, the source must have a high fraction of positron emission from its radioactive decay. Also the amount of overlying material must be thin enough that the positron can escape the source material and interact with the positron moderator. In the ultimate limit a pure radioactive source will be limited to the number of positrons that can escape through the overlying radioactive material. When the thickness is greater than the escape depth for a decay positron no farther increase in positron flux can be achieved. Thus, while it is possible to obtain large amounts of long lived radioactive species such as $^{22}Na$, beams made from these materials have been limited in current.

The solution to this dilemma is to have radioactive species with rapid decay rates so that the amount of overlying material is reduced for a specific positron emission rate. This technique is used to produce high current beams but it requires a close connection between the production of the radioactive materials and the beam experiment. Radioisotopes that have been proposed to produce high current positron beams from reactor irradiation include $^{64}Cu$, $^{58}Co$, and $^{79}Kr$ [Mills 1992].

### *2.1.1 Reactor production of positron sources*

The most well known system based on a high activity, short-lived radioactive sources is the beam based at the High Flux Beam Reactor at Brookhaven National Laboratory. This beam started operation in 1983 and has been the most successful in producing data for atomic and solid state physics. It provided beams for positron research until recently when the reactor was put in an indefinite stand down. With its loss there are no operational high current beams that solely rely on high activity sources derived from nuclear reactors.

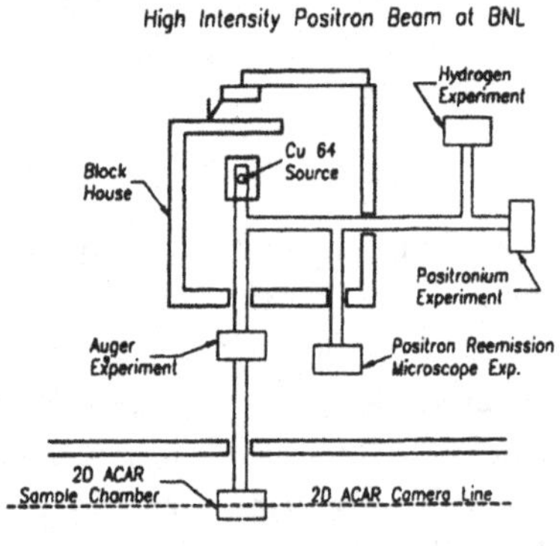

Figure 1 Layout of the Brookhaven intense beam facility [ Lynn 1994 ]

At Brookhaven a small, ~130mg, pellet of natural copper was irradiated in the neutron flux in the core of the High Flux Beam Reactor to produce an intense source of positrons. The reaction $^{63}Cu(n,\gamma)^{64}Cu$ in a neutron flux of 8 $10^{14}$ $ncm^{-2}s^{-1}$ resulted in approximately 100 Curies of short lived $^{64}Cu$ ($T_{1/2}$ = 12.8 h). This pellet was then rapidly transferred to the source-moderator chamber of their high positron current beam. It was then evaporated and deposited on a tungsten crystal at the moderator position. Initially positron moderation was done in the as deposited annealed copper film on the tungsten substrate [Lynn 1987]. Later a solid neon moderator was deposited on the radioactive copper to increase the low energy positron flux [Weber 1992, Lynn 1994]. Moderated positrons were than accelerated and transported to experimental stations. This beam varied in intensity over time with the highest current available immediately after depositing a newly activated source . Beam current would then decrease with the 12.8 hr. half life of $^{64}$Cu. There were also variations in the maximum current obtained form different source deposition episodes due to delays or other variations in depositing the copper source.

Experiments performed with the Brookhaven beam established the value of high current beams for measurements of electron momentum measurements by angular correlation. Both defect depth profiles and the electronic structure of clean metallic surfaces were determined at the Brookhaven facility. They also produced beams of positronium which was then used to measure positronium reflection and surface interactions [Weber 1988]. At the time of the standdown the beam was available for electron momentum by angular correlation experiments, crossed beam atomic physics experiments, positron stimulated Auger experiments and a positron imaging microscope.

A new, reactor based high current beam facility is under development at the university of Delft reactor. This facility will employ a hybrid of $^{64}$Cu production from thermal neutron activation of copper and pair production from neutron capture gamma rays at a positron source region that is located at the reactor core. The production geometry of this facility is similar to the geometry using pair production from neutron capture gamma rays alone. A complete discussion of the Delft design will be made in the reactor based pair production section.

Another facility based on a novel concept for capturing and moderating fast positrons may soon produce a high intensity beam at the Paul Scherrer Institute [Waeber, 1997]. This system confines the initial positron sources in a Tesla magnetic field and dissipates the initial positron energy through repeated passes through a thin carbon foil. Moderated positrons are then extracted with high efficiency into an electrostatic transport system capable of producing high quality beam characteristics. This system is presently operating with a $^{58}$Co source to produce a beam with $310^6$ $e^+s^{-1}$ in a 0.1 mm spot. A high intensity, short-lived source renewed at the PSI reactor is planned to increase the beam strength to high intensity.

Other facilities have been proposed to couple very radioactive positron emitters made in the core of nuclear reactors and the production of high current positron beams but have failed due to lack of support for the general reactor facility. These include irradiation of many copper pellets in the Advanced Neutron Source reactor proposed for installation at Oak Ridge National Laboratory [ Hulett 1994 ] and production of extremely active large area $^{58}$Co at INEEL [ Makowitz 1992 ]. The need to couple the high current positron beam to a high flux nuclear reactor limits the opportunity to develop these facilities and similar reactor based systems described in the next session.

### *2.1.2 Accelerator production of positron emitters*

Irradiation of materials with accelerator ion beams can also produce positron emitting radionuclides[Stein 1974, Itoh, 1997]. Up to now positron beams produced by ion irradiation have been limited to low current levels by the ion beam current. This limitation may be removed in a promising new design to use high current, low energy deuterium ions to irradiate $^{12}$C which undergoes ( d,n ) reactions to produce $^{13}$N. $^{13}$N has a 9.97 minute half life and emits 100% of 1.19 MeV positrons for each radioactive decay. The cross section for ( d,n ) is peaked at 1.2 MeV deuteron energy. Consequently high current deuteron beams of a few MeV energy can be used to produce high concentrations of $^{13}$N. Compact, single stage linear accelerators are now available that can supply milliampere currents of 1.5 MeV deuterons. By rapid cycling of a carbon target between a high current deuteron beam and a positron moderation station a quasi continuous beam of positrons can be made. This technique offers great promise to produce an inexpensive, high current positron beam. First experiments to prove the principle of these devices have been performed and $^{13}$N produced in a thick target by a 1.8 MeV deuteron beam have been moderated in a tungsten single crystal foil resulting in about 5 $10^3$ moderated positrons per microamp of deuterons in a preliminary geometry[ Xie, 1994].

A scaled up design has been proposed [Hughey 1998] that could result in a moderated beam with greater than $10^7$ $e^+s^{-1}$ in a 7 mm spot from a tungsten moderator foil. A first attempt to produce such a beam is under way at Brandeis University. In the Brandeis system solutions to problems of scaling the demonstration experiment into a working beam will be addressed. Some problems that will be addressed include finding methods to efficiently couple the tungsten foil to the $^{13}$N positron source, minimizing positron absorption in the C target, enhancing carbon target stability under the high current deuteron beam implantation and minimizing associated radiation damage in the moderator and other beam elements by the neutrons produced as a byproduct of the activation reaction. Construction of the accelerator to produce the deuteron beam has begun at this time.

### *2.2 Pair production*

Pair production is the dominant method of producing high current positron beams. When a photon having energy greater than the rest energy of an electron and a positron, 1.022 MeV, interacts with the fields surrounding an atomic nucleus that energy can be converted into an electron-positron pair. Probability of pair production grows as the photon energy increases and pair production is the major interaction for photons having high energies. There is no physical limit on the number of pairs that can be produced but photon energy deposition and heating of the pair production target establishes a practical limit.

Photons for use in pair production to produce high current positron beams have been obtained from two primary sources, high energy electron beam interactions in a target or nuclear capture gamma rays in a nuclear reactor core. Both of these sources are capable of producing high photon fluxes that can result in high rates of positron production. The current worlds record positron beam is produced by pair production from bremsstrahlung photons that are made by interactions of a 100 MeV electron beam with a tungsten target. Most intense positron beams are located at electron linacs.

#### *2.2.1 Linacs*

The majority of intense positron beams have been produced by collection of positrons produced during interaction of the energetic electron beam with a target. Intense beams based on this principle include: European beams at Mainz, Giessen and Ghent and new beams at Rossendorf and Aarhus, United States beams at Lawrence Livermore National Laboratory, LLNL, Oak Ridge National Laboratory , ORNL and ATT-Bell Laboratory, and Japanese beams at the University of Tokyo, the University of Kyoto, the University of Osaka, the Electrotechnical Laboratory, ETL, the KEK injector, at KEK-B and in a proposed "Positron Factory" facility. This large number is due to the ease of production of positrons by this method and the availability of linacs that are used for nuclear or high energy physics research.

Producing a positron beam with a linac high energy electron beam is a direct conversion of power from the electrical grid into particle mass in a beam. Electron linacs can produce high currents of electrons with energies of 100 MeV or higher.

When a high energy electron beam interacts in a target it looses its energy by a combination of photon emission and electronic excitation. For electrons over 15 MeV the dominant energy loss mechanism is by bremsstrahlung photon emission. These photons then interact with surrounding material to produce electron-positron pairs. After creating a pair, any excess photon energy appears as kinetic energy of electron-positron pair. If the original photons are high energy then there is a cycle of radiation and pair production until the average energy of the electron and positron is too low to produce photons above the pair rest mass. This can lead to many new electrons and positrons similar to the events found in a cosmic ray shower.

An efficient method of collecting and moderating the positrons from an electron beam was developed at the 100 MeV electron linac at Lawrence Livermore National Laboratory [Howell, 1982]. The geometry of the primary electron beam and accelerated positron beam is shown in figure 2. In early attempts to optimize the connection between the electron beam and positron intensity efficiencies of 1 moderated positron per $10^6$ incident electrons was achieved for a 100 MeV electron beam at LLNL. This corresponds to about 10 pA of positrons per kW power deposited by the electron beam in the stopping target. The efficiency of positron production depends on the electron beam energy by approximately ($E_{electron\ beam}$ - 15 MeV) [Howell, 1982 ]. This leads to a clear preference for linacs with higher energy beams. Since the first beam at LLNL there have been many successful variations of this basic geometry installed at linacs with a range of electron beam energies and currents. Trends in positron production on electron beam energy and electron stopping target thickness have been explored and higher efficiencies have been obtained [Ley, 1997] but up to now there has not been a major improvement from the early designs.

For a particular conversion-moderation geometry and electron energy the positron current is proportional to the linac power and the time structure of the beam is tied to the pulse structure of the linac. The factors that control the current produced from stopping of high energy electron beams are technical. The three most significant issues are the ability of the linac to produce a high energy, high power electron beam, the ability of the target to dissipate the thermal load from the power in the electron beam and the efficiency of production of low energy positrons from high energy electrons.

The best efficiencies so far have been obtained from empirical optimization by testing positron yields from new geometries. A hard upper limit on the efficiency of positron production has not yet been established. Present efficiencies enable this technique to produce the highest positron currents achieved by any method. Higher currents may be possible as details of the complex processes involved in coupling the bremsstrahlung target and positron moderators become better understood.

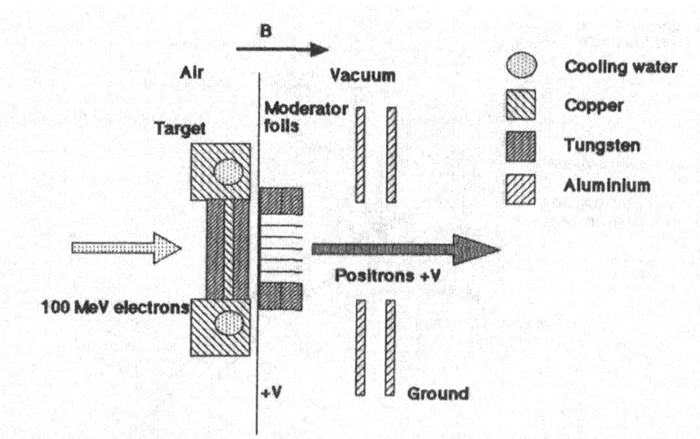


Figure 2 Fixture for production of intense positron beams from high energy electrons.

Up to now the limit on linac power set by accelerator technology has been the most significant factor in determining the beam characteristics of the facility. All operating linac based intense positron beams have been developed at existing linacs. In this case linac characteristics are set by the needs defined by the original facility function. A proposal to build a new, dedicated linac based "Positron Factory" is under consideration in Japan. In this case a new 100 kW linac would be used to produce a beam of more than $10^{10}$ $e^{+}s^{-1}$. This new linac would supply more than twice the power of the LLNL linac with a corresponding increase in positron current. Limits set by the available linac power are constantly improving as better designs for electron linacs are achieved. Target technology that can convert the electron beam into positrons is also under constant development in order to supply positrons for accelerator beam in electron positron colliders and in positron beam based synchrotrons.

Linacs are inherently pulsed devices and most have both a sub nanosecond micropulse structure and a microsecond scale macrostructure. Experimental needs often are best served with a continuous beam of positrons. Conversion of linac beams to quasi-continuous beams is done by trapping the positron pulse from the linac in a magnetic bottle and then slowly releasing the trapped positrons over the period between linac beam pulses. This was first done with the Ghent linac and is now part of the beam transport at LLNL, ORNL and ETL. At LLNL a transition into an electrostatic transport system through a magnetic grid is made following the magnetic bottle providing a nearly continuous, focused beam.

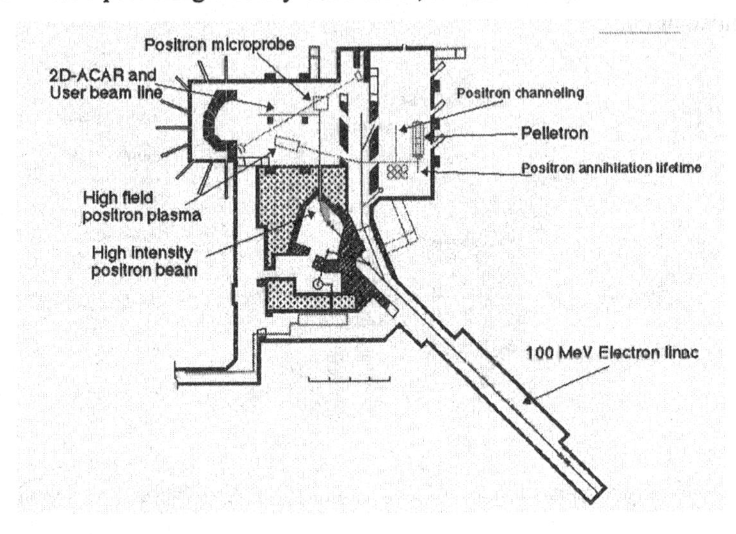


Figure 3. layout of the Lawrence Livermore National Laboratory positron facilities. Both intense and high energy beams are available

The combination of ease of access and reliability have made linac based positron beams very popular and today there are several active facilities. Over longer operation times all of the factors important to beam strength and quality are stable. Linac beams tend to have reproducible characteristics over periods of months. Here are the activities at those with the highest intensities.

The facility at LLNL has been in operation since 1980 and has been the record holder for beam current since that time. Today the primary function of the high power, 45 kW, electron linac is to produce a $10^{10}$ $e^{+}s^{-1}$ primary, pulsed positron beam. This makes the LLNL facility the only dedicated intense positron beam facility in the world. This systems has been used to perform high resolution electron momentum experiments by angular correlation, positronium time of flight experiments for surface analysis and fundamental electron-positron interactions [Howell 1987 ]. Recently beam stretching and extraction into an electrostatic transport have been added to the primary transport so that a pulsed positron microprobe for positron lifetime experiments can be installed. Electrostatic beam switching will also service diffraction, electron momentum spectroscopy, and other positron spectroscopic experiments. The LLNL facility provides beams for solid state and surface analysis using positron spectroscopy and for fundamental physics measurements.

The ORNL positron beam is at the ORELLA accelerator. The linac is primarily used for nuclear physics experiments and the positron beam is produced in a parasitic mode from the nuclear physics target. This secondary access to the beam results in a $10^8$ $e^+s^{-1}$. primary positron beam which is then stretched or bunched to shorter pulses depending on experimental needs. The primary use of the ORNL beam has been the study positron interactions with atoms and molecules.

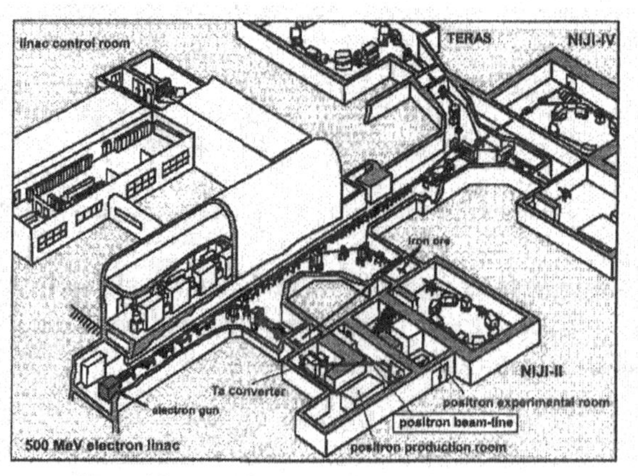

Figure 4 Layout of the ETL intense beam facility

The ElectroTechnology Laboratory in Tsukuba shares linac time at the injector for the electron storage rings NIJI II, NIJI-IV and TERAS. The linac electron beam directly strikes a target and moderator assembly similar to the LLNL design. They obtain $10^8$ $e^+s^{-1}$ in an 8mm spot in from less than 1 kW deposited in the target. This beam is trapped and stretched for use in a fast buncher system [Suzuki 1997]. The ETL facility has been very productive in measuring depth profiles of low energy beam positron lifetimes for defect analysis in thin films and buried interfaces of semiconductors and plastics. It has also been used for time of flight experiments of Auger electrons or of positronium to determine surface properties.

The present operating facility at KEK shares linac time with the injector for the TRISTAN facility. They bring the primary 2.0 GeV electron beam onto their target. A $10^8$ $e^+s^{-1}$ positron beam is produce d from 2 kW deposited into the target. This beam is guided to experimental areas where positronium time of flight and positron reemission microscopy experiments are underway and electron momentum

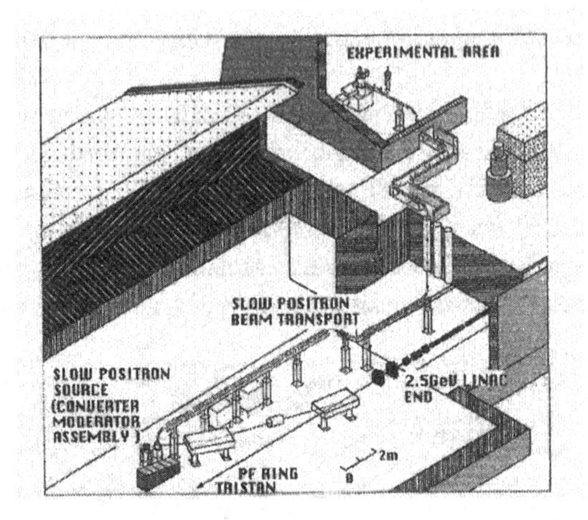


Figure 5. Layout of the KEK intense beam facility [Kanazawa 1997 ]

momentum measurements using angular correlation are planned. A second facility which will use both the KEKB 1.5 GeV injector and have a dedicated linac for positron production is under development at KEKB.

*2.2.2 Synchrotron radiation*

It is now possible to conceive of MeV photon sources derived from wigglers inserted in the beams of very high energy electron accelerators. Several high energy accelerator facilities including ESRF in Grenoble, APS at Argonne National Laboratory, TRISTAN at KEK in Tsukuba, PETRA II, in Hamburg, the B-Factory at the Stanford Linear Accelerator Facility and LEP at CERN all have enough energy to produce photons above the pair production threshold. An insertion device at one of these facilities could produce a high flux of 2-4 MeV photons in a small area. Pair production in a tungsten moderator would then produce an intense, bright positron source. Calculations of wiggler design show that as much as $10^{13}$ $e^{+}s^{-1}$ in a $mm^2$ area are possible at most of the accelerators listed above[Csonka 1995, Barbiellini 1996].

The basic design of such an intense beam consists of a wiggler or undulator inserted into the GeV electron beam of the accelerator to produce a high flux of MeV photons. Characteristics of the photon flux are determined by the details of the insertion design and very high brightness designs have been proposed [Csonka 1995]. Positron production and moderation would be performed using the same basic techniques as described above for production by bremsstrahlung radiation from an electron linac beam. Positron production would have advantages over linac bremsstrahlung systems. The extremely high flux of low energy photons will produce lower energy primary positrons which will moderate with less spreading and higher efficiency than their energetic counterparts. This can result in very bright positron beams when insertion devices such as undulators are used to produce very highly directed photon beams

In principle positron production from an insertion device is attractive due to the high flux and brightness expected in the positron beam. Since 2-4 MeV photons produce few nuclear reactions there is also the attraction of working in a low radiation environment at a facility with a fully supported infrastructure. Unfortunately at this time there is no active program to construct a high energy photon insertion device at any of the available facilities.

*2.2.3 Neutron capture gamma rays*

An alternative source of photons for pair production of positrons is found in the gamma rays produced by capture of the neutrons found in the core of a nuclear reactor. Neutron capture gamma rays from the $^{113}$Cd ( n,$\gamma$) $^{114}$Cd reaction share the neutron binding energy of 9.041 MeV between two or more photons. This results in 2.3 photons/neutron capable of efficiently undergoing pair production. When these photons illuminate a well annealed tungsten foil positrons are produced and efficiently moderated in one step. This process is illustrated in figure 6.

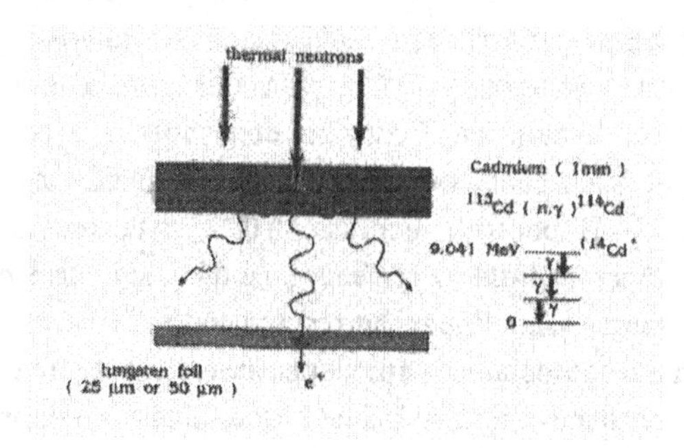


Figure 6 Mechanism of positron production from neutron capture gamma-rays [Triftshauser 1997 ]

Two intense beam facilities under development are based on pair production from capture gamma rays. One is at the Research Reactor Munich at Garching and another is at the Delft 2 MW research reactor. which contains a hybrid of pair production and short lived activation. Both of these share a common geometry. The positron production fixture is inserted near the reactor core through a penetration in the reactor wall and the positrons are transported from the interior of the reactor to experimental stations outside the reactor. These beams are continuous over time since the process of pair production and activation are constantly ongoing. This basic geometry for the Munich system is shown in figure 7. Initial experiments are being performed inside the reactor containment vessel at both facilities. At this time both Munich and Delft have produced preliminary beams and are undergoing design optimization.

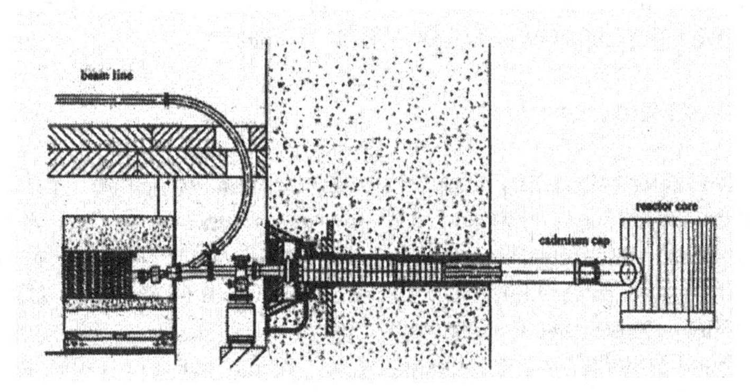


Figure 7 Positron production at the Research Reactor Munich [Triftshauser 1997 ]

The two designs differ as the Munich design depends exclusively on pair production from capture gamma rays while the Delft design uses a combination of pair production and activation of $^{64}$Cu. At this time the Munich group have achieved a 2 $10^8$ $e^+s^{-1}$ beam in a 3 cm diameter spot in a continuous beam. An upgraded version of this beam line will be installed at the new reactor FRM-II in the next year. At FRM-II beams in excess of $10^{10}$ $e^+s^{-1}$ current are expected.

At this time there is little experience in the long term operation of beams produced from systems located near the reactor core. The difficulties of activation and radiation damage associated with experiments inside reactor cores are well known but there are also well established techniques for dealing with them. We must wait for operation at FRM-II to see what effect activation and radiation damage will have on the practical operation of an in core beam.

A similar system is under development at the university of Delft reactor. Positron production from the Delft reactor beam takes place by a combination of pair production from capture gamma rays in the tungsten moderator foils and radioactive decay from $^{64}$Cu. Early tests of this beam have produced 0.8x10$^{8}$ e$^{+}$s$^{-1}$ [van Veen 1998]. An addition to the Delft reactor building will house the experimental area for the beam with stations for electron momentum by angular correlation measurements, defect depth profiling using novel detectors and a positron injected microprobe.

## 3 Summary

There is a critical need for intense beams to do experiments for materials analysis and fundamental physics. Many productive intense beam facilities have been operated to produce new data on fundamental properties and on surfaces, thin films and buried interfaces for materials analysis. Some of these are lost to the positron community due to closure of the host facility. New intense beam facilities are under development at both reactor and linac sites and facilities dedicated to positron production are beginning to appear.

## 4 References

Barbiellini, G, and Petrucci G, 1997, *Appl. Surf. Sci.* **116**, 49

Brandes, GR, Canter, KF, and Mills, AP Jr, 1998, *Phys. Rev. Lett.* **610,** 492

Csonka, PL, 1995, *Nucl. Instum. Meth.* **A352**, 579

Howell RH, Stoeffl W, Kumar A, Sterne PA, Cowan TE, and Hartley J, 1997, *Mat. Sci. Forum* **255** 644

Howell, RH, Rosenberg ,IJ, and Fluss, MJ, 1987, *Appl. Phys.* **A43**

Howell RH, Meyer P, Rosenberg IJ, and Fluss, MJ, 1985, *Phys. Rev. Lett.* **54**, 1698

Howell, RH, Alvarez and RA, Stanek M, 1982, *Appl. Phys. Lett.* **40**, 751

Hulett LD, unpublished proposal 1992

Hughey BJ, Shefer RE, Klimkowstein RE and Canter KF, 1998, *Applications of Accelerators in Research and Industry,* ed. JL Duggan and IL Morgan (New York: AIP Press), p455

Itoh Y, Peng Zl, Lee Kh, Ishii M, Goto A, Nakanishi N, Kase M, and Ito Y, 1997, *Appl. Surf. Sci.* **116**, 68

Kanazawa I, Koizumi T, Iwamoto A, Tashiro M, Komori F, Murata Y, Fukutani K, and Ito Y, 1997, *Mat. Sci. Forum* , **255** , 787

Ley, Richard, 1997, *Hyperfine Interactions* **109**, 167

Lynn KG and Jacobsen FM, 1994, *Hyperfine Interactions* , **89**, 19

Lynn, KG, Roellig, LO, Mills, AP, Moodenbaugh, 1987, *Atomic Physics with Positrons*, p161

Makowitz, H, Abrashoff, JD, Landman, WH, Albano, RK, Tajima, T and Larson, JD, 1993, AIP Conf. 303, Ed. E. Ottewitte, A. Weiss

Mills AP, 1992, Nucl. Sci. Eng. , **110** 165

Ottewitte, EH and Kells, W, 1987, *Intense Positron Beams*, (Singapore: World Scientific)

Rosenberg, IJ, Weiss, AH and Canter, KF, 1980, *Phys. Rev. Lett.* **44**, 1139

Stein, TS, Kauppila, WE and Roellig,LO, 1974, *Rev. Sci. Instrum.* **45**

Suzuki, R, Ohdaira, T, Mikado, T, Ohgaki, H, Chiwaki, M and Yamazaki, T, 1997, *Appl. Surf. Sci.* **116**, 187

Triftshauser, G, Kogel, G, Triftshäuser, W, Springer, M, Strasser, B, Schreckenbach, K, 1997, *Appl. Surf. Sci..* **116**, 45

Waeber, WB and Shi M, 1997, *Appl. Surf. Sci.* **116**, 19

Weiss, A. Mayer, R, Jibaly, M, Lei, C, Mehl, D, and Lynn, KG, 1988, *Phys. Rev. Lett.* **61**, 2245

Weber, M, Schwab, A, Becker, D, Lynn, KG, 1992, *Hyperfine Interactions* **73**, 147,

Weber, M, Tang, s, Berko, S, Brown, BL, Canter, KF, Lynn, KG, Mills, AP Jr, Roellig, LO, and Viescas, AJ, 1988, *Phys Rev Lett.* **61**, 2542

van Veen, A, Labohm,F, Schut, H, de Roode, J, Heijenka, T, Mijnarends, PE, 1997, *Appl. Surf. Sci.* **116**, 39

van Veen, A, 1998, private communication.

Xie, R , Petkov, M, Becker, D, Canter, K, Jacobson, FM, Lynn, KG, Mills, R, and Roellig, LO, 1994, Nucl. Instrum. Methods B**93**, 98

**INDEX**